Computational Methods and Mathematical Modeling in Cyberphysics and Engineering Applications 1

SCIENCES

Mathematics, Field Director – Nikolaos Limnios

Mathematics in Engineering, Subject Head – Dmitri Koroliouk

Computational Methods and Mathematical Modeling in Cyberphysics and Engineering Applications 1

Coordinated by
Dmitri Koroliouk
Sergiy Lyashko
Nikolaos Limnios

iSTE

WILEY

First published 2024 in Great Britain and the United States by ISTE Ltd and John Wiley & Sons, Inc.

ISTE Ltd
27-37 St George's Road
London SW19 4EU
UK

www.iste.co.uk

John Wiley & Sons, Inc.
111 River Street
Hoboken, NJ 07030
USA

www.wiley.com

Library of Congress Control Number: 2023944056

British Library Cataloguing-in-Publication Data
A CIP record for this book is available from the British Library
ISBN 978-1-78945-164-1

ERC code:
PE1 Mathematics
PE1_20 Application of mathematics in sciences
PE1_21 Application of mathematics in industry and society

Contents

Preface

Dmitri KOROLIOUK[1,2]
[1] *National Technical University of Ukraine*
Igor Sikorsky Kyiv Polytechnic Institute, Ukraine
[2] *UNESCO Chair in Biotechnology and Bioethics,*
University of Rome Tor Vergata, Italy

Mathematical methods in engineering are characterized by a wide range of techniques and methods for approaching problems. Moreover, completely different analysis techniques can be applied to the same problem, which is justified by the difference in specific applications.

Therefore, the study of the analysis and solution of specific problems leads the researcher to generate their own techniques and approaches for analyzing similar problems that continuously arise in the process of technical development.

This book provides solutions to specific problems in current areas of computational engineering and cyberphysics, which are considered as examples of ideas and approaches to practical tasks.

October 2023

Computational Methods and Mathematical Modeling in Cyberphysics and Engineering Applications 1,
coordinated by Dmitri KOROLIOUK, Sergiy LYASHKO and Nikolaos LIMNIOS.

1

The Hydrodynamic-type Equations and the Solitary Solutions

Sergiy LYASHKO[1], Valerii SAMOILENKO[1], Yuliia SAMOILENKO[2,3] and Ihor GAPYAK[1]

[1] *Taras Shevchenko National University of Kyiv, Ukraine*

[2] *National Technical University of Ukraine*
Igor Sikorsky Kyiv Polytechnic Institute, Ukraine

[3] *Claude Bernard Lyon 1 University, France*

This chapter analyzes the problem of solitary waves for hydrodynamic-type equations. It mainly focuses on the Korteweg–de Vries (KdV) equation with a small singular perturbation. It should be noted that in the study of various problems associated with the singularly perturbed KdV and Kdv-like equations, linear and nonlinear Wentzel–Kramers–Brillouin (WKB) methods are used. A description of the main idea of this technique is given, and its effectiveness for constructing asymptotic soliton-like solutions for KdV and KdV-like equations with variable coefficients and a singular perturbation is provided.

1.1. Introduction

Nonlinear hydrodynamic-type equations are associated with important models of modern physics and applied mathematics (Whitham 1974; Ablowitz 2011). These

Computational Methods and Mathematical Modeling in Cyberphysics and Engineering Applications 1,
coordinated by Dmitri KOROLIOUK, Sergiy LYASHKO and Nikolaos LIMNIOS.

equations arise in the study of various wave processes and phenomena in hydrodynamics and many other fields of natural science. They are quite difficult to analyze because of their nonlinear nature and the fundamental impossibility of their integration in a closed form. Therefore, it is natural to study either their particular solutions, for example, solutions in the form of a traveling wave, or such equations, using numerical analysis.

These equations include the Boussinesq equation, the Korteweg–de Vries equation (the KdV equation), the modified Korteweg–de Vries equations, the Burgers' equation and the Benjamin–Bona–Mahony equation (Bullough and Caudry 1980; Lamb 1980; Dodd et al. 1982; Newell 1985). The equations have the traveling wave solution $u = f(x - at)$.

The Boussinesq equation:

$$u_{tt} = c_s^2 \frac{\partial^2}{\partial x^2}\left(u + \frac{3}{2}\frac{u^2}{h} + \frac{h^2}{3} u_{xx} \right) \qquad [1.1]$$

and the Korteweg–de Vries equation:

$$u_{xxx} - 6uu_x + u_t = 0 \qquad [1.2]$$

appeared in the 19th century as a result of scientific discussions by many outstanding scientists about the propagation of long waves in rectangular channels and the search for mathematical models to describe the solitary waves that were observed by John Scott Russell in 1834 (Russell 1844).

Equations [1.1] and [1.2] have solutions that describe the waves with profile $A\cosh^{-2}(c(x - vt))$. A characteristic feature of waves is localization in space and time, as well as the dependence of the propagation velocity on their amplitude (Bullough and Caudry 1980).

The Burgers' equation (Burgers 1939)

$$u_t + uu_x = \nu u_{xx}, \quad \nu > 0, \quad x \in R, \; t \geq 0, \qquad [1.3]$$

is one of the simple models describing the evolution of shock waves. It is deduced from the well-known Navier–Stokes system in the one-dimensional case.

The Benjamin–Bona–Mahony equation or regularized long wave equation (Peregrin 1966)

$$u_t + u_x + uu_x - u_{xxt} = 0 \qquad [1.4]$$

was proposed in 1966 because of a search for an alternative to the KdV equation for describing long waves on the surface of a liquid. Although equation [1.4] has one-soliton solutions (Benjamin et al. 1972), it does not have m-soliton $(m \geq 2)$ solutions. Moreover, [1.4] is not an integrable system because it only has three conservation laws.

The integrability of the Korteweg–de Vries (KdV) equation and the KdV-like equations makes it possible to use powerful algebro-geometric approaches (Blackmore et al. 2011) and methods of the Hamiltonian analysis (Faddeev and Takhtajan 1987) to study the properties of the equations and construct a wide set of their exact solutions.

However, the consideration of wave processes in media with variable characteristics and, in particular, with a small dispersion, requires the study of nonlinear equations with variable coefficients and small perturbations. The presence of variable coefficients in nonlinear equations significantly complicates the researching of the corresponding systems. As a result, either the methods of asymptotic analysis are the only mathematical tool for studying such systems or such equations must be studied using numerical analysis (Lyashko 1991, 1995).

1.2. The Korteweg–de Vries equation and the soliton solutions

Most of the above-mentioned equations are models of fluid motion. The Korteweg–de Vries equation is used to mathematically describe the dynamics of solitary waves on a liquid surface. Later, in the middle of the 20th century, it was found that equation [1.2] and the KdV-like equations arise while studying many various phenomena and processes, in particular, in plasma, solid body theory, optic, biology, telecommunication systems, etc. Equation [1.2] is currently one of the fundamental equations of modern physics. If we consider this equation as an example, then we can find many different characteristic phenomena and properties that are inherent in hydrodynamic-type equations.

An in-depth study of the Korteweg–de Vries equation began in the late 1960s after the publication of the paper by Zabusky and Kruskal (1965). The authors pointed out the connection between the equation and the Fermi–Pasta–Ulam problem. Zabusky and Kruskal found that the Korteweg–de Vries equation possesses solutions that have the property of retaining their waveform after collision with waves of the same nature. Such solutions are called *soliton solutions* or *solitons* (Zabusky and Kruskal 1965).

The amazing discovery drew great attention from both mathematicians and physicists to the Korteweg–de Vries equation. This attention increased significantly after the creation of a new technique called the inverse scattering transform, which has been successfully applied to the study of many integrable systems (Gardner et al. 1967). Subsequently, the inverse scattering transform became a powerful tool for studying numerous nonlinear systems of modern theoretical and mathematical physics.

Many monographs are devoted to different mathematical aspects of the theory of nonlinear integrable systems, including the Korteweg–de Vries equation. In particular, one-, two-, multi-soliton solutions, finite-gap solutions, etc. have been constructed for the KdV equation. For this equation and its generalizations, problems on existence, uniqueness and other properties (smoothness, exponential decay, etc.) of its solutions, and solutions to the Cauchy problem for various classes of initial functions were also studied (Sjoberg 1970; Hruslov 1976; Kato 1979; Faminskii 1990).

It has been found that properties of the solutions to the KdV equation essentially depend on the initial conditions (Pokhozhayev 2010; Turbal et al. 2015). In addition to soliton, periodic, finite-gap and shock wave solutions, the KdV equation has solutions with some singularities on the curves (Arkadyev et al. 1984), as well as solutions that are destroyed roughly or destroyed according to the gradient catastrophe scenario (Pokhozhayev 2010).

1.3. The Korteweg–de Vries equation with a small perturbation

The discovery of new classes of solutions to the KdV equation had a great influence on the development of the perturbation theory of nonlinear systems. Such research began with the study of the perturbed Korteweg–de Vries equation by asymptotic methods. In the paper by Miura and Kruskal (1974), the authors generalized the linear Wentzel–Kramers–Brillouin (WKB) method for the nonlinear equations and constructed the leading term of the asymptotic series for the quasi-periodic solution to the KdV equation with a small dispersion:

$$u_t + uu_x + \delta^2 u_{xxx} = 0, \qquad [1.5]$$

where δ is a small parameter. Thus, the nonlinear WKB method was created. It has become a powerful mathematical tool for studying nonlinear systems of the hydrodynamic-type (Maslov 1982; Samoilenko and Samoilenko 2008, 2019; Lyashko et al. 2021) with a singular perturbation.

By means of the Bogoliubov M.M. averaging method (Bogoliubov and Mitropolsky 1961), Flaschka et al. (1980) studied the modulation of nonlinear waves, which are described by the Korteweg–de Vries equation. Although the equation under consideration does not contain a small parameter δ explicitly, the dependence of its finite-gap solutions on the small parameter manifests through a small deformation of the associated Riemann surfaces.

Furthermore, Lax and Levermore (1983a, 1983b, 1983c) considered the problem on the limit of the solution to the Cauchy problem for the singularly perturbed Korteweg–de Vries equation [1.5], as a small parameter tends to zero. While solving the problem they used the inverse scattering transform. In particular, the authors constructed eigenfunctions for the Lax operator with a small perturbation by the linear WKB method. Subsequently, this problem under other initial conditions was considered in many papers and studied mainly by numerical methods.

Asymptotic methods turned out to be an effective means of investigating other problems for the Korteweg–de Vries equation. So, for example, the equation with regular perturbation was considered by de Kerf (1988), who constructed asymptotic solutions to the Cauchy problem of the following form for the non-homogeneous Korteweg–de Vries equation

$$u_t - 6uu_x + u_{xxx} = \varepsilon f[u], \quad u(x,0) = u_0(x), \qquad [1.6]$$

where $f[u]$ is a functional and the initial function $u_0(x)$ is a soliton-like one.

Ilyin and Kalyakin studied the problem [1.6] when the initial condition described the perturbation of multisoliton solutions. The authors constructed perturbed soliton solutions to the Korteweg–de Vries equation (Kalyakin 1992) using the Krylov–Bogoliubov–Mitropolsky method (Bogoliubov and Mitropolsky 1961).

Later Glebov et al. (2005) and Glebov and Kiselev (2005) considered the Korteweg–de Vries equation with forced perturbation of the following form:

$$\varepsilon^2 f(\varepsilon x)\cos\left(\frac{S(\varepsilon^2 x, \varepsilon^2 t)}{\varepsilon^2}\right).$$

They found asymptotic solutions using the nonlinear WKB method. This made it possible to study the phenomenon of passage through resonance.

In the papers mentioned above, the Korteweg–de Vries equation with a small parameter was considered. It is important that under a regular perturbation, the unperturbed equation is the classical Korteweg–de Vries equation, while under a singular perturbation it is an equation with constant coefficients.

The presence of singular perturbations in such problems requires the study of challenging mathematical problems. The first papers on singularly perturbed nonlinear partial differential equations with variable coefficients are due to Maslov and his students, who considered equations that are integrable in particular cases. We should also mention the paper by Maslov and Omel'yanov (1981), in which an asymptotic solution of a differential equation with variable coefficients was constructed. They considered an equation describing the propagation of waves in shallow water with variable depth. It was the KdV-like equation of the following form:

$$u_t + (\rho_1 + 3\rho_2 u)u_x + \varepsilon^2 \rho_3 u_{xxx} + \rho_4 u = 0, \qquad [1.7]$$

where the coefficients of the equation depend on the depth function $H(x) > 0$ of the non-perturbed liquid and have the form:

$$\rho_1 = \sqrt{gH(x)}, \quad \rho_2 = \sqrt{gH^{-1}(x)}\big/2, \quad \rho_3 = \sqrt{gH^5(x)}\big/6, \quad \rho_4 = \rho_{2x}/2,$$

where g is the acceleration of gravity, and ε is a small parameter that characterizes the value of the dispersion.

Using the nonlinear WKB method, the authors constructed asymptotic solutions to equation [1.7]. In structure, the solutions are close to solutions of the soliton type. These solutions are called *asymptotic soliton-like solutions* (Maslov and Omel'yanov 1981, 2001).

Furthermore, the problem of constructing asymptotic soliton-like solutions was considered in a more general statement when the coefficients of the KdV equation are variable and have a rather general form (Samoilenko and Samoilenko 2012a, 2012b).

Finally, we note that singularly perturbed partial differential equations with variable coefficients naturally arise in the study of mathematical models of wave processes and phenomena in media whose characteristics depend on time and spatial variables, and singular perturbation. These problems are generally non-integrable, although they have soliton solutions in the case of constant coefficients. Hence, the problem of constructing asymptotic soliton-like solutions for a singularly perturbed partial differential equation with variable coefficients is of great interest (Whitham 1974; Ablowitz 2011; Samoilenko and Samoilenko 2019).

1.4. The linear WKB technique and its generalization

To construct an asymptotic soliton-like solution to the Korteweg–de Vries equation with a singular perturbation, the nonlinear WKB method (Miura and Kruskal 1974) is used. This technique is generalization of the WKB method.

The WKB method is named after the German scientists Wentzel, Kramers and Brillouin. At the beginning of 1926, in fact, at the same time, they proposed this method (Brillouin 1926; Kramers 1926; Wentzel 1926) for the construction of approximate solutions to the one-dimensional stationary Schrödinger equation of quantum mechanics (Schrödinger 1926)

$$-\frac{\hbar^2}{2m}\frac{d^2\Psi(x)}{dx^2} + V(x)\Psi(x) = E\Psi(x). \qquad [1.8]$$

This was an efficient approach to studying the equation, since [1.8] is a non-integrable differential equation in general.

Here, the Planck constant $\hbar$ is supposed to be a small parameter, and $\Psi = \Psi(x)$ is a wave (complex-valued) function which, according to its physical content, satisfies the conditions of regularity, i.e. the wave function is finite, single-valued and continuous with its first derivatives. These assumptions are satisfied, for example, under condition $E - V(x) > 0$.

The main idea of the WKB method is that a solution of equation [1.8] is represented in the special form $\Psi(x) = e^{\Phi(x)}$, where the complex-valued function $\Phi(x)$ according to equation [1.8] satisfies the second-order nonlinear differential equation

$$\Phi''(x) + \Phi'^2(x) = \frac{2m}{\hbar^2}\left(V(x) - E\right). \qquad [1.9]$$

It is natural to consider the real-valued functions $A(x) = \operatorname{Re}\Phi'(x)$, $B(x) = \operatorname{Im}\Phi'(x)$, characterized amplitude and phase of the oscillating solutions. From relation [1.8], it follows that these functions satisfy the first-order nonlinear differential equation

$$A' + A^2 - B^2 = \frac{2m}{\hbar^2}\left(V(x) - E\right), \quad B' + 2AB = 0, \qquad [1.10]$$

the solutions of which can be constructed as asymptotic expansions in the value $\hbar$:

$$A(x) = \frac{1}{\hbar}\sum_{k=0}^{\infty} \hbar^k A_k(x), \quad B(x) = \frac{1}{\hbar}\sum_{k=0}^{\infty} \hbar^k B_k(x). \qquad [1.11]$$

The coefficients of asymptotic series [1.11] are determined recursively from the equations found in the standard way from equation [1.10]. In particular, the main terms $A_0(x)$ and $B_0(x)$ in equation [1.11] have to satisfy the differential equations

$$A_0^2(x) - B_0^2(x) = 2m\left(V(x) - E\right), \quad A_0(x)B_0(x) = 0, \qquad [1.12]$$

and the terms $A_1(x)$ and $B_1(x)$ satisfy the following differential equations:

$$A_0'(x) + 2A_0A_1 - 2B_0B_1 = 0, \qquad [1.13]$$

$$B_0'(x) + 2A_0B_1 + 2A_1B_0 = 0. \qquad [1.14]$$

Obviously, to find the coefficients $A_0(x)$, $B_0(x)$, $A_1(x)$ and $B_1(x)$ from relations [1.12]–[1.14], both cases $A_0(x) = 0$ and $B_0(x) = 0$ must be considered separately.

Taking into account the condition of regularization, we first consider the case $A_0(x)=0$. This equality means that the amplitude of the oscillation changes more slowly than its phase. Therefore, from equations [1.12] to [1.14], we have

$$B_0(x)=\pm\sqrt{2m(E-V(x))},$$

$$A_1(x)=-\frac{1}{4}\frac{d}{dx}\ln\left(2m(E-V(x))\right),\quad B_1(x)=0.$$

As a result, taking into account the first two terms of the asymptotic for the functions $A(x)$, $B(x)$, the asymptotic approximation for the solution to equation [1.8] is given in the form

$$\Psi(x)=\frac{C_+}{\sqrt[4]{2m(E-V(x))}}\exp\left(\frac{i}{\hbar}\int\sqrt{2m(E-V(x))}\,dx\right)+$$

$$+\frac{C_-}{\sqrt[4]{2m(E-V(x))}}\exp\left(-\frac{i}{\hbar}\int\sqrt{2m(E-V(x))}\,dx\right),\qquad [1.15]$$

where C_+, C_- are arbitrary constants.

If $B_0(x)=0$, i.e. the phase changes more slowly than the amplitude, then from equations [1.12]–[1.14], we have

$$A_0(x)=\pm\sqrt{2m(E-V(x))},\quad B_0(x)=B_1(x)=0,$$

and the asymptotic approximation for the solution to equation [1.8] is written as

$$\Psi_1(x)=\frac{C_+}{\sqrt[4]{2m(E-V(x))}}\exp\left(\frac{1}{\hbar}\int\sqrt{2m(E-V(x))}\,dx\right)+$$

$$+\frac{C_-}{\sqrt[4]{2m(E-V(x))}}\exp\left(-\frac{1}{\hbar}\int\sqrt{2m(E-V(x))}\,dx\right),\qquad [1.16]$$

where C_+, C_- are arbitrary complex constants.

Note that while applying the WKB method, it is enough to construct only the first two terms of the asymptotics.

Formulas [1.15] and [1.16] asymptotically approximate the solutions to equation [1.8] for all real arguments x except for the neighborhoods of the turning points, where the equality $E - V(x) = 0$ holds.

In the neighborhoods of the turning points, the asymptotic solution to equation [1.8] is searched for in the other way, using the expansion of its right-hand side in a Taylor series in the neighborhood of the corresponding point. The main term of the asymptotic solution to equation [1.8] is defined from the Bessel equation and written as a linear combination of the Bessel functions.

While using the linear WKB method, the solution of equation [1.8] is sought as an exponential function in some asymptotic expansion. Miura and Kruskal (1974) proposed to construct a solution to the nonlinear differential equation with a singular perturbation as an asymptotic series in a small parameter with quickly oscillating coefficients. They called this technique the nonlinear WKB method. This approach has proven to be quite effective (Samoilenko and Samoilenko 2005, 2008, 2019, 2012a, 2012b; Lyashko et al. 2021) for constructing the asymptotic soliton-like solutions to the following equation:

$$\varepsilon^n u_{xxx} = a(x,t,\varepsilon)u_t + b(x,t,\varepsilon)uu_x, \quad n \in N, \qquad [1.17]$$

where $\varepsilon > 0$ is a small parameter, the coefficients $a(x,t,\varepsilon)$, $b(x,t,\varepsilon)$, $(x,t) \in R \times [0;T]$ are infinitely differentiable and are given with asymptotic (according to Poincaré) series

$$a(x,t,\varepsilon) = \sum_{k=0}^{N} \varepsilon^k a_k(x,t) + O\left(\varepsilon^{N+1}\right),$$

$$b(x,t,\varepsilon) = \sum_{k=0}^{N} \varepsilon^k b_k(x,t) + O\left(\varepsilon^{N+1}\right),$$

and the condition $a_0(x,t)b_0(x,t) \neq 0$, $(x,t) \in R \times [0;T]$, takes place.

Note that due to the presence of the variable coefficients in equation [1.17], the problem of finding its asymptotic solutions becomes much more complicated, and at the same time, a number of new interesting mathematical problems appear. In addition, the degree of the singularity in the equation [1.17] also has a significant impact on the structure of the constructed asymptotic soliton-like solution, the algorithm for finding it and the properties of the solutions.

1.5. Acknowledgments

Samoilenko Yuliia was supported by Programme PAUSE and the ANR, project no. ANR-22-PAUK-0038-01, in Claude Bernard Lyon 1 University, and she is very appreciative and grateful for the support and the hospitality.

1.6. References

Ablowitz, M.J. (2011). *Nonlinear Dispersive Waves. Asymptotic Analysis and Solitons.* Cambridge University Press, Cambridge.

Arkadyev, V.A., Pogrebkov, A.K., Polivanov, M.K. (1984). Singular solutions to the Korteveg–de Vries equation and inverse scattering transform. *Zapiski nauchnych seminarov Leningradskogo otdeleniya Matematicheskogo instituta*, 133, 17–37 [in Russian].

Benjamin, T.B., Bona, J.L., Mahony, J.J. (1972). Model equations for long waves in nonlinear dispersive systems. *Philosoph. Trans. Royal Society of London. Series A, Mathematical and Physical Sciences*, 272(1220), 47–78.

Blackmore, D., Prykarpatsky, A.K., Samoylenko, V.H. (2011). *Nonlinear Dynamical Systems of Mathematical Physics. Spectral and Integrability Analysis.* World Scientific, Singapore.

Bogoliubov, N.N. and Mitropolsky, Y.A. (1961). *Asymptotic Methods in the Theory of Nonlinear Oscillations.* Gordon and Breach, New York.

Brillouin, L. (1926). Remarques sur la mécanique ondulatoire. *Journ. Phys. Radium.*, 7, 353–368.

Bullough, R.K. and Caudrey, P.J. (1980). Further remarks on John Scott Russel and on the early history of his solitary wave. In *Solitons*, Bullough, R.K. and Caudrey, P.J. (eds). Springer, Berlin, Heidelberg.

Burgers, J.M. (1939). Mathematical examples illustrating relations occurring in the theory of turbulent fluid motion. *Verhandelingen der Koninklijke Nederlandsche Akademie van Wetenschappen, Afdeeling natuurkunde*, Eerste sectie, Deel XVII (2), 1–53.

Dodd, R.K., Eilbeck, J.G., Gibbon, J.D., Morris, H.C. (1982). *Solitons and Nonlinear Wave Equations.* Academic Press, London.

Faddeev, L.D. and Takhtajan, L. (1987). *Hamiltonian Methods in the Theory of Solitons.* Springer, Berlin, Heidelberg.

Faminskii, A.V. (1990). Cauchy problem for the Korteweg–de Vries equation and its generalizations. *J. Sov. Math.,* 50(1), 1381–1420.

Flaschka, H., Forest, M.G., McLaughlin, D.W. (1980). Multiphase averaging and the inverse spectral solution of the Korteweg–de Vries equation. *Comm. Pure Appl. Math.*, 33(6), 739–784.

Gardner, C.S., Green, J.M., Kruskal, M.D., Miura, R.M. (1967). Method for solving the Korteweg-de Vries equation. *Phys. Rev. Lett.*, 19, 1095–1097.

Glebov, S.G. and Kiselev, O.M. (2005). The stimulated scattering of solitons on a resonance. *J. Nonlin. Math. Phys*., 12(3), 330–341.

Glebov, S.G., Kiselev, O.M., Lazarev, V.A. (2003). Birth of solitons during passage through local resonance. *Proc. Steklov Inst. Math. Suppl*., 1, 84–90.

Glebov, S.G., Kiselev, O.M., Lazarev, V.A. (2005). Slow passage through resonance for a weakly nonlinear dispersive waves. *SIAM J. Appl. Math*., 65(6), 2158–2177.

Hruslov, E.J. (1976). Asymptotics of the solutions of the Cauchy problem for the Korteweg–de Vries equation with initial data of step type. *Math. USSR-Sbornik*, 28(2), 229–248.

Kalyakin, L.A. (1992). Perturbation of the Korteweg–de Vries soliton. *Theor. Math. Phys.*, 92(1), 736–747.

Kato, T. (1979). On the Korteweg–de Vries equation. *Manuscripta Math*., 28, 89–99.

de Kerf, F. (1988). Asymptotic analysis of a class of perturbed Korteweg–de Vries initial value problems. *Centrum voor Wiskunde en Informatica*, Amsterdam.

Kramers, H.A. (1926). Wellenmechanik und halbzahlige Quantisierung. *Z. Phys.*, 39, 828–840.

Lamb Jr., G.R. (1980). *Elements of Soliton Theory*. John Wiley & Sons, New York.

Lax, P.D. and Levermore, C.D. (1983a). The small dispersion limit of the Korteweg–de Vries equation I. *Comm. Pure Appl. Math.*, 36(3), 253–290.

Lax, P.D. and Levermore, C.D. (1983b). The small dispersion limit of the Korteweg–de Vries equation II. *Comm. Pure Appl. Math.*, 36(5), 571–593.

Lax, P.D. and Levermore, C.D. (1983c). The small dispersion limit of the Korteweg–de Vries equation III. *Comm. Pure Appl. Math.*, 36(6), 809–829.

Lyashko, S.I. (1991). The approximate solution of a pseudoparabolic equation. *Comput. Math. Math. Phys.*, 31(12), 107–111.

Lyashko, S.I. (1995). Numerical solution of pseudoparabolic equations. *Cybernet. Systems. Anal.*, 31(5), 718–722.

Lyashko, S.I., Samoilenko, V.H., Samoilenko, Y.I., Lyashko, N.I. (2021). Asymptotic analysis of the Korteweg–de Vries equation by the nonlinear WKB method. *Math. Model. and Comput.*, 8(3), 368–378.

Maslov, V.P. and Omel'yanov, G.A. (1981). Asymptotic soliton-like solutions of equations with small dispersion. *Russ. Math. Surveys*, 36(3), 73–149.

Maslov, V.P. and Omel'yanov, G.A. (2001). *Geometric Asymptotics for PDE. I.* American Mathematical Society, Providence.

Miura, R.M. and Kruskal, M.D. (1974). Application of nonlinear WKB-method to the Korteweg–de Vries equation. *SIAM Appl. Math.*, 26(2), 376–395.

Newell, A.C. (1985). *Solitons in Mathematics and Physics*. SIAM, Philadelphia.

Peregrin, D.H. (1966). Calculations of the development of an undular bore. *J. Fluid Mech.*, 25(2), 321–330.

Pokhozhayev, S.I. (2010). On the singular solutions of the Korteweg–de Vries equation. *Math. Notes*, 88(5), 741–747.

Russell, J.S. (1844). Report on waves. Report, Fourteenth Meeting of the British Association, John Murray, London.

Samoilenko, V.H. and Samoilenko, Y.I. (2005). Asymptotic expansions for one-phase soliton-type solutions of the Korteweg–de Vries equation with variable coefficients. *Ukr. Math. J.*, 57(1), 132–148.

Samoilenko, V.H. and Samoilenko, Y.I. (2007). Asymptotic solutions of the Cauchy problem for the singularly perturbed Korteweg-de Vries equation with variable coefficients. *Ukr. Math. J.*, 59(1), 126–139.

Samoilenko, V.H. and Samoilenko, Y.I. (2008). Asymptotic two-phase soliton-like solutions of the singularly perturbed Korteweg–de Vries equation with variable coefficients. *Ukr. Math. J.*, 60(3), 449–461.

Samoilenko, V.H. and Samoilenko, Y.I. (2012a). Asymptotic *m*-phase soliton-type solutions of a singularly perturbed Korteweg–de Vries equation with variable coefficients. *Ukr. Math. J.*, 64(7), 1109–1127.

Samoilenko, V.H. and Samoilenko, Y.I. (2012b). Asymptotic *m*-phase soliton-type solutions of a singularly perturbed Korteweg–de Vries equation with variable coefficients II. *Ukr. Math. J.*, 64(8), 1241–1259.

Samoilenko, V.H. and Samoilenko, Y.I. (2019). Asymptotic soliton-like solutions to the singularly perturbed Benjamin-Bona-Mahony equation with variable coefficients. *J. Math. Phys.*, 60(1), 011501-1–011501-13.

Schrödinger, E. (1926). Quantisierung als Eigenwertproblem. *Ann. Phys.*, 384(4), 361–376.

Sjoberg, A. (1970). On the Korteweg–de Vries equation: Existence and uniqueness. *J. Math. Anal. Appl.*, 29(3), 569–579.

Turbal, Y., Turbal, M., Bomba, A., Sokh, A. (2015). *T*-transformation method for studying the multi-soliton solutions of the Korteweg–de Vries type equations. *J. Math. System Sci.*, 4, 164–169.

Wentzel, G. (1926). Eine Verallgemeinerung der Quantenbedingung fur die Zwecke der Wallenmechanik. *Z. Phys.*, 38, 518–529.

Whitham, G.B. (1974). *Linear and Nonlinear Waves*. John Wiley & Sons, New York.

Zabusky, N.J. and Kruskal, M.D. (1965). Interaction of solitons in a collisionless plasma and recurrence of initial states. *Phys. Rev. Lett.*, 15, 240–243.

2

The Nonlinear WKB Technique and Asymptotic Soliton-like Solutions to the Korteweg–de Vries Equation with Variable Coefficients and Singular Perturbation

Sergiy LYASHKO[1], Valerii SAMOILENKO[1], Yuliia SAMOILENKO[2,3] and Evgen VAKAL[1]

[1] *Taras Shevchenko National University of Kyiv, Ukraine*

[2] *National Technical University of Ukraine*
Igor Sikorsky Kyiv Polytechnic Institute, Ukraine

[3] *Claude Bernard Lyon 1 University, France*

Here we consider the problem of the mathematical description of soliton-like solutions of hydrodynamic models, such as the Korteweg–de Vries (KdV) equation and its generalizations, describing wave processes in inhomogeneous media with variable characteristics and a small dispersion. Mainly, attention is paid to the development of an algorithm for constructing approximate (asymptotic) solutions of soliton type. These solutions contain a regular part, which is a background function, and a singular part, which reflects the soliton properties of the solutions. They are

Computational Methods and Mathematical Modeling in Cyberphysics and Engineering Applications 1,
coordinated by Dmitri KOROLIOUK, Sergiy LYASHKO and Nikolaos LIMNIOS.

constructed using the nonlinear WKB method. The construction of the regular and singular parts of the searched asymptotic solution is described in detail.

Problems that arise while using the algorithm as well as while its justification are discussed. Statements about the accuracy with which the constructed asymptotic solution satisfies the original equation are proved.

This approach provides us with effective tools for studying the influence of model parameters on the properties of the dynamical system under consideration. In particular, the obtained results can be used for the subsequent analysis of these solutions in different ways, for example, by using computer simulation.

2.1. Introduction

In the papers by Dobrokhotov and Maslov (1981), and Maslov and Omel'yanov (1981), the notion of an asymptotic soliton-like solution to equations of integrable type with a small singular perturbation is proposed. In accordance with the kind of small parameter, the nonlinear WKB method (Miura and Kruskal 1974) was used. Later, these results were detailed in the monograph (Maslov and Omel'yanov 2001). Much attention is also paid to other issues related to the properties of solutions to equations with a singular perturbation. In particular, the problem regarding discontinuous solutions to the unperturbed problems is considered. These solutions can be obtained from the formulas for the exact or asymptotic solutions in the limiting case as a small parameter tends to zero (Lax and Levermore 1983a, 1983b, 1983c; Samoilenko and Samoilenko 2010).

In the work on this subject (Maslov and Omel'yanov 1981), the equation of hydrodynamic KdV-like equation

$$u_t + (\rho_1 + 3\rho_2 u)u_x + \varepsilon^2 \rho_3 u_{xxx} + \rho_4 u = 0, \qquad [2.1]$$

was considered. Here, the coefficients of the equation depend on the depth function $H(x) > 0$ of the non-perturbed liquid and have the form

$$\rho_1 = \sqrt{gH(x)}, \quad \rho_2 = \sqrt{gH^{-1}(x)}\Big/2, \quad \rho_3 = \sqrt{gH^5(x)}\Big/6, \quad \rho_4 = \rho_{2x}/2,$$

where g is the acceleration of gravity, and ε is a small parameter characterized value of dispersion.

Subsequently, this line of research was developed in many studies. In particular, for the KdV equation with variable coefficients and a singular perturbation, the problem regarding the existence of asymptotic one-, two- and multi-phase soliton-like solutions (Samoilenko and Samoilenko 2005, 2008, 2012a, 2012b) as well as the corresponding Cauchy problems (Samoilenko and Samoilenko 2007) was studied. The concept of the asymptotic Σ-solution similar to multiphase soliton solutions was proposed (Samoilenko and Samoilenko 2014).

This study deals with constructing and researching asymptotic soliton-like solutions to the singularly perturbed KdV equation with variable coefficients of the following form:

$$\varepsilon^n u_{xxx} = a(x,t,\varepsilon)u_t + b(x,t,\varepsilon)uu_x, \quad n \in N, \qquad [2.2]$$

where $\varepsilon > 0$ is a small parameter, the functions $a(x,t,\varepsilon)$, $b(x,t,\varepsilon)$, $(x,t) \in R \times [0;T]$, are smooth or infinitely differentiable, and are written as asymptotic (according to Poincaré) expansions

$$a(x,t,\varepsilon) = \sum_{k=0}^{N} \varepsilon^k a_k(x,t) + O\left(\varepsilon^{N+1}\right), \qquad [2.3]$$

$$b(x,t,\varepsilon) = \sum_{k=0}^{N} \varepsilon^k b_k(x,t) + O\left(\varepsilon^{N+1}\right).$$

Condition $a_0(x,t)b_0(x,t) \neq 0$, $(x,t) \in R \times [0;T]$, is supposed to be true.

Next, the structure of an asymptotic series for soliton-like solutions to equation [2.2] is given, a general scheme for their construction is described, an algorithm for constructing such solutions is proposed and justified, the differential equations for the terms of asymptotic series are obtained and the problems on the existence of their solutions are analyzed. The accuracy with which the asymptotic solutions satisfy this equation is also found.

The approach described here is successfully realized for the long-wave equation (Samoilenko and Samoilenko 2019) and can be used for the other equations of the hydrodynamic type.

2.2. Main notations and definitions

First, we give the main notations and definitions used in what follows. Denote by $S = S(R)$ the space of quickly decreasing functions, i.e. such infinitely differentiable on the set R functions that for any integers $m,n >= 0$, the following condition

$$\sup_{x\in R}\left|x^m \frac{d^n}{dx^n}u(x)\right| < +\infty$$

holds.

Denote by $C^\infty(0,T;S)$ the space of such infinitely differentiable on the set $R\times[0;T]$ functions $u(x,t)$, that for any integers $m,k >= 0$, the following condition

$$\int_{-\infty}^{+\infty}\left(D_x^m D_t^k u\right)^2 dx + \int_{-\infty}^{+\infty}\left(1+x^2\right)^m \left(D_t^k u\right)^2 dx < +\infty$$

takes place.

Analogously to Maslov and Omel'yanov (2001) and Samoilenko and Samoilenko (2005), denote by $G_1 = G_1(R\times[0;T]\times R)$ a linear space of infinitely differentiable functions $f = f(x,t,\tau)$, $(x,t,\tau)\in R\times[0;T]\times R$, such that for any non-negative integers n, p, q, r uniformly with respect to $(x;t)$ on every compact set $K\subset R\times[0;T]$, the following conditions are fulfilled:

1) the relation

$$\lim_{\tau\to+\infty} \tau^n \frac{\partial^p}{\partial x^p}\frac{\partial^q}{\partial t^q}\frac{\partial^r}{\partial \tau^r} f(x,t,\tau) = 0, \quad (x,t)\in K,$$

is true;

2) there exists such an infinitely differentiable function $f^-(x,t)$ that

$$\lim_{\tau\to-\infty} \tau^n \frac{\partial^p}{\partial x^p}\frac{\partial^q}{\partial t^q}\frac{\partial^r}{\partial \tau^r}(f(x,t,\tau) - f^-(x,t)) = 0, \quad (x,t)\in K.$$

Let $G_1^0(R\times[0;T]\times R)\subset G_1$ be a space of functions $f=f(x,t,\tau)$, $(x,t,\tau)\in R\times[0;T]\times R$, such that uniformly with respect to variables $(x;t)$ on any compact $K\subset R\times[0;T]$, the following condition

$$\lim_{\tau\to-\infty} f(x,t,\tau)=0,\quad (x,t)\in K,$$

is satisfied.

DEFINITION 2.1 (Maslov and Omel'yanov 1981; Samoilenko and Samoilenko 2005).– *Function $u=u(x,t,\varepsilon)$, where $\varepsilon>0$ is a small parameter, is called an asymptotic soliton-like function, if for any integer $N\geq 0$, the one is represented as*

$$u(x,t,\varepsilon)=\sum_{j=0}^{N}\varepsilon^j\left[u_j(x,t)+V_j(x,t,\tau)\right]+O\left(\varepsilon^{N+1}\right),\quad \tau=\frac{x-\varphi(t)}{\varepsilon},\qquad [2.4]$$

where $\varphi(t)\in C^\infty([0;T])$ is a scalar real-valued function, the functions $u_j(x,t)\in C^\infty(R\times[0;T])$, $j=\overline{0,N}$, are infinitely differentiable (in the points $t=0$, $t=T$, they are considered to be the left and right derivatives correspondingly); functions $V_0(x,t,\tau)\in G_1^0$, $V_j(x,t,\tau)\in G_1$, $j=\overline{1,N}$.

The value $x-\varphi(t)$ is called a phase function of the asymptotic soliton-like function $u(x,t,\varepsilon)$.

Here and below, we use the standard notation of asymptotic analysis. That is, the relation $\Psi(x,t,\varepsilon)=O\left(\varepsilon^N\right)$, $\varepsilon\to 0$, means that there exist such a value $\varepsilon_0>0$ and a constant $C>0$ depending on the integer N and the compact $K\subset R\times[0;T]$ that $|\Psi(x,t,\varepsilon)|\leq C\varepsilon^N$ for any $\varepsilon\in(0;\varepsilon_0)$ and all $(x,t)\in K$.

2.3. The structure of the asymptotic one-phase soliton-like solution

Consider the problem on constructing the asymptotic one-phase soliton-like solutions to equation [2.2]. In a sense, these solutions are deformation of soliton solutions to the KdV equation with constant coefficients. First, we describe the structure of the asymptotic solutions to equation [2.2] and represent the algorithm for finding them.

The form of the asymptotic expansion for the asymptotic one-phase soliton-like solutions to equation [2.2] depends on integer n in [2.2], i.e. on the degree of singularity. Moreover, it is necessary to differ a case when n is even and a case when n is odd. That is, the asymptotic expansion for the one-phase soliton-like solution is written as

$$u(x,t,\varepsilon) = \sum_{j=0}^{N} \varepsilon^j u_j(x,t) + \sum_{j=0}^{N} \varepsilon^j V_j(x,t,\tau) + O\left(\varepsilon^{N+1}\right), \quad \tau = \frac{x-\varphi(t)}{\varepsilon^{n/2}}, \quad [2.5]$$

if n is even, and

$$u(x,t,\varepsilon) = \sum_{j=0}^{k} \varepsilon^j u_j(x,t) + \sum_{j=0}^{k} \varepsilon^j V_j(x,t,\tau) + \varepsilon^k \sum_{j=1}^{2N-k} \varepsilon^{j/2} u_j(x,t) +$$

$$+\varepsilon^k \sum_{j=1}^{2N-k} \varepsilon^{j/2} V_j(x,t,\tau) + O\left(\varepsilon^{N+1/2}\right), \quad \tau = \frac{x-\varphi(t)}{\varepsilon^{n+1/2}}, \quad [2.6]$$

if n is odd, and $n = 2k+1$, where $k \in N \cup \{0\}$.

Formulas [2.5] and [2.6] are deduced (Samoilenko and Samoilenko 2005) in the standard way by taking into account equation [2.2], forms for its coefficients [2.3] and searched solution [2.4]. Next, we demonstrate the main ideas for constructing the asymptotic one-phase soliton-like solution to equation [2.2] for the case $n = 2$. Then, we present asymptotic expansion for the one-phase soliton-like solution to equation [2.2] for the case of $n = 1$.

2.4. The KdV equation with quadratic singularity

In this section, we describe in detail the algorithm for constructing the asymptotic soliton-like solution to the KdV equation [2.2] for n=2. In this case, the asymptotic solution to equation [2.2] is searched as the following series:

$$u(x,t,\varepsilon) = Y_N(x,t,\tau,\varepsilon) + O\left(\varepsilon^{N+1}\right), \quad [2.7]$$

where

$$Y_N(x,t,\tau,\varepsilon) = \sum_{j=0}^{N} \varepsilon^j \left[u_j(x,t) + V_j(x,t,\tau)\right] + O\left(\varepsilon^{N+1}\right), \quad \tau = \frac{x-\varphi(t)}{\varepsilon}.$$

The function

$$U_N(x,t,\varepsilon) = \sum_{j=0}^{N} \varepsilon^j u_j(x,t),$$

is called a regular part of asymptotics [2.5], and the function

$$V_N(x,t,\tau,\varepsilon) = \sum_{j=0}^{N} \varepsilon^j V_j(x,t,\tau)$$

is called a singular part of asymptotics [2.5].

Obviously, $Y_N = U_N + V_N$. We note also that the regular part of asymptotics [2.5] is a background function, while its singular part reflects soliton properties of the searched solution.

While constructing the asymptotic solution of form [2.5], we first define the regular part and then we find the singular part. This is done as follows. Having calculated the derivatives $u_t(x,t,\varepsilon)$, $u_x(x,t,\varepsilon)$, $u_{xxx}(x,t,\varepsilon)$, substituting them into equation [2.2] and multiplying obtained equality by ε, we obtain:

$$\varepsilon^3\left(\frac{\partial^3 Y_N}{\partial x^3} + \frac{3}{\varepsilon}\frac{\partial^3 Y_N}{\partial x^2 \partial \tau} + \frac{3}{\varepsilon^2}\frac{\partial^3 Y_N}{\partial x\, \partial \tau^2} + \frac{3}{\varepsilon^3}\frac{\partial^3 Y_N}{\partial \tau^3}\right) = \qquad [2.8]$$

$$= \varepsilon a(x,t,\varepsilon)\left(\frac{\partial Y_N}{\partial t} - \frac{1}{\varepsilon}\varphi'(t)\frac{\partial Y_N}{\partial \tau}\right) + \varepsilon b(x,t,\varepsilon)\left(\frac{\partial Y_N}{\partial x} + \frac{1}{\varepsilon}\frac{\partial Y_N}{\partial \tau}\right) Y_N +$$

$$+\varepsilon g_N(x,t,\tau,\varepsilon),$$

where $g_N(x,t,\tau,\varepsilon)$ is an infinitely differentiable function that is defined recursively in functions Y_j, $j = \overline{1, N-1}$.

Thus, relation [2.8] yields

$$\varepsilon^3\left(\frac{\partial^3 U_N}{\partial x^3} + \frac{\partial^3 V_N}{\partial x^3} + \frac{3}{\varepsilon}\frac{\partial^3 V_N}{\partial x^2 \partial \tau} + \frac{3}{\varepsilon^2}\frac{\partial^3 V_N}{\partial x\, \partial \tau^2} + \frac{3}{\varepsilon^3}\frac{\partial^3 V_N}{\partial \tau^3}\right) = \qquad [2.9]$$

$$= \varepsilon a(x,t,\varepsilon)\left(\frac{\partial U_N}{\partial t} + \frac{\partial V_N}{\partial t} - \frac{1}{\varepsilon}\varphi'(t)\frac{\partial Y_N}{\partial \tau}\right) +$$

$$+\varepsilon b(x,t,\varepsilon)\left(\frac{\partial U_N}{\partial x} + \frac{\partial V_N}{\partial x} + \frac{1}{\varepsilon}\frac{\partial V_N}{\partial \tau}\right)(U_N + V_N) + \varepsilon g_N(x,t,\tau,\varepsilon).$$

From equation [2.9], we deduce equations for terms of the regular and the singular parts of asymptotics [2.5].

2.5. Equations for the regular part of the asymptotics and their analysis

The regular terms $u_j(x,t)$, $j = \overline{0,N}$, of the asymptotics are defined from the following system of the first-order partial differential equations:

$$a_0(x,t)\frac{\partial u_0}{\partial t} + b_0(x,t)u_0\frac{\partial u_0}{\partial t} = 0, \qquad [2.10]$$

$$a_0(x,t)\frac{\partial u_j}{\partial t} + b_0(x,t)\left(u_0\frac{\partial u_j}{\partial x} + u_j\frac{\partial u_0}{\partial x}\right) =$$

$$= f_j(x,t,u_0,u_1,\ldots,u_{j-1}), \quad j = \overline{1,N}, \qquad [2.11]$$

where the functions $f_j(x,t,u_0,u_1,\ldots,u_{j-1})$, $j = \overline{1,N}$, on the right side of equation [2.11] are found recurrently. In particular,

$$f_1(x,t,u_0) = -a_1(x,t)\frac{\partial u_0}{\partial t} - b_1(x,t)u_0\frac{\partial u_0}{\partial x},$$

$$f_2(x,t,u_0,u_1) = -a_1(x,t)\frac{\partial u_1}{\partial t} - a_2(x,t)\frac{\partial u_0}{\partial t} - b_1(x,t)u_1\frac{\partial u_0}{\partial x} -$$

$$-b_0(x,t)u_1\frac{\partial u_1}{\partial x} - b_2(x,t)u_0\frac{\partial u_0}{\partial x} - b_1(x,t)u_0\frac{\partial u_1}{\partial x}.$$

Equation [2.10] for the main term of the asymptotics is quasi-linear, while equations [2.11] for the other terms of the asymptotics are linear inhomogeneous equations. Thus, the solutions $u_j(x,t)$, $j = \overline{0,N}$, to equations [2.10] and [2.11] can be found by the direct integration of these equations, for example, through the method of characteristics. Thus, the terms of the regular part of the asymptotics can be considered known.

On the contrary, the problem has certain features related to the domain of determining these coefficients. That is, even though the coefficients of equation [2.2] are smooth functions and the terms of the regular part of the asymptotics are defined for all values of their arguments, the regular part can exist only in a bounded domain. To make sure, let us consider the case, when the functions $a_0(x,t)$, $b_0(x,t)$ are constant and quasi-linear equation [2.10] is equivalent to the Hopf equation

$$u_t + uu_x = 0 . \qquad [2.12]$$

The Hopf equation is associated with the Burgers equation of hydrodynamics. Solutions of the Hopf equation exhibit many interesting properties among which the well-known phenomenon of gradient catastrophe should be mentioned (Sagdeev et al. 1988). The phenomenon is connected with the properties of the initial function

$$u(x,0) = u_0(x), \;\; x \in R . \qquad [2.13]$$

Indeed, the formula $u = u_0(x - ut)$ gives us an implicit form of the solution to the Cauchy problem [2.12] and [2.13]. It is easy to see that under certain conditions, for the initial function, the partial derivatives of the solution

$$u_x = \frac{u_0'}{1 + t u_0'}, \qquad u_t = \frac{-u\, u_0'}{1 + t u_0'} \qquad [2.14]$$

can infinitely grow as the arguments tend to some finite point. In other words, the solution of equation [2.12] collapses according to the gradient catastrophe scenario.

It follows that in the case of variable coefficients in equation [2.2], equation [2.10] under certain initial conditions for the main term of the regular part of the asymptotics can have a solution only in a bounded domain. This feature of equation [2.10] has to be taken into account while constructing an asymptotic one-phase soliton-like solution to the Cauchy problem (Samoilenko and Samoilenko 2007) for equation [2.2].

2.6. Equations for the singular part of the asymptotics and their analysis

From relation [2.9], we deduce differential equations for the terms of the singular part of asymptotics [2.5]:

$$\frac{\partial^3 V_0}{\partial \tau^3} + a_0(x,t)\frac{\partial V_0}{\partial \tau}\varphi'(t) - b_0(x,t)\left(u_0\frac{\partial V_0}{\partial \tau} + V_0\frac{\partial V_0}{\partial \tau}\right) = 0, \qquad [2.15]$$

$$\frac{\partial^3 V_j}{\partial \tau^3} + a_0(x,t)\frac{\partial V_j}{\partial \tau}\varphi'(t) - b_0(x,t)\left(u_0\frac{\partial V_j}{\partial \tau} + \frac{\partial}{\partial \tau}(V_0 V_j)\right) = F_j(x,t,\tau), \qquad [2.16]$$

where the functions

$$F_j(t,\tau) =$$

$$= F_j(t, V_0(x,t,\tau), V_1(x,t,\tau), \ldots, V_{j-1}(x,t,\tau), u_0(x,t), u_1(x,t), \ldots, u_j(x,t))\Big|_{x=\varphi(t)}$$

are recurrently defined after determining the functions $u_0(x,t)$, $u_1(x,t)$, …, $u_j(x,t)$, $V_0(x,t,\tau)$, $V_1(x,t,\tau)$, …, $V_{j-1}(x,t,\tau)$, $j = \overline{1,N}$.

The problem of searching the terms of the singular part is more complicated than the problem of defining the regular part of the asymptotics, because equations [2.10] and [2.11] have higher order, and these solutions have to belong to some functional spaces (see Definition 2.1). Moreover, it is also necessary to find the phase function $\varphi = \varphi(t)$ that determines the so-called discontinuity curve $\Gamma = \{(x;t) \in R \times [0;T] : x = \varphi(t)\}$.

The problem is solved in the following way. First, equations [2.15] and [2.16] are considered on the discontinuity curve Γ, which is assumed to be known a priori. Then, we find the functions $v_j = v_j(t,\tau) = V_j(x,t,\tau)\big|_{\Gamma}$, $j = \overline{0,N}$. Furthermore, we deduce the differential equation for the function $\varphi = \varphi(t)$. Finally, the prolongation of the functions $v_j = v_j(t,\tau)$, $j = \overline{0,N}$, from the discontinuity curve Γ is constructed in such a way that the found functions belong to the corresponding function spaces.

Now, let us move on to the problem of constructing the functions $v_j = v_j(t,\tau)$, $j = \overline{0,N}$, satisfying the following differential equations:

$$\frac{\partial^3 v_0}{\partial \tau^3} + a_0(\varphi,t)\varphi'(t)\frac{\partial v_0}{\partial \tau} - b_0(\varphi,t)\left(u_0(\varphi,t)\frac{\partial v_0}{\partial \tau} + v_0\frac{\partial v_0}{\partial \tau}\right) = 0, \qquad [2.17]$$

$$\frac{\partial^3 v_j}{\partial \tau^3} + a_0(\varphi,t)\frac{\partial v_j}{\partial \tau}\varphi'(t) - b_0(\varphi,t)\left(u_0(\varphi,t)\frac{\partial v_j}{\partial \tau} + \frac{\partial}{\partial \tau}(v_0 v_j)\right) =$$

$$= F_j(t,\tau), \qquad [2.18]$$

where $F_j(t,\tau) = F_j(x,t,\tau)\big|_{x=\varphi(t)}$ and $j = \overline{1,N}$.

2.6.1. *The main term of the singular part*

First, we consider equation [2.17] and write its solution explicitly. Foremost, we integrate equation [2.17] with respect to the variable τ considering the value t to be a parameter. We obtain

$$\frac{d^2 v_0}{d\tau^2} = -[a_0(\varphi,t)\varphi'(t) - b_0(\varphi,t)u_0(\varphi,t)]v_0(t,\tau) + \frac{1}{2}b_0(\varphi,t)v_0^2(t,\tau) + c_1(t), \qquad [2.19]$$

where $c_1(t)$ is the integration constant.

Accordingly, the condition $V_0(x,t,\tau) \in G_1^0$, we put $c_1(t) \equiv 0$.

After multiplying equation [2.19] by $\dfrac{dv_0}{d\tau}$, we integrate the obtained relation in the variable τ. We have

$$\left(\frac{dv_0}{d\tau}\right)^2 = -[a_0(\varphi,t)\varphi'(t) - b_0(\varphi,t)u_0(\varphi,t)]v_0^2(t,\tau) + \frac{1}{3}b_0(\varphi,t)v_0^3(t,\tau) + c_2(t) \qquad [2.20]$$

From the condition $V_0(x,t,\tau) \in G_1^0$, it is deduced that the integration constant $c_2(t) \equiv 0$. Then, a solution to equation [2.17] in the space G_1^0 is written in the following form:

$$v_0(t,\tau) = -\frac{3A(\varphi,t)}{b_0(\varphi,t)} \cosh^{-2}\left(\frac{\sqrt{A(\varphi,t)}}{2}(\tau + C_0(t))\right), \qquad [2.21]$$

where $\varphi = \varphi(t)$, and $C_0 = C_0(t)$ is an integration constant.

We use the following note:

$$A(\varphi,t) = -a_0(\varphi,t)\varphi'(t) + b_0(\varphi,t)u_0(\varphi,t), \quad t \in [0;T], \qquad [2.22]$$

and we assume that the condition $A(\varphi(t),t) > 0$ is fulfilled for all $t \in [0;T]$.

The obtained result can be summarized by the following lemma.

LEMMA 2.1.– *If the condition* $A(\varphi(t),t) > 0$ *is satisfied for all* $t \in [0;T]$, *then function [2.21] is a solution to equation [2.17] in the space* G_1^0.

2.6.2. The higher terms of the singular part and the orthogonality condition

We now proceed to the consideration of differential equations [2.18]. Although these equations are linear inhomogeneous equations, we need to obtain a solution of these equations belonging to the space G_1. This means that the solutions have certain asymptotic properties with respect to the variable τ. Therefore, we previously deduce the necessary and sufficient conditions on the right-hand side of equations [2.18] under which these equations will have solutions in space G_1. Note that this problem is related to the problem on the existence of solutions to a differential inhomogeneous equation with the Schrödinger operator

$$Lv(x) = f(x), \text{ where } L = \frac{d^2}{dx^2} + q(x), \ x \in R, \qquad [2.23]$$

in the space of quickly decreasing functions.

Furthermore, we use the following result (Samoilenko and Samoilenko 2012c).

THEOREM 2.1.– *Let the following conditions hold:*

1) $q(x) = q_0 + q_1(x)$, *where the constant* $q_0 < 0$, *and the function* $q_1(x) \in S(R)$;

2) *the function* $f(x) \in S(R)$.

If the kernel of the operator $L : S(R) \to S(R)$ *is trivial, then equation [2.23] has a solution in the space* $S(R)$ *for any function* $f(x) \in S(R)$.

If the kernel of the operator $L : S(R) \to S(R)$ *is non-trivial, then equation [2.23] has a solution in the space* $S(R)$ *if and only if the orthogonality condition*

$$\int\limits_{-\infty}^{+\infty} f(x)\, v_0(x)\, dx = 0 \qquad [2.24]$$

is satisfied, where $v_0(x) \in S(R)$ *is a non-trivial solution to the equation*

$$\frac{d^2 v(x)}{dx^2} + q(x)v(x) = 0\,.$$

Consider equation [2.18] and rewrite it in the operator form

$$L_0 v_j = F_j, \quad j = \overline{1,N}\,, \qquad [2.25]$$

where

$$L_0 = \frac{\partial^3}{\partial\tau^3} + \left(a_0(\varphi,t)\varphi'(t) - b_0(\varphi,t)v_0 - b_0(\varphi,t)u_0(\varphi,t)\right)\frac{\partial}{\partial\tau} - \frac{\partial v_0}{\partial\tau} b_0(\varphi,t)\,.$$

Let us prove the statement.

LEMMA 2.2.– *Let* $F_j(t,\tau) \in G_1^0$, $j = \overline{1,N}$. *Then, a solution* $v_j(t,\tau)$, $j = \overline{1,N}$, *to equation [2.25] exists in the space* G_1 *if and only if the orthogonality condition*

$$\int\limits_{-\infty}^{+\infty} F_j(t,\tau)\, v_0(t,\tau)\, d\tau = 0, \quad j = \overline{1,N}\,, \qquad [2.26]$$

is fulfilled. The function $v_0(t,\tau)$ *is written by formula [2.21].*

PROOF.– Let us represent a solution to equation [2.25] in the following form:

$$v_j(t,\tau) = v_j(t)\eta_j(t,\tau) + \psi_j(t,\tau), \quad j = \overline{1,N}, \tag{2.27}$$

where

$$v_j(t) = \left[a_0(\varphi,t)\varphi' - b_0(\varphi,t)u_0(\varphi,t)\right]^{-1} \lim_{\tau\to-\infty} \Phi_j(t,\tau),$$

$$\Phi_j(t,\tau) = \int_{-\infty}^{\tau} F_j(t,\zeta)d\zeta + E_j(t).$$

Here, an integration constant $E_j(t)$ fulfills the condition $\lim_{\tau\to+\infty} \Phi_j(t,\tau) = 0$, the function $\eta_j(t,\tau) \in G_1$ satisfies the relation $\lim_{\tau\to-\infty} \eta_j(t,\tau) = 1$, and the value t is a parameter.

It is clear that from formula [2.27], it follows that the belonging $v_j(t,\tau) \in G_1$ is equivalent to the inclusion $\psi_j(t,\tau) \in G_1^0$, $j = \overline{1,N}$. Therefore, for proving Lemma 2.2, it is necessary to set the property $\psi_j(t,\tau) \in G_1^0$, $j = \overline{1,N}$.

With this purpose, we consider the auxiliary equation deduced from [2.25] by integrating it with respect to the variable τ (from $-\infty$ to τ):

$$L_1 v_j = \Phi_j(t,\tau), \tag{2.28}$$

where

$$L_1 = \frac{d^2}{d\tau^2} + a_0(\varphi(t),t)\varphi'(t) - b_0(\varphi(t),t)u_0(\varphi(t),t) - b_0(\varphi(t),t)v_0(t,\tau).$$

From equations [2.27] and [2.28], it follows that the function $\psi_j(t,\tau)$, $j = \overline{1,N}$, satisfies the inhomogeneous equation

$$L_1\psi_j = \Phi_j - v_j L_1\eta. \tag{2.29}$$

Since the kernel of the operator $L_1 : S(R) \to S(R)$ is non-trivial because of $\ker L_1 = \{v_{0\tau}\}$, then accordingly Theorem 2.1's solution to equation [2.29] exists in the space G_1^0 if and only if the orthogonality condition

$$\int_{-\infty}^{+\infty} (\Phi_j - v_j L_1 \eta) v_{0\tau} d\tau = 0, \quad j = \overline{1,N},$$

is fulfilled. The last relation provides us with the orthogonality condition [2.26].

Lemma 2.2 is proved.

Thus, the functions $v_j = v_j(t,\tau) = V_j(x,t,\tau)\big|_{\Gamma}$, $j = \overline{1,N}$, belong to the space G_1 if and only if condition [2.26] is satisfied.

2.6.3. *The orthogonality condition and the discontinuity curve*

The orthogonality condition [2.26] for $j = 1$ allows us to deduce the differential equation for the function $\varphi = \varphi(t)$ which defined the discontinuity curve Γ. For this purpose, we consider the following function

$$F_1(t,\tau) = a_0(\varphi,t)\frac{\partial v_0}{\partial t} + b_0(\varphi,t)u_{0x}(\varphi,t)v_0 +$$

$$+\left[-a_1(\varphi,t)\varphi'\frac{\partial v_0}{\partial t} + b_1(\varphi,t)u_0(\varphi,t) + b_0(\varphi,t)u_1(\varphi,t)\right]\frac{\partial v_0}{\partial \tau} +$$

$$+\tau\left[-a_{0x}(\varphi,t)\varphi'\frac{\partial v_0}{\partial t} + b_{0x}(\varphi,t)u_0(\varphi,t) + b_0(\varphi,t)u_{0x}(\varphi,t)\right]\frac{\partial v_0}{\partial \tau} +$$

$$+b_1(\varphi,t)\frac{\partial v_0}{\partial \tau}v_0 + \tau b_{0x}(\varphi,t)\frac{\partial v_0}{\partial \tau}v_0 \qquad [2.30]$$

satisfying the condition of Lemma 2.2.

Furthermore, we consider the orthogonality condition for $j = 1$ and calculate the expression

$$\int_{-\infty}^{+\infty} F_1(t,\tau)v_0(t,\tau)d\tau = \left(\frac{1}{2}a_0(\varphi,t)\frac{\partial}{\partial t} + b_0(\varphi,t)u_{0x}(\varphi,t)\right)\int_{-\infty}^{+\infty} v_0^2 d\tau +$$

$$+\frac{1}{2}(-a_1(\varphi,t)\varphi' + b_1(\varphi,t)u_0(\varphi,t) + b_0(\varphi,t)u_1(\varphi,t))\int_{-\infty}^{+\infty}\frac{\partial v_0^2}{\partial \tau}d\tau +$$

$$+\frac{1}{2}\left(-a_{0x}(\varphi,t)\varphi'\frac{\partial v_0}{\partial t} + b_{0x}(\varphi,t)u_0(\varphi,t) + b_0(\varphi,t)u_{0x}(\varphi,t)\right)\int_{-\infty}^{+\infty}\tau\frac{\partial v_0^2}{\partial \tau}d\tau +$$

$$+\frac{1}{3}b_1(\varphi,t)\int_{-\infty}^{+\infty}\frac{\partial v_0^3}{\partial \tau}d\tau + \frac{1}{3}b_{0x}(\varphi,t)\int_{-\infty}^{+\infty}\tau\frac{\partial v_0^3}{\partial \tau}d\tau.$$

According to the properties of the function $v_0(t,\tau)$, we find

$$\int_{-\infty}^{+\infty} F_1(t,\tau)v_0(t,\tau)d\tau = 12a_0(\varphi,t)\frac{d}{dt}\left(\frac{A^2(\varphi,t)}{b_0^2(\varphi,t)\sqrt{A(\varphi,t)}}\right) +$$

$$+12\left[a_{0x}(\varphi,t)\varphi' - b_{0x}(\varphi,t)u_0(\varphi,t) + b_0(\varphi,t)u_{0x}(\varphi,t)\right]\times$$

$$\times\frac{A^2(\varphi,t)}{b_0^2(\varphi,t)\sqrt{A(\varphi,t)}} + \frac{96}{5}b_{0x}(\varphi,t)\frac{A^3(\varphi,t)}{b_0^3(\varphi,t)\sqrt{A(\varphi,t)}},$$

which gives the differential equation for the function $\varphi(t)$ in the following form:

$$15a_0(\varphi,t)b_0(\varphi,t)\frac{d}{dt}A(\varphi,t) + \left[(10a_{0x}(\varphi,t)b_0(\varphi,t) - 36a_0(\varphi,t)b_{0x}(\varphi,t))\varphi' +\right.$$

$$\left.+10b_0^2(\varphi,t)u_{0x}(\varphi,t) + 3\left(b_0^2(\varphi,t)\right)_x u_0(\varphi,t) - 20a_0(\varphi,t)b_{0t}(\varphi,t)\right]A(\varphi,t) = 0. \quad [2.31]$$

Relation [2.31] is the nonlinear second-order differential equation with variable coefficients, which is not explicitly integrable in general. The problem regarding the existence and uniqueness of its solution, as well as the existence of its global solutions, needs to be studied in each case with the given functions $a_0(x,t)$, $b_0(x,t)$. It should be noted that since the coefficients of this equation are smooth functions, the equation satisfies conditions of the theorem on the existence and uniqueness of a solution to the Cauchy problem. Consequently, its solution exists in some neighborhood of the initial point under enough general conditions on $a_0(x,t)$, $b_0(x,t)$. Thus, equation [2.31] is considered to have a solution $\varphi = \varphi(t)$, defined on an interval $[0;T]$, where $T > 0$ is real.

In addition to orthogonality conditions [2.26], given necessary and sufficient conditions for the existence of a solution to equations [2.25] in the space G_1, there are also other necessary and sufficient conditions that ensure the existence of a solution to these equations in the space G_1^0. This is an important aspect, since belonging terms of the singular part to the space G_1^0 produces the soliton properties of the asymptotic solution being constructed.

The following lemma is true.

LEMMA 2.3.– *If* $F_j(t,\tau) \in G_1^0$, $j = \overline{1,N}$, *and condition [2.26] takes place, then the solution* $v_j(t,\tau)$, $j = \overline{1,N}$, *to equation [2.25] belongs to the space* G_1^0 *if and only if the following condition is satisfied:*

$$\lim_{\tau \to -\infty} \Phi_j(t,\tau) = 0, \quad j = \overline{1,N}. \qquad [2.32]$$

The proof of Lemma 2.3 is carried out according to the properties of the spaces G_1, G_1^0 in a similar way to proving Lemma 2.2.

Now, let us find out the conditions for the coefficients of equation [2.2] under which the solution $v_1(t,\tau)$ of equation [2.25] belongs to the space G_1^0. To do this, consider the expression for the function $\Phi_1(t,\tau)$. We have

$$\Phi_1(t,\tau) = 6\left[a_0(\varphi,t)\frac{d}{dt}\left(\frac{\sqrt{A(\varphi,t)}}{b_0(\varphi,t)}\right) + \sqrt{A(\varphi,t)}u_{0x} + \right.$$

$$\left. + \frac{\sqrt{A(\varphi,t)}}{b_0(\varphi,t)}\left[a_{0x}(\varphi,t)\varphi' - \left(b_0(\varphi,t)u_0(\varphi,t)\right)_x \right] + \right.$$

$$+\frac{A(\varphi,t)\sqrt{A(\varphi,t)}}{b_0^2(\varphi,t)}b_{0x}(\varphi,t)\Bigg]\times\left(1-\tanh\left(\frac{\sqrt{A(\varphi,t)}}{2}(\tau+c_0(t))\right)\right)+$$

$$-\left[\frac{3}{2}\frac{a_0(\varphi,t)}{b_0(\varphi,t)}(c_0(t)+\tau)\frac{d}{dt}A(\varphi,t)+3\frac{A(\varphi,t)}{b_0(\varphi,t)}a_0(\varphi,t)c_0'(t)-\right.$$

$$-\frac{3A(\varphi,t)}{b_0(\varphi,t)}\left[a_1(\varphi,t)\varphi'-\left(b_0(\varphi,t)u_0(\varphi,t)\right)_x\right]+$$

$$\left.+3\tau\frac{A(\varphi,t)}{b_0(\varphi,t)}\left[a_{0x}(\varphi,t)\varphi'-\left(b_0(\varphi,t)u_0(\varphi,t)\right)_x\right]\right]\times$$

$$\times\cosh^{-2}\left(\frac{\sqrt{A(\varphi,t)}}{2}(\tau+c_0(t))\right)+\frac{9}{2}\frac{A^2(\varphi,t)}{b_0^2(\varphi,t)}\left[b_{0x}(\varphi,t)\tau+b_1(\varphi,t)\right]\times$$

$$\times\cosh^{-4}\left(\frac{\sqrt{A(\varphi,t)}}{2}(\tau+c_0(t))\right)-3\frac{A(\varphi,t)\sqrt{A(\varphi,t)}}{b_0^2(\varphi,t)}b_{0x}(\varphi,t)\times$$

$$\times\tanh\left(\frac{\sqrt{A(\varphi,t)}}{2}(\tau+c_0(t))\right)\cosh^{-2}\left(\frac{\sqrt{A(\varphi,t)}}{2}(\tau+c_0(t))\right),$$

where $\varphi=\varphi(t)$.

It is obvious that the condition $\lim_{\tau\to-\infty}\Phi_1(t,\tau)=0$ can be rewritten as

$$a_0(\varphi,t)\frac{d}{dt}\left(\frac{\sqrt{A(\varphi,t)}}{b_0(\varphi,t)}\right)+\sqrt{A(\varphi,t)}\,u_{0x}+$$

$$+\frac{\sqrt{A(\varphi,t)}}{b_0(\varphi,t)}\left[a_{0x}(\varphi,t)\varphi'-\left(b_0(\varphi,t)u_0(\varphi,t)\right)_x\right]+\frac{A(\varphi,t)\sqrt{A(\varphi,t)}}{b_0^2(\varphi,t)}b_{0x}(\varphi,t)=0,\qquad[2.33]$$

that yields certain relations for the coefficients $a_0(x,t)$, $b_0(x,t)$.

These conditions have a rather general form. In particular, for a rather interesting and meaningful zero background case $u_0(x,t) \equiv 0$ under assumptions $a_0(x,t) = a_0(x),\ b_0(x,t) = b_0(x)$, condition [2.33] is written as

$$a_0(\varphi)\frac{d}{dt}\left(\frac{\sqrt{A(\varphi,t)}}{b_0(\varphi)}\right) + \frac{\sqrt{A(\varphi,t)}}{b_0(\varphi)}a_0'(\varphi)\varphi' + \frac{A(\varphi,t)\sqrt{A(\varphi,t)}}{b_0^2(\varphi)}b_0'(\varphi) = 0\,.$$

After simplification, this relation takes the form

$$a_0(\varphi)b_0(\varphi)\frac{d}{dt}A(\varphi,t) + 2A(\varphi,t)\left[2a_0(\varphi)b_0'(\varphi) - a_0'(\varphi)b_0(\varphi)\right]\varphi' = 0. \qquad [2.34]$$

Conditions [2.31] and [2.34] are compatible, for example, in a particular case

$$12a_0(x)b_{0x}(x) = 5a_{0x}(x)b_0(x)\,.$$

The last equality gives us the condition

$$a_0^5(x) = c_1 b_0^{12}(x)\,, \qquad [2.35]$$

where $c_1 \neq 0$ is a real.

In this case, the differential equation for the function $\varphi = \varphi(t)$ takes the following simple form:

$$\left(a_0(\varphi)\right)^{2/3}\frac{d\varphi}{dt} = c_1\,. \qquad [2.36]$$

It can be analyzed using the qualitative theory of differential equations (Samoilenko and Kaplun 2003).

As a result, we can formulate the following lemma.

LEMMA 2.4.– *Let the coefficients* $a_0(x,t) = a_0(x)$, $b_0(x,t) = b_0(x)$ *of equation [2.2] satisfy condition [2.35] and the function* $\varphi = \varphi(t)$, $t \in [0;T]$, *be a solution to equation [2.36] for a real* $c_1 \neq 0$. *Then, the solution* $v_1(t,\tau)$ *to equation [2.25] for* $u_0(x,t) \equiv 0$ *belongs to the space* G_1^0.

Finally, we write the general solution to equation [2.18] in explicit form. Since $\ker L_1 = \{v_{0\tau}\}$, where the function $v_0(t,\tau)$ is given by formula [2.21], and then another solution to the homogeneous equation $L_1 v = 0$ linearly independent with $v_{0\tau}(t,\tau)$ is represented by the following expression:

$$v(t,\tau) = v_{0\tau} \int_{\tau_0}^{\tau} v_{0\tau}^{-2}(t,\tau)d\tau .$$

Thus, a general solution to inhomogeneous equation [2.18] can be found via the arbitrary constant variation method, which is given by the formula

$$v_j(t,\tau) = \left(\int_{\tau_0}^{\tau} \Phi_j(t,\tau_1) v_{0\tau}(t,\tau_1)d\tau_1 + c_1 \right) v_{0\tau}(t,\tau) \int_{\tau_0}^{\tau} v_0^{-2}(t,\tau_1)d\tau_1 -$$

$$- \left(\int_{\tau_0}^{\tau} \Phi_j(t,\tau_1) v_{0\tau}(t,\tau_1)d\tau_1 \int_{\tau_0}^{\tau_1} v_{0\tau}^{-2}(t,\xi)d\xi + c_2 \right) v_{0\tau}(t,\tau), \qquad [2.37]$$

where c_1 and c_2 are arbitrary constants.

2.6.4. *Prolongation of the singular terms from the discontinuity curve*

Now, let us move on to the problem of defining the functions $V_j(x,t,\tau)$, $j = \overline{0,N}$. They are obtained through the analytical prolongation of the functions $v_j(t,\tau)$, $j = \overline{0,N}$, from the curve Γ. The prolongation is done in the following way. Since the function $v_0(t,\tau)$ belongs to the space G_1^0 according to Definition 2.1, we can put

$$V_0(x,t,\tau) = v_0(t,\tau) . \qquad [2.38]$$

With presentation [2.27], the functions $V_j(x,t,\tau)$, $j = \overline{1,N}$, can be written as

$$V_j(x,t,\tau) = u_j^-(x,t)\eta_j(t,\tau) + \psi_j(t,\tau), \qquad [2.39]$$

where $u_j^-(x,t)$, $j=\overline{1,N}$, are solutions to the Cauchy problems

$$\Lambda u_j^-(x,t) = f_j^-(x,t), \quad j=\overline{1,N}, \tag{2.40}$$

$$u_j^-(x,t)\Big|_{\Gamma} = v_j(t), \quad j=\overline{1,N}. \tag{2.41}$$

Equation [2.40] is deduced from asymptotic representation [2.8] after substituting formula [2.39] into [2.8] and calculating limit as $\tau \to -\infty$ as well as condition [2.41], found from relations [2.39] and [2.27] by calculating limit as $\tau \to -\infty$.

The differential operator Λ is given by the formula

$$\Lambda = a_0(x,t)\frac{\partial}{\partial t} + b_0(x,t)u_0(x,t)\frac{\partial}{\partial x} + b_0(x,t)u_{0x}(x,t). \tag{2.42}$$

For example, for the right-side functions, we have

$$f_1^-(x,t) = 0,$$

$$f_2^-(x,t) = -a_1(x,t)\frac{\partial u_1^-}{\partial t} - b_1(x,t)u_1^-\frac{\partial u_0}{\partial x} - b_0(x,t)u_1^-\frac{\partial u_1}{\partial x} -$$

$$-b_0(x,t)u_1\frac{\partial u_1^-}{\partial x} - b_2(x,t)u_1^-\frac{\partial u_1^-}{\partial x} - b_1(x,t)u_0\frac{\partial u_1^-}{\partial x}.$$

Let us consider the Cauchy problem [2.40] and [2.41] in detail. Since the curve Γ is transversal to the characteristics of the operator Λ accordingly condition $A(\varphi,t) > 0$ for all $t \in [0;T]$, then the Cauchy problem is well-posed. Thus, for enough small μ in domain $\Omega_\mu(\Gamma) = \{(x;t) \in R \times [0;T] : |x-\varphi(t)| < 2\mu\}$, the Cauchy problem [2.40] and [2.41] has a solution $u_j^-(x,t) \in C^{(\infty)}(\Omega_\mu(\Gamma))$, $j=\overline{1,N}$. Moreover, this solution can be found by means of the method of characteristics. This procedure is given later.

Therefore, we assume that the functions $u_j^-(x,t)$, $j=\overline{1,N}$, are known, and prolongation of the functions $v_j(t,\tau)$, $j=\overline{0,N}$, from the curve Γ into the domain $\Omega_\mu(\Gamma)$ is constructed.

This prolongation provides us with the asymptotic one-phase soliton-like solution [2.5] to equation [2.2]. It satisfies asymptotic relation [2.8] as $\tau \to -\infty$ with better accuracy. For example, the first constructed asymptotic approximation $Y_1(x,t,\varepsilon)$ satisfies equality [2.8] with an accuracy $O\left(\varepsilon^2\right)$, while it satisfies [2.8] with an accuracy $O\left(\varepsilon^3\right)$ as $\tau \to -\infty$.

Now, we will find the explicit solution of the Cauchy problem [2.40] and [2.41] under the assumption

$$|a_0(x,t)| + |u_0(x,t)b_0(x,t)| \neq 0, \quad (x,t) \in R \times [0;T].$$

At the beginning, we construct a general solution to equation [2.40] by applying the method of characteristics. We need to find the first integrals to the system of differential equations

$$\frac{dt}{a_0(x,t)} = \frac{dx}{b_0(x,t)u_0(x,t)} = \frac{du_j^-}{f_j^-(x,t) - b_0(x,t)u_{0x}(x,t)}. \qquad [2.43]$$

A solution to equation [2.40] is implicitly written as

$$\Phi_j(\xi(x,t), \zeta(x,t,u_j^-)) = 0, j = \overline{1, N}, \qquad [2.44]$$

where $\xi(x,t)$ is the first integral to the nonlinear equation

$$\frac{dx}{dt} = \frac{b_0(x,t)u_0(x,t)}{a_0(x,t)}, \text{w}$$

and $\zeta(x,t,u_j^-)$ is the first integral to the linear equation

$$\frac{du_j^-}{dx} = \frac{f_j^-(x,t) - b_0(x,t)u_{0x}(x,t)}{b_0(x,t)u_0(x,t)}.$$

The function $\zeta(x,t,u_j^-)$ is explicitly determined by the following formula:

$$\zeta(x,t,u_j^-) = \left[u_j^- - B(x,t)\right]e^{A(x,t)}$$

with

$$A(x,t) = \int_{x_0}^{x} \frac{u_{0x}(\tau, g(\tau,\xi(x,t)))}{u_0(\tau, g(\tau,\xi(x,t)))} d\tau \ ,$$

$$B(x,t) = \int_{x_0}^{x} \frac{e^{A(\tau,t)-A(x,t)} f_j^-(\tau, g(\tau,\xi(x,t)))}{u_0(\tau, g(\tau,\xi(x,t)))b_0(\tau, g(\tau,\xi(x,t)))} d\tau \ .$$

Here, the function $g(x,c_1)$ is such that $\xi(x,g(x,c_1)) \equiv c_1$, i.e. it is implicitly defined from equality $\xi(x,t) = c_1$ with a real c_1.

The function $\Phi_j(\varphi,\psi) \in C^\infty(\Xi;\mathrm{R})$ in formula [2.44] is arbitrary. Its whole derivative with respect to the variable u does not equal zero for all (x,t,u_j^-) from the domain of definition of the mapping $(x,t,u_j^-) \to (\varphi,\psi)$.

We use the notation $\varphi = \varphi(x,t)$, $\psi = \psi(x,t,u_j^-)$ and assume that there exists at least a point $(\varphi_0,\psi_0) \in \Xi$ such that $\Phi_j(\varphi_0,\psi_0) = 0$.

To obtain a solution to the Cauchy problem [2.40] and [2.41], we need to find such a function $\Phi_j(\varphi,\psi) \in C^\infty(\Xi;\mathrm{R})$ that implicitly defines a solution to equation [2.40] that satisfies the Cauchy condition [2.41]. To do this, we consider the first integrals of the system [2.43] $\xi(x,t) = c_1$, $\zeta(x,t,u_j^-) = c_2$ under the Cauchy condition [2.41]. Thus, we come to the relations

$$\xi(\varphi(t),t) = c_1, \quad \left[v_j(t) - B(\varphi(t),t)\right]e^{A(\varphi(t),t)} = c_2 .$$

Using the inverse mappings $t = h_1^{-1}(c_1)$, $t = h_2^{-1}(c_2)$, where according to the definition

$$h_1(t) = \xi(\varphi(t),t)\,, \ h_2(t) = \left[v_j - B(\varphi(t),t)\right]e^{A(\varphi(t),t)},$$

the solution $u_j^-(x,t)$, $j=\overline{1,N}$, to the Cauchy problem [2.40] and [2.41] is implicitly represented by the formula

$$\xi(x,t) = h_1\left(h_2^{-1}\left(\left[u_j^- - B(x,t)\right]e^{A(x,t)}\right)\right). \qquad [2.45]$$

It follows from relation [2.45] that the domain of existence of the solution to the Cauchy problem [2.40] and [2.41] depends on the domains of definitions of inverse mappings $\theta \to h_1^{-1}(\theta)$, $\theta_2 \to h_2^{-1}(\theta)$.

Thus, we have completed constructing the asymptotic one-phase soliton-like solution [2.5].

REMARK 2.1.– *The procedure for constructing the solution* $u_j^-(x,t)$, $j=\overline{1,N}$, *to the Cauchy problem [2.40] and [2.41] described above looks somewhat complicated and cumbersome, but in the case of specifically given coefficients of equation [2.2] and some regular part, the corresponding formulas are greatly simplified. For example, if the main term of the regular part of the asymptotic [2.5] has a form* $u_0(x,t) \equiv 0$, *then the Cauchy problem [2.40] and [2.41] for the function* $u_1^-(x,t)$ *is written as*

$$\frac{du_1^-}{dt} = 0, \qquad [2.46]$$

$$u_1^-(x,t)\Big|_{\Gamma} = v_1(t). \qquad [2.47]$$

Thus, instead of [2.45] as $j=1$, *its solution is given by a simple formula*

$$u_1^-(x,t) = \rho(x), \quad x \in R$$

where $\rho = \rho(x)$ *is an arbitrary function satisfying relation* $\rho(\varphi(t)) = v_1(t)$, $t \in [0;T]$.

2.7. Justification of the algorithm

Now, we consider the problem of the justification of the constructed asymptotics. To do this, we demonstrate that the obtained function $Y_N(x,t,\varepsilon)$, $N \geq 0$, is indeed an asymptotic solution to problem [2.2], that is, the function satisfies this equation with a certain asymptotic accuracy. The following statement is true.

THEOREM 2.2.– *Let the following assumptions be true:*

1) *the functions* $a_k(x,t), b_k(x,t), c_k(x,t) \in C^\infty(R\times[0;T])$, $k=\overline{0,N}$, *and* $a_0(x,t)b_0(x,t) \neq 0$ *for all* $(x,t)\in R\times[0;T]$;

2) *the inequality* $A(\varphi,t)>0$ *holds for the function* $\varphi(t)$, $t\in[0;T]$, *which is a solution to equation [2.31];*

3) *the functions* $F_j(t,\tau)\in G_1^0$, $j=\overline{1,N}$, *and they satisfy conditions [2.26];*

4) *the functions* $\Phi_j(t,\tau)$, $j=\overline{1,N}$, *satisfy conditions [2.32].*

Then, the asymptotic one-phase soliton-like solution to equation [2.2] is written as

$$u_N(x,t,\varepsilon)=\begin{cases} Y_N^-(x,t,\varepsilon), & (x,t)\in D^- / \Omega_\mu(\Gamma), \\ Y_N(x,t,\varepsilon), & (x,t)\in \Omega_\mu(\Gamma), \\ Y_N^+(x,t,\varepsilon), & (x,t)\in D^+ / \Omega_\mu(\Gamma), \end{cases} \qquad [2.48]$$

where

$$Y_N^-(x,t,\varepsilon)=\sum_{j=0}^{N}\varepsilon^j u_j(x,t),\quad (x,t)\in D^-=\{(x,t)\in R\times[0;T]: x-\varphi(t)<0\},$$

$$Y_N(x,t,\varepsilon)=\sum_{j=0}^{N}\varepsilon^j\left[u_j(x,t)+v_j(t,\tau)\right],\quad \tau=\frac{x-\varphi(t)}{\varepsilon},\quad (x,t)\in\Omega_\mu(\Gamma),$$

$$Y_N^+(x,t,\varepsilon)=\sum_{j=0}^{N}\varepsilon^j u_j(x,t),\quad (x,t)\in D^+=\{(x,t)\in R\times[0;T]: x-\varphi(t)>0\}.$$

Asymptotic solution [2.48] satisfies equation [2.2] with an accuracy $O\left(\varepsilon^N\right)$ *on the set* $R\times[0;T]$. *Moreover, function [2.48] satisfies equation [2.2] with an accuracy* $O\left(\varepsilon^{N+1}\right)$ *as* $\tau\to\pm\infty$.

PROOF.– First, we consider the domain $\Omega_{\mu}(\Gamma)$. After substituting the function [2.48] into equation [2.2] in this domain, we have

$$\varepsilon^2\left(\sum_{j=0}^{N}\varepsilon^j\frac{\partial^3 u_j}{\partial x^3}+\sum_{j=0}^{N}\varepsilon^j\frac{\partial^3 V_j}{\partial x^3}+\frac{3}{\varepsilon}\sum_{j=0}^{N}\varepsilon^j\frac{\partial^3 V_j}{\partial x^2\partial\tau}+\right.$$

$$\left.+\frac{3}{\varepsilon^2}\sum_{j=0}^{N}\varepsilon^j\frac{\partial^3 V_j}{\partial x\partial\tau^2}+\frac{1}{\varepsilon^3}\sum_{j=0}^{N}\varepsilon^j\frac{\partial^3 V_j}{\partial\tau^3}\right)=$$

$$=\sum_{k=0}^{N}\varepsilon^k a_k(x,t)\left(\sum_{j=0}^{N}\varepsilon^j\frac{\partial u_j}{\partial t}+\sum_{j=0}^{N}\varepsilon^j\frac{\partial V_j}{\partial t}-\frac{1}{\varepsilon}\varphi'(t)\sum_{j=0}^{N}\varepsilon^j\frac{\partial V_j}{\partial\tau}\right)+$$

$$+\sum_{k=0}^{N}\varepsilon^k b_k(x,t)\left(\sum_{j=0}^{N}\varepsilon^j u_j+\sum_{j=0}^{N}\varepsilon^j V_j\right)\times$$

$$\times\left(\sum_{j=0}^{N}\varepsilon^j\frac{\partial u_j}{\partial x}+\sum_{j=0}^{N}\varepsilon^j\frac{\partial V_j}{\partial x}+\frac{1}{\varepsilon}\sum_{j=0}^{N}\varepsilon^j\frac{\partial V_j}{\partial\tau}\right). \qquad [2.49]$$

Taking into account relation [2.49] in μ-neighborhood of the curve Γ and equations for the terms of the regular and the singular parts of the asymptotics, we conclude that in order to prove Theorem 2.2, it is necessary to estimate the following function:

$$g_N(x,t,\varepsilon)=\left[a_0(x,t)-\sum_{k=0}^{N}\frac{(\varepsilon\tau)^k}{k!}\frac{\partial^k a_0(\varphi,t)}{\partial x^k}+\right.$$

$$\left.+\varepsilon\left(a_1(x,t)-\sum_{k=0}^{N-1}\frac{(\varepsilon\tau)^k}{k!}\frac{\partial^k a_1(\varphi,t)}{\partial x^k}\right)+\ldots+\varepsilon^N\left(a_N(x,t)-a_N(\varphi,t)\right)\right]\times$$

$$\times\left(\sum_{j=0}^{N}\varepsilon^j\frac{\partial V_j}{\partial t}-\frac{1}{\varepsilon}\varphi'(t)\sum_{j=0}^{N}\varepsilon^j\frac{\partial V_j}{\partial\tau}\right)+\left[b_0(x,t)-\sum_{k=0}^{N}\frac{(\varepsilon\tau)^k}{k!}\frac{\partial^k b_0(\varphi,t)}{\partial x^k}+\right.$$

$$+\varepsilon\left(b_1(x,t)-\sum_{k=0}^{N-1}\frac{(\varepsilon\tau)^k}{k!}\frac{\partial^k b_1(\varphi,t)}{\partial x^k}\right)+\ldots+\varepsilon^N\left(b_N(x,t)-b_N(\varphi,t)\right)\Bigg]\times$$

$$\times\left[u_0(x,t)-\sum_{k=0}^{N}\frac{(\varepsilon\tau)^k}{k!}\frac{\partial^k u_0(\varphi,t)}{\partial x^k}+\varepsilon\left(u_1(x,t)-\sum_{k=0}^{N-1}\frac{(\varepsilon\tau)^k}{k!}\frac{\partial^k u_1(\varphi,t)}{\partial x^k}\right)+\right.$$

$$\left.+\ldots+\varepsilon^N\left(u_N(x,t)-u_N(\varphi,t)\right)\right]\times\left(\sum_{j=0}^{N}\varepsilon^j\frac{\partial V_j}{\partial x}+\frac{1}{\varepsilon}\sum_{j=0}^{N}\varepsilon^j\frac{\partial V_j}{\partial\tau}\right)+$$

$$+\left[b_0(x,t)-\sum_{k=0}^{N}\frac{(\varepsilon\tau)^k}{k!}\frac{\partial^k b_0(\varphi,t)}{\partial x^k}+\varepsilon\left(b_1(x,t)-\sum_{k=0}^{N-1}\frac{(\varepsilon\tau)^k}{k!}\frac{\partial^k b_1(\varphi,t)}{\partial x^k}\right)+\right.$$

$$\left.+\ldots+\varepsilon^N\left(b_N(x,t)-b_N(\varphi,t)\right)\right]\times\left[\frac{\partial u_0(x,t)}{\partial x}-\sum_{k=0}^{N}\frac{(\varepsilon\tau)^k}{k!}\frac{\partial^{k+1}u_0(\varphi,t)}{\partial x^{k+1}}+\right.$$

$$\left.+\varepsilon\left(\frac{\partial u_1(x,t)}{\partial x}-\sum_{k=0}^{N-1}\frac{(\varepsilon\tau)^k}{k!}\frac{\partial^{k+1}u_1(\varphi,t)}{\partial x^{k+1}}\right)+\ldots+\varepsilon^{N+1}\frac{\partial u_{N+1}(x,t)}{\partial x}\right]\times$$

$$\times\sum_{j=0}^{N}\varepsilon^j V_j+\left[b_0(x,t)-\sum_{k=0}^{N}\frac{(\varepsilon\tau)^k}{k!}\frac{\partial^k b_0(\varphi,t)}{\partial x^k}+\right.$$

$$+\varepsilon\left(b_1(x,t)-\sum_{k=0}^{N-1}\frac{(\varepsilon\tau)^k}{k!}\frac{\partial^k b_1(\varphi,t)}{\partial x^k}\right)+$$

$$\left.+\ldots+\varepsilon^N\left(b_N(x,t)-b_N(\varphi,t)\right)\right]\times\sum_{j=0}^{N}\varepsilon^j V_j\left(\sum_{j=0}^{N}\varepsilon^j\frac{\partial V_j}{\partial x}+\frac{1}{\varepsilon}\sum_{j=0}^{N}\varepsilon^j\frac{\partial V_j}{\partial\tau}\right),$$

where $\varphi=\varphi(t)$, because the other terms of this equality have the order $O\left(\varepsilon^{N+1}\right)$ or higher with respect to a small parameter.

Since the functions $a_j(x,t)$, $j=\overline{0,N}$, are infinitely differentiable, for all $(x,t)\in\Omega_\mu(\Gamma)$ the following inequality

$$\left|a_j(x,t)-\sum_{k=0}^{N-j}\frac{(\varepsilon\tau)^k}{k!}\frac{\partial^k a_j(\varphi,t)}{\partial x^k}\right|\le C_j\left|(\varepsilon\tau)^{N+1-j}\right|,\quad \varphi=\varphi(t),$$

is true. Here, the constant C_j, $j=\overline{0,N}$, depends on a compact $K\subset R\times[0;T]$.

So, taking into account infinite differentiability of the functions $b_j(x,t)$, $V_j(x,t,\tau)$, $u_j(x,t)$, $j=\overline{0,N}$, we can write the analogical inequality for these functions. It follows that the function $g_N(x,t,\varepsilon)$ satisfies asymptotic relation $g_N(x,t,\varepsilon)=O\left(\varepsilon^N\right)$, $\varepsilon\to 0$, for all $(x,t)\in\Omega_\mu(\Gamma)$.

Now, we consider the set $D^+ / \Omega_\mu(\Gamma)$. Since for every $k\in N$, any function $f(x,t,\tau)\in G_1$ satisfies the equality

$$\left|(\varepsilon\tau)^k f(x,t,\tau)\right|\le\frac{C_k(f)}{\tau^k},\quad k\ge 1,\quad \tau\ge\tau_0,$$

where the constant $C_k(f)$, $k\in N$, depends on the compact $K\subset R\times[0;T]$, and $\tau_0>0$ is a value, then for all $k\in N$ and $\tau>0$, we have

$$\left|(\varepsilon\tau)^k f(x,t,\tau)\right|\le\varepsilon^k C_k(f).$$

Hence, we obtain the statement of Theorem 2.2 for the domain $D^+ / \Omega_\mu(\Gamma)$.

Analogously, we can prove the statement of Theorem 2.2 for the case of the domain $D^- / \Omega_\mu(\Gamma)$, when $u_N(x,t,\varepsilon)=Y_N^-(x,t,\varepsilon)$.

Theorem 2.2 is proved.

Theorem 2.2 provides us with sufficient conditions of existence of the asymptotic one-phase soliton-like solutions to equation [2.2]. All terms of the singular part of this asymptotics are quickly decreasing functions with respect to the phase variable τ, i.e. all singular terms belong to the space G_1^0.

The procedure for constructing asymptotic solutions described above is rather general. This makes it possible to obtain asymptotic solutions of equations [2.2] with singular terms having different properties, as the phase variable τ tends to infinity. In particular, these terms may be step-like, when ones belong to the space G_1 but do not belong to the space G_1^0. These solutions are called the asymptotic step-wise solutions (Lyashko et al. 2020a, 2020b). The justification for constructing this kind of asymptotic solution is given by the following theorem.

THEOREM 2.3.– *Let the Assumptions 1–3 of Theorem 2.2 be satisfied and the Cauchy problem [2.40] and [2.41] have a solution on the set* $\{(x,t) \in R \times [0;T] : x - \varphi(t) \le 0\}$. *Then, the asymptotic solution to equation [2.2] is written as*

$$u_N(x,t,\varepsilon) = \begin{cases} Y_N^-(x,t,\varepsilon), & (x,t) \in D^- / \Omega_\mu(\Gamma), \\ Y_N(x,t,\varepsilon), & (x,t) \in \Omega_\mu(\Gamma), \\ Y_N^+(x,t,\varepsilon), & (x,t) \in D^+ / \Omega_\mu(\Gamma), \end{cases} \qquad [2.50]$$

where

$$Y_N^-(x,t,\varepsilon) = u_0(x,t) + \sum_{j=1}^{N} \varepsilon^j \left[u_j(x,t) + u_j^-(x,t) \right], \quad (x,t) \in D^-,$$

$$Y_N(x,t,\varepsilon) = \sum_{j=0}^{N} \varepsilon^j \left[u_j(x,t) + V_j(x,t,\tau) \right], \quad \tau = \frac{x - \varphi(t)}{\varepsilon}, \quad (x,t) \in \Omega_\mu(\Gamma),$$

$$Y_N^+(x,t,\varepsilon) = \sum_{j=0}^{N} \varepsilon^j u_j(x,t), \quad (x,t) \in D^+$$

Asymptotic solution [2.50] satisfies equation [2.2] with an accuracy $O\left(\varepsilon^N\right)$ *on the set* $R \times [0;T]$. *Moreover, function [2.50] satisfies equation [2.2] with an accuracy* $O\left(\varepsilon^{N+1}\right)$ *as* $\tau \to \pm\infty$.

PROOF.– Since asymptotic solutions [2.48] and [2.50] coincide on the sets $\Omega_\mu(\Gamma)$ and $D^+ / \Omega_\mu(\Gamma)$, we just need to consider the set $D^- / \Omega_\mu(\Gamma)$. In this case, the functions $V_j(x,t,\tau)$, $j=\overline{1,N}$, are written as

$$V_j(x,t,\tau) = u_j^-(x,t)\eta_j(t,\tau) + \psi_j(t,\tau),$$

where $\eta_j(t,\tau) \in G_1$, $\lim\limits_{\tau\to-\infty} \eta_j(t,\tau) = 1$, $\psi_j(t,\tau) \in G_1^0$.

For any $k \in N$ and $\tau \le -\tau_0$, we have

$$\left|V_j(x,t,\tau) - u_j^-(x,t)\right| \le \frac{C_{1k}}{|\tau|^k}, \quad \left|\frac{\partial}{\partial \tau} V_j(x,t,\tau)\right| \le \frac{C_k(V_j)}{|\tau|^k},$$

where C_{1k} and $C_k(V_j)$ are constants that depend only on the compact set $K \subset R \times [0;T]$.

Thus, analogously to proving Theorem 2.2 for the set $D^+ / \Omega_\mu(\Gamma)$, we obtain that asymptotic solution [2.50] satisfies equation [2.2] with an accuracy $O\left(\varepsilon^{N+1}\right)$ as $\tau \to \pm\infty$.

Theorem 2.3 is proved.

2.8. Discussion and conclusion

While studying wave processes in media with variable characteristics, equations of the hydrodynamic type with variable coefficients are used. Among them, the Korteweg–de Vries equation should be mentioned. It has many different solutions with important properties. For the Korteweg–de Vries equation with variable coefficients, the problem of the deformation of its soliton solutions arises. The problem is successfully solved if the medium is characterized by a small dispersion. In this case, we obtain the KdV equation with a singular perturbation.

In this section, using the WKB technique, the asymptotic one-phase soliton-like solution to the KdV equation with variable coefficients and singular perturbation is constructed. The procedure for constructing asymptotic solutions is rather general. It makes it possible to obtain asymptotic solutions of equations [2.2] with singular terms having different properties as the phase variable τ tends to infinity. These terms may be step-wise, when ones belong to the space G_1 but do not belong to the space G_1^0. These solutions are called the asymptotic step-wise solutions and are important for studying various applied problems.

The algorithm for finding the asymptotic one-phase soliton-like solution is presented in detail. Theorems on its justification are proved.

The effectiveness of the method for constructing the asymptotic solutions is demonstrated in Chapter 3 while studying the singular perturbed Korteweg–de Vries equation with certain variable coefficients.

Note that it is advisable to compare the obtained asymptotic soliton-like solutions to the KdV equation with its numerical solutions found through the finite difference method by applying effective iterative methods for solving systems of nonlinear algebraic equations (Vakai et al. 1991).

2.9. Acknowledgments

Samoilenko Yuliia was supported by Programme PAUSE and the ANR, project N. ANR-22-PAUK-0038-01, in Claude Bernard Lyon I University, and she is very appreciative and grateful for the support and hospitality.

2.10. References

Dobrokhotov, S.Y. and Maslov, V.P. (1981). Finite-zone, almost-periodic solutions in WKB approximations, *J. Math. Sci.*, 16, 1433–1487.

Lax, P.D. and Levermore, C.D. (1983a). The small dispersion limit of the Korteweg–de Vries equation I. *Comm. Pure Appl. Math.*, 36(3), 253–290.

Lax, P.D. and Levermore, C.D. (1983b). The small dispersion limit of the Korteweg–de Vries equation II. *Comm. Pure Appl. Math.*, 36(5), 571–593.

Lax, P.D. and Levermore, C.D. (1983c). The small dispersion limit of the Korteweg–de Vries equation III. *Comm. Pure Appl. Math.*, 36(6), 809–829.

Lyashko, S.I., Samoilenko, V.H., Samoilenko, Yu.I., Lyashko, N.I. (2020a). Step-wise asymptotic solutions to the Korteweg–de Vries equation with variable coefficients and a small parameter at the higher order derivative. *Cyberne. Syst. Anal.*, 56(6), 934–942.

Lyashko, S.I., Samoilenko, V.H., Samoilenko, Yu.I., Gapyak, I.V., Lyashko, N.I., Orlova, M.S. (2020b). Global asymptotic step-type solutions to singularly perturbed Korteweg–de Vries equation with variable coefficients. *J. Autom. Inform. Sci.*, 52(9), 27–38.

Maslov, V.P. and Omel'yanov, G.A. (1981). Asymptotic soliton-like solutions of equations with small dispersion. *Russ. Math. Surveys*, 36(3), 73–149.

Maslov, V.P. and Omel'yanov, G.A. (2001). *Geometric Asymptotics for PDE I*. American Mathematical Society, Providence.

Miura, R.M. and Kruskal, M.D. (1974). Application of nonlinear WKB-method to the Korteweg–de Vries equation. *SIAM Appl. Math.*, 26(2), 376–395.

Sagdeev, R.Z., Usikov, D.A., Zaslavsky, G.M. (1988). *Nonlinear Physics. From the Pendulum to Turbulence and Chaos*. Harwood Academic Publishers, Chur.

Samoilenko, V.H. and Kaplun, Yu.I. (2003). Extendability of solutions of differential equations to a singular set. *Ukr. Math. J.*, 55(3), 450–455.

Samoilenko, V.H. and Samoilenko, Yu.I. (2005). Asymptotic expansions for one-phase soliton-type solutions of the Korteweg–de Vries equation with variable coefficients. *Ukr. Math. J.*, 57(1), 132–148.

Samoilenko, V.H. and Samoilenko, Yu.I. (2007). Asymptotic solutions of the Cauchy problem for the singularly perturbed Korteweg–de Vries equation with variable coefficients. *Ukr. Math. J.*, 59(1), 126–139.

Samoilenko, V.H. and Samoilenko, Yu.I. (2008). Asymptotic two-phase soliton-like solutions of the singularly perturbed Korteweg–de Vries equation with variable coefficients. *Ukr. Math. J.*, 60(3), 449–461.

Samoilenko, V.H. and Samoilenko, Yu.I. (2010). Boundary layer method and the Hugoniot type condition for the Korteweg–de Vries equation. *Vesnik of Brest University. Series 4. Phys.. Math.*, 2, 111–129.

Samoilenko, V.H. and Samoilenko, Yu.I. (2012a). Asymptotic m-phase soliton-type solutions of a singularly perturbed Korteweg–de Vries equation with variable coefficients. *Ukr. Math. J.*, 64(7), 1109–1127.

Samoilenko, V.H. and Samoilenko, Yu.I. (2012b). Asymptotic m-phase soliton-type solutions of a singularly perturbed Korteweg–de Vries equation with variable coefficients II. *Ukr. Math. J.*, 64(8), 1241–1259.

Samoilenko, V.H. and Samoilenko, Yu.I. (2012c). Existence of a solution to the inhomogeneous equation with the one-dimensional Schrödinger operator in the space of quickly decreasing functions. *J. Math. Sci.*, 187(1), 70–76.

Samoilenko, V.H. and Samoilenko, Yu.I. (2014). Asymptotic multiphase Σ-solutions to the singularly perturbed Korteweg–de Vries equation with variable coefficients. *J. Math. Sci.*, 200(3), 358–373.

Samoilenko, V.H. and Samoilenko, Yu.I. (2019). Asymptotic soliton-like solutions to the singularly perturbed Benjamin–Bona–Mahony equation with variable coefficients. *J. Math. Phys.*, 60(1), 011501-1–011501-13.

Vakai, E.S., Kivva, S.L., Mistetskii, G.E., Stelya, O.B. (1991). Solution method for nonlinear parabolic equations. *J. Sov. Math.*, 54(2), 781–786.

3

Asymptotic Analysis of the vcKdV Equation with Weak Singularity

Sergiy LYASHKO[1], Valerii SAMOILENKO[1], Yuliia SAMOILENKO[2,3] and Nataliia LYASHKO[4]

[1] *Taras Shevchenko National University of Kyiv, Ukraine*

[2] *National Technical University of Ukraine Igor Sikorsky Kyiv Polytechnic Institute, Ukraine*

[3] *Claude Bernard Lyon 1 University, France*

[4] *V.M. Glushkov Institute of Cybernetics, National Academy of Sciences of Ukraine, Kyiv, Ukraine*

In this chapter, we analyze the asymptotic soliton-like solutions of the variable coefficients Korteweg–de Vries (vcKdV) equation with weak singularity, describing wave processes in inhomogeneous media with variable characteristics and a small dispersion. Attention is paid to the description of an algorithm for constructing asymptotic solutions and the subsequent analysis of these solutions. This approach provides us with effective tools for studying the influence of model parameters on the properties of the dynamical system under consideration.

Computational Methods and Mathematical Modeling in Cyberphysics and Engineering Applications 1, coordinated by Dmitri KOROLIOUK, Sergiy LYASHKO and Nikolaos LIMNIOS.

3.1. Introduction

In the previous chapters, the problem of constructing the asymptotic one-phase soliton-like solutions to the Korteweg–de Vries equation with variable coefficients and singular perturbation

$$\varepsilon^n u_{xxx} = a(x,t,\varepsilon)u_t + b(x,t,\varepsilon)uu_x, \quad n \in N, \qquad [3.1]$$

is considered in detail. Here, $\varepsilon > 0$ is a small parameter, and the functions $a(x,t,\varepsilon)$, $b(x,t,\varepsilon)$, $(x,t) \in R\times[0;T]$, are smooth and written as asymptotic (according to Poincaré) expansions

$$a(x,t,\varepsilon) = \sum_{k=0}^{N} \varepsilon^k a_k(x,t) + O\left(\varepsilon^{N+1}\right), \; b(x,t,\varepsilon) = \sum_{k=0}^{N} \varepsilon^k b_k(x,t) + O\left(\varepsilon^{N+1}\right), \qquad [3.2]$$

where the condition $a_0(x,t)b_0(x,t) \neq 0, \; (x,t) \in R\times[0;T]$ holds.

To solve this problem using the nonlinear WKB method, an algorithm for constructing these solutions is proposed and justified. While implementing this algorithm, new rather interesting mathematical problems arise. The main problem is related to the existence of the global asymptotic one-phase soliton-like solutions to equation [3.1]. It is connected with the problem of global solvability of the second-order ordinary differential equation for the phase function $x = \varphi(t)$ defining a discontinuity curve.

The problem of global asymptotic soliton-like solutions is closely related to the physical interpretation of the constructed solutions, so let us look at this in more detail. The equation for the function $x = \varphi(t)$ is nonlinear and has enough general form. As a rule, its solution exists in some neighborhood of the initial point and does not exist globally. On the other hand, the interval of existence of the solution depends both on its coefficients and on the initial data. Below, we will consider examples of an equation of the form [3.1] with some given coefficients, for which the equation for the discontinuity curve has a global solution.

Below, we study the vcKdV equation with weak singularity of the following form:

$$\varepsilon u_{xxx} = a(x,t,\varepsilon)u_t + b(x,t,\varepsilon)uu_x, \qquad [3.3]$$

where the functions $a(x,t,\varepsilon)$, $b(x,t,\varepsilon)$ are given by equation [3.2].

We construct the asymptotic one-phase soliton-like solutions to equation [3.3] and consider their properties. It should be mentioned that the corresponding expansions contain fraction power of a small parameter ε that is a specific feature of the problem under consideration.

3.2. The asymptotic soliton-like solutions

Here, as in the previous chapter, we use notation $S = S(R)$ for the space of quickly decreasing functions, definitions of the asymptotic soliton-like solution and the spaces $G_1 = G_1(R\times[0;T]\times R)$ and $G_1^0(R\times[0;T]\times R) \subset G_1$ (Maslov and Omel'yanov 2001; Samoilenko and Samoilenko 2005).

In Samoilenko and Samoilenko (2005), it was found that the degree of a small parameter at the highest derivative significantly affects the algorithm for constructing an asymptotic solution. In particular, the structure of the equation for the terms of the asymptotic expansions has been modified.

This leads to new qualitative properties of the singular part of the asymptotics, in particular, to a change in the accuracy with which the constructed asymptotic solution satisfies the equation under consideration. At the same time, the basic ideas and steps of constructing the asymptotic soliton-like solutions in the case of a weak singularity for equation [3.3] are similar to the case $n = 2$ considered in the previous chapter. Below, we first present general results on the structure, construction and properties of the asymptotic soliton-like solution of the Korteweg–de Vries equation [3.3], and then demonstrate these results for equation [3.1] with certain coefficients, for which we obtained the first-order asymptotic step-wise solutions (Lyashko et al. 2021b), as well as the first-order asymptotic soliton type solutions (Samoilenko et al. 2019).

The asymptotic one-phase soliton-like solution to equation [3.3] is written as

$$u(x,t,\tau,\varepsilon) = \sum_{j=0}^{2N} \varepsilon^{\frac{j}{2}} \left[u_j(x,t) + V_j(x,t,\tau) \right] + O\left(\varepsilon^{N+\frac{1}{2}} \right), \quad \tau = \frac{x - \varphi(t)}{\sqrt{\varepsilon}}. \qquad [3.4]$$

Without resorting to a detailed description of the algorithm for constructing an asymptotic soliton-like solution for the Korteweg–de Vries equation [3.3], we note that the regular part of the asymptotic solution [3.4] is determined from some system of the first-order partial differential equations. Its main term is determined from a quasi-linear homogeneous equation with variable coefficients, while the other terms are determined from linear inhomogeneous equations with variable coefficients. These equations can be integrated by the method of characteristics (Fritz 1982). Therefore, without loss of generality, it is assumed that the regular terms $u_j(x,t)$, $j = 0,1,\ldots$, of the asymptotic solution are already defined.

Let us pay more attention to the problem on constructing the singular part of the asymptotics since this part reflects the soliton properties of the searched solution. We outline only the general scheme of its solution, since it is similar to the case $n = 2$. We focus on its features that arise for equations with a weak singular perturbation. As in the case $n = 2$, solving the problem contains two steps. First, we consider equations for the terms of the singular part of the asymptotic on the discontinuity curve. These equations are given as

$$\frac{\partial^3 v_0}{\partial \tau^3} + a_0(\varphi,t)\varphi'(t)\frac{\partial v_0}{\partial \tau} - b_0(\varphi,t)\left(u_0(\varphi,t)\frac{\partial v_0}{\partial \tau} + v_0\frac{\partial v_0}{\partial \tau}\right) = 0, \qquad [3.5]$$

$$\frac{\partial^3 v_j}{\partial \tau^3} + a_0(\varphi,t)\frac{\partial v_j}{\partial \tau}\varphi'(t) - b_0(\varphi,t)\left(u_0(\varphi,t)\frac{\partial v_j}{\partial \tau} + \frac{\partial}{\partial \tau}(v_0 v_j)\right) = F_j(t,\tau), \qquad [3.6]$$

where $F_j(t,\tau) = F_j(x,t,\tau)\big|_{x=\varphi(t)}$, $j = \overline{1,N}$.

In particular, the function $F_1(t,\tau)$ has the following form:

$$F_1(t,\tau) = a_0(\varphi,t)\frac{\partial v_0}{\partial t} + \Big[-\tau a_{0x}(\varphi,t)\varphi'\frac{\partial v_0}{\partial t} + \tau b_{0x}(\varphi,t)u_0(\varphi,t) +$$

$$+\tau b_0(\varphi,t)u_{0x}(\varphi,t) + \tau b_{0x}(\varphi,t)v_0 + b_0(\varphi,t)u_1(\varphi,t)\Big]\frac{\partial v_0}{\partial \tau} + b_0(\varphi,t)u_{0x}(\varphi,t)v_0.$$

where $v_0(t,\tau)$ is explicitly defined from equation [3.5] and is given by the formula

$$v_0(t,\tau) = -\frac{3A(\varphi,t)}{b_0(\varphi,t)}\cosh^{-2}\left(\frac{\sqrt{A(\varphi,t)}}{2}(\tau + C_0(t))\right). \qquad [3.7]$$

Here, we use the notation

$$A(\varphi,t) = -a_0(\varphi,t)\varphi'(t) + b_0(\varphi,t)u_0(\varphi,t), \quad t \in [0;T],$$

$C_0 = C_0(t)$ is a constant of integration, the function $\varphi = \varphi(t)$ defines the discontinuity curve and satisfies the equation

$$15a_0(\varphi,t)b_0(\varphi,t)\frac{d}{dt}A(\varphi,t) + \Big[\big(10a_{0x}(\varphi,t)b_0(\varphi,t) - 36a_0(\varphi,t)b_{0x}(\varphi,t)\big)\varphi' +$$

$$+10b_0^2(\varphi,t)u_{0x}(\varphi,t) + 3\left(b_0^2(\varphi,t)\right)_x u_0(\varphi,t) - 20a_0(\varphi,t)b_{0t}(\varphi,t)\Big]A(\varphi,t) = 0. \qquad [3.8]$$

Let us denote

$$\Phi_j(t,\tau) = \int_{-\infty}^{\tau} F_j(t,\zeta)d\zeta + E_j(t), \qquad [3.9]$$

where the constant of integration $E_j(t)$ satisfies the condition

$$\lim_{\tau \to +\infty} \Phi_j(t,\tau) = 0.$$

The following statement is true.

THEOREM 3.1.– *Let the following assumptions be satisfied:*

1) *the functions* $a_k(x,t), b_k(x,t), c_k(x,t) \in C^\infty(R \times [0;T])$, $k = \overline{0,N}$, *and* $a_0(x,t)b_0(x,t) \neq 0$ *for all* $(x,t) \in R \times [0;T]$;

2) *the inequality* $A(\varphi,t) > 0$ *is true for the function* $\varphi(t)$, $t \in [0;T]$, *that satisfies equation [3.8]*;

3) *the functions* $F_j(t,\tau) \in G_1^0$, $j = \overline{1,N}$, *and the conditions*

$$\int_{-\infty}^{+\infty} F_j(t,\tau)\, v_0(t,\tau)\, d\tau = 0, \quad j = \overline{1,2N}, \qquad [3.10]$$

are true;

4) *the functions* $\Phi_j(t,\tau)$, $j=\overline{1,2N}$, *satisfy conditions*

$$\lim_{\tau\to-\infty}\Phi_j(t,\tau)=0,\quad j=\overline{1,2N}. \qquad [3.11]$$

Then, the asymptotic one-phase soliton-like solution to equation [3.3] can be written as

$$u_N(x,t,\varepsilon)=\begin{cases} Y_N^-(x,t,\varepsilon), & (x,t)\in D^-/\Omega_\mu(\Gamma),\\ Y_N(x,t,\varepsilon), & (x,t)\in \Omega_\mu(\Gamma),\\ Y_N^+(x,t,\varepsilon), & (x,t)\in D^+/\Omega_\mu(\Gamma),\end{cases} \qquad [3.12]$$

where

$$Y_N^-(x,t,\varepsilon)=\sum_{j=0}^{2N}\varepsilon^{\frac{j}{2}}u_j(x,t),\quad (x,t)\in D^-=\{(x,t)\in R\times[0;T]:x-\varphi(t)<0\},$$

$$Y_N(x,t,\varepsilon)=\sum_{j=0}^{2N}\varepsilon^{\frac{j}{2}}\left[u_j(x,t)+v_j(t,\tau)\right],\quad \tau=\frac{x-\varphi(t)}{\sqrt{\varepsilon}},\quad (x,t)\in\Omega_\mu(\Gamma),$$

$$Y_N^+(x,t,\varepsilon)=\sum_{j=0}^{2N}\varepsilon^{\frac{j}{2}}u_j(x,t),\quad (x,t)\in D^+=\{(x,t)\in R\times[0;T]:x-\varphi(t)>0\},$$

and satisfies equation [3.3] with an accuracy $O\left(\varepsilon^N\right)$ *on the set* $R\times[0;T]$. *Moreover, function [3.12] satisfies equation [3.3] with an accuracy* $O\left(\varepsilon^{N+\frac{1}{2}}\right)$ *as* $\tau\to\pm\infty$.

Theorem 3.1 is obtained under conditions [3.11] for the function $\Phi_j(t,\tau)$, $j=\overline{1,2N}$. If the conditions are not fulfilled, then the structure of the singular part of the asymptotics changes. To construct it, we need to prolong the functions $v_j(t,\tau)$, $j=\overline{1,2N}$, from the discontinuity curve in a special way.

Let us denote

$$v_j(t) = \left[a_0(\varphi,t)\varphi' - b_0(\varphi,t)u_0(\varphi,t)\right]^{-1} \lim_{\tau\to-\infty} \Phi_j(t,\tau).$$

While extending the functions $v_j(t,\tau)$, $j=\overline{1,2N}$, we use the representation of a solution of equation [3.6] in the space G_1 according to the following formula (Samoilenko and Samoilenko 2005):

$$v_j(t,\tau) = v_j(t)\eta_j(t,\tau) + \psi_j(t,\tau), \quad j=\overline{1,2N}, \qquad [3.13]$$

where $\psi_j(t,\tau)\in G_1^0$, $j=\overline{1,2N}$, the function $\eta_j(t,\tau)\in G_1$ satisfies the relation $\lim_{\tau\to-\infty} \eta_j(t,\tau)=1$, and value t is a parameter.

Prolongation of the functions $v_j(t,\tau)$, $j=\overline{0,2N}$, is realized in such a way

$$V_0(x,t,\tau) = v_0(t,\tau), \qquad [3.14]$$

$$V_j(x,t,\tau) = u_j^-(x,t)\eta(t,\tau) + \psi_j(t,\tau), \qquad [3.15]$$

where $u_j^-(x,t)$, $j=\overline{1,2N}$, is a solution to the Cauchy problem

$$\Lambda u_j^-(x,t) = f_j^-(x,t), \quad j=\overline{1,N}, \qquad [3.16]$$

$$\left. u_j^-(x,t)\right|_{\Gamma} = v_j(t), \quad j=\overline{1,N}. \qquad [3.17]$$

Equation [3.16] is deduced from equation [3.3] after substituting representation [3.15] into equation [3.3] and calculating the limit as $\tau\to-\infty$. Condition [3.17] is obtained from relations [3.15], [3.13] by means of calculating the limit as $\tau\to-\infty$.

Thus, we have the following statement.

THEOREM 3.2.– *Let assumptions 1–3 of Theorem 3.1 hold and the Cauchy problem [3.16], [3.17] have a solution on the set* $\{(x,t)\in R\times[0;T] : x-\varphi(t)\le 0\}$. *Then, the asymptotic solution to equation [3.3] is written as*

$$u_N(x,t,\varepsilon)=\begin{cases} Y_N^-(x,t,\varepsilon), & (x,t)\in D^- / \Omega_\mu(\Gamma), \\ Y_N(x,t,\varepsilon), & (x,t)\in \Omega_\mu(\Gamma), \\ Y_N^+(x,t,\varepsilon), & (x,t)\in D^+ / \Omega_\mu(\Gamma), \end{cases} \quad [3.18]$$

where

$$Y_N^-(x,t,\varepsilon)=u_0(x,t)+\sum_{j=1}^{2N}\varepsilon^{\frac{j}{2}}\left[u_j(x,t)+u_j^-(x,t)\right],\quad (x,t)\in D^-,$$

$$Y_N(x,t,\varepsilon)=\sum_{j=0}^{2N}\varepsilon^{\frac{j}{2}}\left[u_j(x,t)+V_j(x,t,\tau)\right],\quad \tau=\frac{x-\varphi(t)}{\sqrt{\varepsilon}},\quad (x,t)\in \Omega_\mu(\Gamma),$$

$$Y_N^+(x,t,\varepsilon)=\sum_{j=0}^{2N}\varepsilon^{\frac{j}{2}}u_j(x,t),\quad (x,t)\in D^+,$$

and satisfies the equation with an accuracy $O\left(\varepsilon^N\right)$ *on the set* $R\times[0;T]$.

Moreover, function [3.18] satisfies equation [3.3] with an accuracy $O\left(\varepsilon^{N+\frac{1}{2}}\right)$ *as* $\tau\to\pm\infty$.

The proofs of Theorems 3.1 and 3.2 are like those of Theorems 2.2 and 2.3 in the previous chapter.

Theorem 3.1 gives justification of the algorithm of constructing the asymptotic soliton-like solutions to equation [3.3], while Theorem 3.2 provides us with justification of the algorithm of constructing the asymptotic step-wise solutions.

3.3. The examples of the asymptotic soliton-like solutions

As mentioned above, two types of the asymptotic soliton-like solutions can be distinguished, namely solutions of soliton type and step-wise solutions. Recall that solutions of the first kind are solutions, where all terms of the singular part are quickly decreasing functions with respect to the phase variable τ. The solutions of the other kind are the solutions, where the main singular term is a quickly decreasing function with respect to the variable τ and the other singular terms are step-wise solutions (Lyashko et al. 2021b).

3.3.1. *The asymptotic step-wise solutions*

Now, we consider the example of an asymptotic step-wise solution for the vc Korteweg–de Vries equation [3.1] of the following form:

$$\varepsilon u_{xxx} = \left(t^2 + 1\right) u_t + \left(x^2 + 1\right)^{5/12} u u_x. \qquad [3.19]$$

Thus, coefficients of equation [3.3] are written as

$$a(x,t,\varepsilon) = a_0(x,t) = t^2 + 1, \; b(x,t,\varepsilon) = b_0(x,t) = \left(x^2 + 1\right)^{5/12}.$$

Let us suppose a zero background. In this case

$$A(\varphi(t),t) = -\left(t^2 + 1\right)\varphi',$$

and equation [3.8] for the phase function $\varphi = \varphi(t)$ is given by means of

$$\left(\varphi^2 + 1\right)\frac{d}{dt}\left(\left(t^2 + 1\right)\varphi'\right) = 2\left(t^2 + 1\right)\varphi'^2\varphi. \qquad [3.20]$$

It yields relation

$$\frac{1}{\left(t^2 + 1\right)\varphi'}\frac{d}{dt}\left(\left(t^2 + 1\right)\varphi'\right) = \frac{1}{\left(\varphi^2 + 1\right)}\frac{d}{d\varphi}\left(\varphi^2 + 1\right),$$

that is equivalent to equation

$$\frac{d}{dt}\ln\left(\left(t^2 + 1\right)\varphi'\right) = \frac{d}{d\varphi}\ln\left(\varphi^2 + 1\right). \qquad [3.21]$$

By integrating equation [3.21], we get

$$\left(t^2+1\right)\varphi' = -\alpha\left(\varphi^2+1\right) \tag{3.22}$$

i.e. $\arctan\varphi = -\alpha \arctan t + \beta$, where α, β are real constants.

From the inequality $A(\varphi,t) > 0$ and relation [3.22], we get condition $\alpha > 0$. Considering the initial condition $\varphi(0) = 0$, we find $\beta = 0$. Then, for the phase function $\varphi = \varphi(t)$, the equality $\varphi = -\tan(\alpha \arctan t)$ is true. Hence, for $\alpha > 1$, the phase function is defined only in a neighborhood of the point $t = 0$, i.e. on a finite interval, and for $\alpha \in (0;1]$, the phase function is defined for all $t \in R$, i.e. the solution to equation [3.20] is global.

Let us put $\alpha = 1$. Then, the solution to equation [3.20] is function $\varphi = -t$ and the main term of the singular part of the asymptotics is written as

$$V_0(t,\tau) = -3\frac{A(\varphi,t)}{b_0(\varphi,t)}\cosh^{-2}\left(\frac{\sqrt{A(\varphi,t)}}{2}\tau\right) =$$

$$= -3\left(t^2+1\right)^{7/12}\cosh^{-2}\left(\frac{\sqrt{t^2+1}}{2}\frac{x+t}{\sqrt{\varepsilon}}\right). \tag{3.23}$$

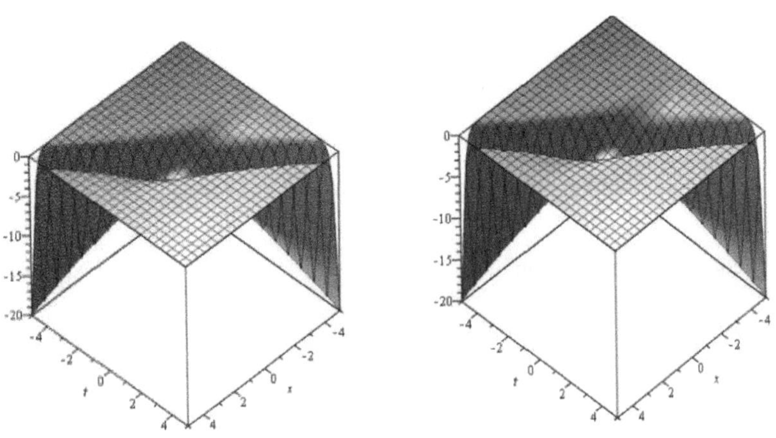

Figure 3.1. *Solution [3.23] as $\varepsilon = 0.5$ and $\varepsilon = 0.25$. For a color version of this figure, see www.iste.co.uk/koroliouk/computational.zip*

To find the term $V_1(x,t,\tau)$, consider the functions $F_1(t,\tau)$ and $\Phi_1(t,\tau)$. We have

$$F_1(t,\tau) = \left(t^2+1\right)\frac{\partial V_0}{\partial t} + \frac{5}{6}\tau\varphi(t)\left(\varphi^2+1\right)^{-7/12} V_0 \frac{\partial V_0}{\partial \tau},$$

where the function $V_0(t,\tau)$ is given by formula [3.23], and

$$\Phi_1(t,\tau) = t\left(t^2+1\right)^{1/12}\Big[4\left(\tanh\vartheta(t,\tau)-1\right)-$$

$$-\left(6t\vartheta(t,\tau) - \frac{5}{2}\tanh\vartheta(t,\tau) + \frac{15}{2}\vartheta(t,\tau)\cosh^{-2}\vartheta(t,\tau)\right)\cosh^{-2}\vartheta(t,\tau)\Big].$$

Here, we use the following notation:

$$\vartheta(t,\tau) = \frac{\sqrt{t^2+1}}{2}\tau.$$

From formula [2.37] in section 2.6 for $j=1$, we obtain

$$v_1(t,\tau) = \Bigg[\frac{t}{\left(t^2+1\right)^{5/12}}\left(\frac{83}{4} - \frac{20}{3}\cosh^{-2}\vartheta(t,\tau) -\right.$$

$$\left.-\frac{105}{4}\cosh^{-4}\vartheta(t,\tau) + 6\tanh\vartheta(t,\tau)\right)\tau - \frac{t}{\left(t^2+1\right)^{11/12}}\times$$

$$\times\left[12 - 15\cosh^{-2}\vartheta(t,\tau) + \left(\frac{23}{4} - \frac{105}{4}\cosh^{-2}\vartheta(t,\tau) + \frac{1}{16}\cosh^{-4}\vartheta(t,\tau)\right)\right]\times$$

$$\times\tanh\vartheta(t,\tau) - t\left(t^2+1\right)^{1/12}\left(\frac{3}{4} + \frac{315}{16}\tau^2\cosh^{-4}\vartheta(t,\tau)\right)\times\tanh\vartheta(t,\tau) -$$

$$-\frac{105t}{\left(t^2+1\right)^{11/12}}\tanh\vartheta(t,\tau)\ln\vartheta(t,\tau)\Bigg]\cosh^{-2}\vartheta(t,\tau)-$$

$$-\frac{4t}{\left(t^2+1\right)^{11/12}}\left(\tanh\vartheta(t,\tau)-1\right).$$

Using this expression for the function $v_1(t,\tau)$, we can specify the coefficients of the function in the form [3.13]. We have

$$v_1(t,\tau)=-\frac{1}{t^2+1}\lim_{\tau\to-\infty}\Phi_1(t,\tau)=8t\left(t^2+1\right)^{-11/12}, \quad [3.24]$$

$$\eta_1(t,\tau)=-\frac{1}{2}\tanh\vartheta(t,\tau)+\frac{1}{2}, \quad [3.25]$$

$$\psi_1(t,\tau)=v_1(t,\tau)-v_1(t,\tau)\eta_1(t,\tau). \quad [3.26]$$

Since the function $v_1(t,\tau)\notin G_1^0$, we apply Theorem 3.2. With this purpose, consider the Cauchy problem [3.16] and [3.17], which is written as

$$\left(t^2+1\right)\frac{\partial u_1^-}{\partial t}=0,$$

$$u_1^-(x,t)\Big|_{\Gamma}=v_1(t),$$

where the discontinuity curve Γ is given by the equation $x=-t$.

Considering formula [3.25] and the function $\eta_1(t,\tau)$ in the initial condition, we find the solution to the Cauchy problem as

$$u_1^-(x,t)=-8x\left(x^2+1\right)^{-11/12}.$$

Thus, the first-order approximate solution to equation [3.19] is written as

$$Y_1(x,t,\tau) = V_0(t,\tau) + \sqrt{\varepsilon} V_1(x,t,\tau),$$

where $V_0(t,\tau)$ is given by formula [3.23] and function $V_1(x,t,\tau)$ is written as

$$V_1(x,t,\tau) = u_1^-(x,t)\eta_1(t,\tau) + \psi_1(t,\tau).$$

A comparison of the corresponding graphs for the main and first terms demonstrates that for a suitable description of the qualitative properties of the asymptotic step-wise solutions of the considered Korteweg–de Vries equation, it is necessary to construct at least the first asymptotic approximation.

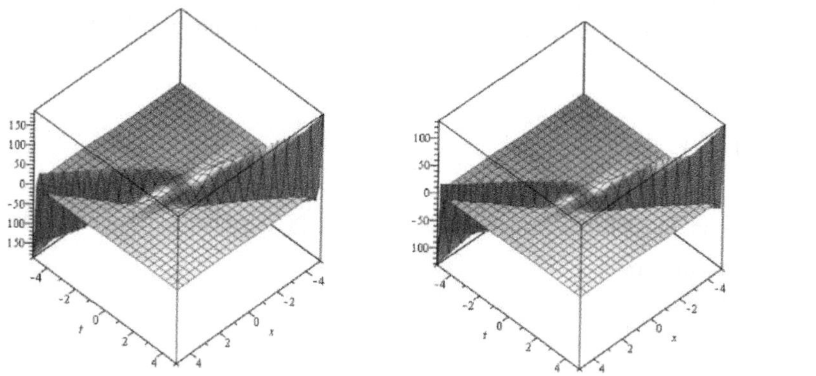

Figure 3.2. *The first-order approximations as $\varepsilon = 0.5$ and $\varepsilon = 0.25$. For a color version of this figure, see www.iste.co.uk/koroliouk/computational.zip*

3.3.2. *The asymptotic solutions of soliton type*

Let equation [3.1] be written as follows (Samoilenko et al. 2019):

$$\varepsilon^2 u_{xxx} = -\left(x^2+1\right)^{3/2} u_t + \left(x^2+1\right)^{5/8} u u_x. \qquad [3.27]$$

In this case, the degree of the singularity $n = 2$ and coefficients of the equation contain only the main terms given as

$$a(x,t,\varepsilon) = a_0(x,t) = -\left(x^2+1\right)^{3/2},$$

$$b(x,t,\varepsilon) = b_0(x,t) = \left(x^2+1\right)^{5/8}. \qquad [3.28]$$

The asymptotic soliton-like solution to equation [3.27] is written as (Samoilenko and Samoilenko 2005)

$$u(x,t,\varepsilon) = \sum_{j=0}^{N} \varepsilon^j u_j(x,t) + \sum_{j=0}^{N} \varepsilon^j V_j(x,t,\tau) + O\left(\varepsilon^{N+1}\right),$$

$$\tau = \frac{x-\varphi(t)}{\varepsilon}. \qquad [3.29]$$

Since the soliton properties of the sought asymptotic solutions are reflected by its singular terms, to simplify the calculation, we consider the zero background case. So, we have

$$A(\varphi,t) = \left(\varphi^2+1\right)^{3/2} \varphi'$$

and equation [3.8] for the phase function $\varphi = \varphi(t)$ is reduced to the first-order differential equation

$$\left(\varphi^2+1\right)\frac{d\varphi}{dt} = \rho, \qquad [3.30]$$

where $\rho \neq 0$ is an arbitrary real.

Then, $A(\varphi,t) = \rho\sqrt{\varphi^2+1}$ and the condition $A(\varphi,t) > 0$ results in the constant ρ being positive.

Consider equation [3.30] for $\rho > 0$ and the interval of existence of its solution. The general solution to the equation is implicitly given by the formula

$$\frac{1}{3}\varphi^3 + \varphi + C = \rho t,$$

that provides a one-to-one correspondence between values $\varphi \in R$ and $t \in R$ for all constants of integration C.

This means that the solution to equation [3.30] for $\rho > 0$ exists for all values $t \in R$, i.e. it is a global solution (Samoilenko et al. 2019).

The solution to equation [3.30] for $\rho = 1$ and the initial condition $\varphi(0) = 0$ is written through the formula

$$\varphi(t) = \sqrt[3]{\frac{3}{2}t + \sqrt{1 + \frac{9}{4}t^2}} + \sqrt[3]{\frac{3}{2}t - \sqrt{1 + \frac{9}{4}t^2}}, \qquad [3.31]$$

where $t \in R$.

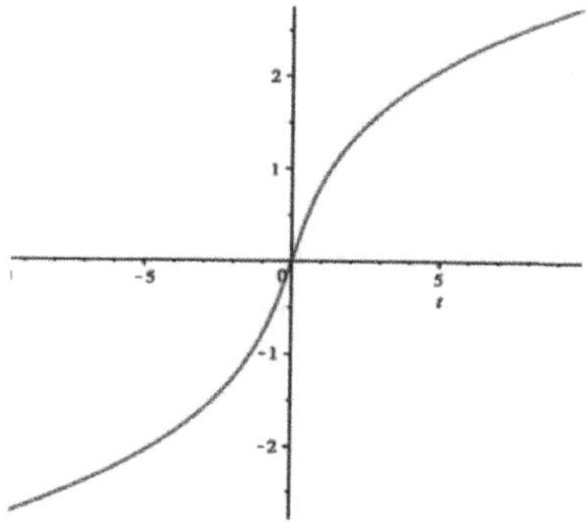

Figure 3.3. *The discontinuity curve $\varphi = \varphi(t)$. For a color version of this figure, see www.iste.co.uk/koroliouk/computational.zip*

According to the above-described technique, the main term of the singular part of the asymptotic solution is found as follows (Samoilenko et al. 2019):

$$V_0(t,\tau) = -3\left(\varphi^2(t) + 1\right)^{-1/8} \cosh^{-2}\left(\sqrt{\left(\varphi^2(t) + 1\right)^{3/2}} \varphi'(t) \frac{\tau}{2}\right). \qquad [3.32]$$

As a result, we finally get the first-order approximation of solution [3.29] to equation [3.19] in the form

$$Y_0(x,t,\varepsilon) = -3\left(\varphi^2(t) + 1\right)^{-1/8} \cosh^{-2}\left(\left(\varphi^2(t) + 1\right)^{-1/4} \frac{x - \varphi(t)}{\varepsilon}\right), \qquad [3.33]$$

where function $\varphi = \varphi(t)$ is given with expression [3.31].

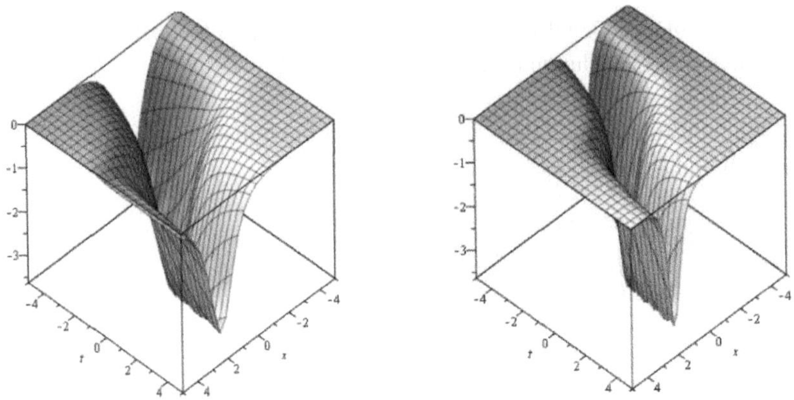

Figure 3.4. *Solution [3.33] as $\varepsilon = 1$ and $\varepsilon = 0.5$. For a color version of this figure, see www.iste.co.uk/korolIouk/computational.zip*

Considering the coefficients to equation [3.19], we obtain that the function $\Phi_1(t,\tau)$ satisfies condition [3.11] and can be represented in the following form:

$$\Phi_1(t,\tau) = \frac{\varphi(t)}{\left(\varphi^2(t)+1\right)^{5/8}}\left[\frac{21}{2}\tau + \frac{45}{8}\tau\cosh^{-2}\left(\left(\varphi^2(t)+1\right)^{1/4}\frac{\tau}{2}\right) - \right.$$

$$\left. -\frac{15}{4}\left(\varphi^2(t)+1\right)^{-1/4}\tanh\left(\left(\varphi^2(t)+1\right)^{1/4}\frac{\tau}{2}\right)\right]\cosh^{-2}\left(\left(\varphi^2(t)+1\right)^{1/4}\frac{\tau}{2}\right).$$

It is clear that $\Phi_1(t,\tau)$ is a quickly decreasing function with respect to phase variable τ, i.e. $\Phi_1(t,\tau)\in G_1^0$. Then, $v_1(t,\tau)\in G_1^0$ and accordingly, Lemma 2 (Lyashko et al. 2021b), result in $V_1(x,t,\tau) = v_1(t,\tau)$. So, in the case under consideration from formula [2.37] in section 2.6 as $j=1$, we find the first term of the singular part of the searched asymptotic solution as follows:

$$V_1(x,t,\tau) = \frac{15}{\left(\varphi^2(t)+1\right)^{9/8}}\frac{\tau}{2}\cosh^{-4}\psi(t,\tau)\times$$

$$\times\left[\frac{37}{4}\left(\cosh^2\psi(t,\tau) - \cosh^{-2}\psi(t,\tau)\right) - \frac{1}{2}\right]+$$

$$+\frac{3\varphi(t)}{\left(\varphi^2(t)+1\right)^{11/8}}\tanh\psi(t,\tau)\cosh^{-4}\psi(t,\tau)\times$$

$$\times\left[\frac{3}{8}+\frac{155}{8}\cosh^{-2}\psi(t,\tau)+\frac{1}{32}\cosh^{-4}\psi(t,\tau)+\frac{165}{2}\ln|\cosh\psi(t,\tau)|\right]-$$

$$-\frac{3\varphi(t)}{\left(\varphi^2(t)+1\right)^{7/8}}\left[\frac{525}{32}\tau^2\cosh^{-4}\psi(t,\tau)+\frac{5}{8}\right]\tanh\psi(t,\tau)\cosh^{-2}\psi(t,\tau),\qquad [3.34]$$

where the following notation $\psi(t,\tau)=\left(\varphi^2(t)+1\right)^{1/4}\frac{\tau}{2}$ is used.

It should be noted that the term $V_1(x,t,\tau)$ is a quickly decreasing function with respect to the phase variable τ, i.e. $V_1(x,t,\tau)\in G_1^0$.

Thus, the first-order approximate solution to equation [3.19] is written as

$$Y_1(x,t,\tau)=V_0(t,\tau)+\varepsilon V_1(x,t,\tau),\qquad [3.35]$$

where $V_0(t,\tau)$ is given by formula [3.32] and $V_1(x,t,\tau)=v_1(t,\tau)$ is represented by equation [3.34].

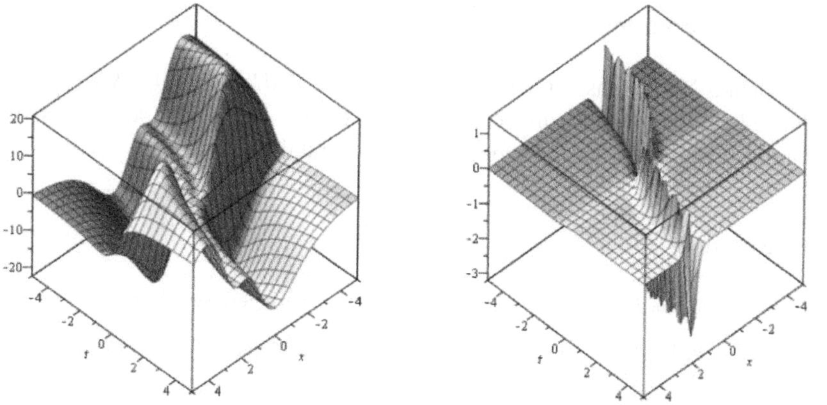

Figure 3.5. *The first-order approximation [3.35] as $\varepsilon=1$ and $\varepsilon=0.5$. For a color version of this figure, see www.iste.co.uk/koroliouk/computational.zip*

Using formulae [3.32] and [3.34], it is also possible to calculate the second-order approximation for asymptotic solution [3.19] exactly. In a similar way, we can find the higher terms of the singular part of the asymptotic solution explicitly. But to obtain terms, it is necessary to apply special packets of applied programs because of tedious formulas.

3.4. Discussion and conclusion

The present section deals with the consideration of the vcKdV equation with a weak singularity. The results regarding the structure, construction and properties of the asymptotic soliton-like solution are presented. The theorems on accuracy of the constructed asymptotic solution are given. The problem of searching for global asymptotic soliton-like solutions is pointed out.

Examples of constructing global asymptotic soliton-like solutions to the equation of the given type are also presented. The first asymptotic approximation for soliton-type solution and step-wise solution is found explicitly.

A comparison of the corresponding graphs (Figures 3.1 to 3.3, and 3.4 to 3.5) for the main and first terms of the asymptotic solution demonstrates that for a suitable description of the qualitative properties of the asymptotic soliton-like solutions of the given equation, it is necessary to construct at least the first asymptotic approximation.

3.5. Acknowledgments

Samoilenko Yuliia was supported by Programme PAUSE and the ANR, project no. ANR-22-PAUK-0038-01, in Claude Bernard Lyon 1 University, and she is very appreciative and grateful for the support and the hospitality.

3.6. References

Fritz, J. (1982). *Partial Differential Equations*, 4th edition. Springer, New York.

Lyashko, S.I. (1991). The approximate solution of a pseudoparabolic equation. *Comput. Math. Math. Phys.*, 31(12), 107–111.

Lyashko, S.I. (1995). Numerical solution of pseudoparabolic equations. *Cybernet. Systems. Anal.*, 31(5), 718–722.

Lyashko, S.I., Samoilenko, V.H., Samoilenko, Yu.I., Lyashko, N.I. (2021a). Asymptotic analysis of the Korteweg–de Vries equation by the nonlinear WKB method. *Math. Model. Comput.*, 8(3), 368–378.

Lyashko, S.I., Samoilenko, V.H., Samoilenko, Yu.I., Gapyak, I.V., Orlova, M.S. (2021b). Asymptotic step-wise solutions of the Korteweg–de Vries equation with a singular perturbation and their accuracy. *Math. Model. Comput.*, 8(3), 410–421.

Maslov, V.P. and Omel'yanov, G.A. (2001). *Geometric Asymptotics for PDE. I.* American Mathematical Society, Providence.

Samoilenko, V.H. and Samoilenko, Y.I. (2005). Asymptotic expansions for one-phase soliton-type solutions of the Korteweg–de Vries equation with variable coefficients. *Ukr. Math. J.*, 57(1), 132–148.

Samoilenko, V.H. and Samoilenko, Y.I. (2007). Asymptotic solutions of the Cauchy problem for the singularly perturbed Korteweg–de Vries equation with variable coefficients. *Ukr. Math. J.*, 59(1), 126–139.

Samoilenko, V.H. and Samoilenko, Y.I. (2008). Asymptotic two-phase soliton-like solutions of the singularly perturbed Korteweg–de Vries equation with variable coefficients. *Ukr. Math. J.*, 60(3), 449–461.

Samoilenko, V.H. and Samoilenko, Y.I. (2012a). Asymptotic *m*-phase soliton-type solutions of a singularly perturbed Korteweg–de Vries equation with variable coefficients. *Ukr. Math. J.*, 64(7), 1109–1127.

Samoilenko, V.H. and Samoilenko, Y.I. (2012b). Asymptotic *m*-phase soliton-type solutions of a singularly perturbed Korteweg-de Vries equation with variable coefficients II. *Ukr. Math. J.*, 64(8), 1241–1259.

Samoilenko, V.H. and Samoilenko, Y.I. (2019). Asymptotic soliton-like solutions to the singularly perturbed Benjamin–Bona–Mahony equation with variable coefficients. *J. Math. Phys.*, 60(1), 011501-1–011501-13.

Samoilenko, V.H., Samoilenko, Y.I., Limarchenko, V.O., Vovk, V.S., Zaitseva, K.S. (2019). Asymptotic solutions of soliton type of the Korteweg–de Vries equation with variable coefficients and singular perturbation. *Math. Model. Comput.*, 6(2), 374–385.

4

Modeling of Heterogeneous Fluid Dynamics with Phase Transitions and Porous Media

Gennadiy V. SANDRAKOV

Taras Shevchenko National University of Kyiv, Ukraine

4.1. Introduction

Mathematical modeling is an essential modern tool for solving scientific and engineering problems. Mathematical and numerical modeling plays a significant role in understanding the phenomena and processes that are investigated in modern physics and chemistry. The modeling can help to interpret and even predict new phenomena. When researching the physical and chemical processes which are in heterogeneous media at large speeds, high pressures and energies in the field of phase transitions, it is required to have accurate values of pressures, energies and velocities arising at various points of these substances at suitable times. Such values are significant in order to understand the state of media and the phase transitions that have taken place. The determination of such values is possible on the basis of modeling of nonlinear hydrodynamic processes in heterogeneous fluids.

Actually, there are no real models for processes describing the behavior of heterogeneous materials at high pressures and energies in the field of phase

Computational Methods and Mathematical Modeling in Cyberphysics and Engineering Applications 1,
coordinated by Dmitri KOROLIOUK, Sergiy LYASHKO and Nikolaos LIMNIOS.

transitions in the available literature. The problem of the models, taking into account the basic provisions of the phase transition theory, was posed in Yudovich (2003) as one of the main problems of mathematical physics. This problem is not solved here, but a possible approach will be presented for calculating the parameters of heterogeneous media at large speeds, high pressures and energies in the field of phase transitions. The approach is based on an approximation of conservation laws for masses, momentums and energies of fluids and performing volume-conserving simulations on relevant time intervals. Initially, such an approach to this problem was proposed in Sandrakov and Boyko (2011).

The significance of this problem is due to the intensive development of the branches of physics and chemistry associated with the study of dynamic processes occurring during pulsed loading and the passage of shock waves in metals, polymers and heterogeneous composites (Nigmatulin 1991). The study of such processes is necessary for the development of new technologies that use the methods of pulsed and shock loading, which make it possible to synthesize new substances. The analysis of these processes requires the development of suitable mathematical models and the construction of numerical methods and effective computational algorithms.

An alternative approach to this problem is known, which is based on the concept of multiphase continuous media. In this case, the assumptions of a multi-speed continuum and interpenetrating motions of the components are accepted when deriving multiphase equations for suitable models. In a sense, this concept means the simultaneous presence of several materials at each point of the space under consideration; further details can be found in Nigmatulin (1991). In addition, the general multiphase equations for such multi-speed continuous media are not closed and contain indefinite terms. In such equations, the terms of describing the redistribution of energies between phases during the development of the dynamics of multiphase media are not definite, for example.

In order to define the terms in specific cases, it is likely that additional physical and physicochemical experiments will be required, the implementation of which is difficult in practice. In addition, in multiphase models that take into account phase transitions in the sense of Nigmatulin (1991), it is assumed that some of the considered phases initially have zero density, which can increase at some points in space with the development of process dynamics. However, when using known computational algorithms to find equation solutions of such multiphase models, it becomes necessary to divide by density in the numerical calculation of the relevant parameters. Thus, the need to divide by a small or even zero density can lead to instability and possible incorrectness of a suitable computational algorithm. The fact of calculation instability is well known when modeling processes in rarefied media.

For the problems of diffusion and heat transfer in heterogeneous composite media, there is a rigorous mathematical approach leading to multiphase homogenization models, which was originally presented in Sandrakov (1997, 2004a, 2004b). For dynamic elasticity problems, this approach was developed in Sandrakov (1998, 2006). General results on this topic were obtained in Sandrakov (2007). In the multiphase models, the media are not considered as several materials at each point of the space simultaneously. In these results, the solutions to the relevant multiphase equations for each phase are restrictions on the corresponding phase that can interact through the interchange terms. Such interchange terms are defined explicitly, and so the problem of describing the redistribution of energies between phases does not arise when considering the dynamics of the multiphase media.

This approach is also used in the method for mathematical modeling of structurally heterogeneous materials as heterogeneous fluid dynamics, which is presented here. The problem of describing energy redistribution between phases does not arise when this method is implemented. Energy redistribution in this method is performed through the assumption that the local pressure in each phase is the same. The assumption is natural, because the coincidence of pressures is typical for mixtures of liquids and the properties of solids are close to the properties of liquids at high pressures according to Nigmatulin (1991, p. 242). The fluids are assumed to be compressible and nonviscous. Liquid inhomogeneities are considered as small particles or droplets of one liquid distributed in another (main) liquid. Thus, simulations of the main fluid with transited particle dynamics are considered, where the total number of the particles may be large enough. The particle dynamics will be modeled as in the particle-in-cell method (Harlow 1964), and in the main fluid as in the large particle method (Belotserkovskij and Davidov 1982).

Other combinations of these methods were also presented in Belotserkovskij and Davidov (1982) for modeling the problems of contact and free boundaries. However, for the energy redistribution in the book, it was assumed that the increments of the internal energies coincide. Particle methods are often used to simulate the dynamics of homogeneous fluids (Grigoriev et al. 2002). Also, these methods are relevant for modeling complex processes of plasma dynamics (Boyko et al. 2007; Bacchini et al. 2017; Dikalyuk and Kuratov 2018). Recent simulation results by the methods for complex problems of plasma dynamics are presented in Olshevsky et al. (2019), Kormann and Sonnendrücker (2021) and Wang et al.

(2021). These methods have proven useful not only for modeling complex processes, but also for visualizing the simulation results (Parreiras et al. 2021).

The particle methods have been generalized to the smoothed (or smeared) particle method to simulate a variety of fluid dynamics problems (Liu and Liu 2003; Li and Liu 2004). An overview of the developments of this method are given in Liu and Liu (2010). This method has also been applied to multiphase flow modeling in Wang et al. (2016) and developed further, for example, in Koukouvinis et al. (2018). However, in this approach, the terms of redistribution between phases are postulated from physical considerations. In this situation, it is not entirely clear how to apply this approach when modeling processes in which phase transitions are permissible.

The method presented here is a generalization of the method discussed in Sandrakov and Boyko (2011, 2012). Such a generalization was announced in Sandrakov (2019, 2020) to obtain more accurate and stable calculations.

Initially, the method was developed with the prospect of simulating processes in which phase transitions are possible. Some results of modeling by the method are presented in Boyko and Sandrakov (2013). The method is a combination of particle-in-cell and large particle methods. Therefore, it is natural to discuss the methods briefly in the next two sections. And then, we describe the presented method in the next two sections. Through this method, fluid particles move along Lagrangian trajectories in the main fluid, the motion of which is determined by the conservation laws for masses, momentums and energy. Thus, by freezing the motion of these particles, it is possible to create a method for simulating the motion of a fluid in a porous medium based on conservation laws, which is presented in the last section. The changes in the method for modeling viscous fluids are also discussed there.

4.2. The large particle method

The methods (Belotserkovskij and Davidov 1982) are based on an approximation of conservation laws for masses, momentums and energy in the integral forms

$$\int_{o(t)} \rho_t' \, d\tau = - \int_{s(t)} (\rho \mathrm{v}) \cdot \eta \, ds,$$

$$\frac{d}{dt} \int_{o(t)} \rho \mathrm{v} \, d\tau = - \int_{s(t)} p \eta \, ds, \qquad [4.1]$$

$$\frac{d}{dt}\int_{o(t)} \rho\varepsilon d\tau = -\int_{s(t)} (p\mathrm{v})\cdot\eta\, ds,$$

where $o(t)$ and $s(t)$ are the volume and surface of some Lagrangian domain in the fluid, η is an exterior normal to the domain, $p = p(\rho,\varepsilon)$ is a given equation of state and $\rho, \mathrm{v}, \varepsilon$ are the unknown density, velocity and full energy, respectively. For example, the case of a three-dimensional space may be discussed and therefore, by definition, ones have $\mathrm{v} = (u, v, w)$ for the velocity vector function of the fluid under consideration.

Euler and Lagrange approaches are used simultaneously in the large particle method for homogeneous fluid. Therefore, conservation laws for masses, momentums and energy will be used, which are equivalent (e.g. Belotserkovskij and Davidov 1982) in a certain sense (equation [4.1]), in the following differential forms:

$$\frac{\partial\rho}{\partial t} + \mathrm{div}(\rho\mathrm{v}) = 0,$$

$$\frac{\partial\rho\mathrm{v}}{\partial t} + \mathrm{div}(\rho\mathrm{v}\otimes\mathrm{v}) + \nabla p = 0, \qquad [4.2]$$

$$\frac{\partial\rho\varepsilon}{\partial t} + \mathrm{div}(\rho\mathrm{v}\varepsilon) + \mathrm{div}(p\mathrm{v}) = 0,$$

where $\mathrm{v}\otimes\mathrm{v}$ is the tensor square of vector function $\mathrm{v} = \mathrm{v}(t,x,y,z)$ and x, y, z is a point of some domain Ω of a three-dimensional space, which is filled by the fluid under consideration at time $t \in [0,T]$ for some T that is fixed.

Spatial discretization in the large particle method is defined as follows. The domain Ω that is occupied by the liquid is divided into small cells, for example, with uniform steps $\Delta x, \Delta y, \Delta z$ in the spatial variables. Each such cell, as usual, is assigned the number i, j, k, and the values of density, velocities, pressure and energy refer to the center of the cell and are denoted, respectively, as

$$\rho_{ijk}^{n},\ u_{ijk}^{n},\ v_{ijk}^{n},\ w_{ijk}^{n},\ p_{ijk}^{n},\ E_{ijk}^{n}. \qquad [4.3]$$

The subscript n in these notations indicates the number of the time step, which corresponds to the natural time discretization, since the calculations are carried out step by step with a sufficiently small time interval Δt, starting from a given initial configuration of the fluid. Thus, the values of n in the course of calculations change from 0 to $N = T / \Delta t$ under the assumption that N is an integer.

The corresponding time step from the time layer n to the time layer $n+1$ to simulate the homogeneous fluid dynamics is divided into three stages: Euler, Lagrangian and final stages in accordance with Belotserkovskij and Davidov (1982). The division is also called the splitting into physical processes of the step or the splitting of the time step into fractional steps of the transition from the layer n to the layer $n+1$. Thus, the total mass of large particles is saved at every time step of such simulation, which is mass and volume conserving. The Euler stage is the calculation of the following intermediate values of velocities and energy

$$u_{ijk}^{n+1/2}, \quad v_{ijk}^{n+1/2}, \quad w_{ijk}^{n+1/2}, \quad E_{ijk}^{n+1/2} \qquad [4.4]$$

for a cell with number i, j, k by the following formulas:

$$u_{ijk}^{n+1/2} = u_{ijk}^{n} - \frac{p_{i+1,j,k}^{n} - p_{i-1,j,k}^{n}}{2\Delta x} \frac{\Delta t}{\rho_{ijk}^{n}},$$

$$v_{ijk}^{n+1/2} = v_{ijk}^{n} - \frac{p_{i,j+1,k}^{n} - p_{i,j-1,k}^{n}}{2\Delta x} \frac{\Delta t}{\rho_{ijk}^{n}},$$

$$w_{ijk}^{n+1/2} = w_{ijk}^{n} - \frac{p_{i,j,k+1}^{n} - p_{i,j,k-1}^{n}}{2\Delta x} \frac{\Delta t}{\rho_{ijk}^{n}},$$

$$E_{ijk}^{n+1/2} = E_{ijk}^{n} - \frac{p_{i+1/2,j,k}^{n} u_{i+1/2,j,k}^{n} - p_{i-1/2,j,k}^{n} u_{i-1/2,j,k}^{n}}{\Delta x} \frac{\Delta t}{\rho_{ijk}^{n}} - \qquad [4.5]$$

$$- \frac{p_{i,j+1/2,k}^{n} v_{i,j+1/2,k}^{n} - p_{i,j-1/2,k}^{n} v_{i,j-1/2,k}^{n}}{\Delta x} \frac{\Delta t}{\rho_{ijk}^{n}} -$$

$$- \frac{p_{i,j,k+1/2}^{n} w_{i,j,k+1/2}^{n} - p_{i,j,k-1/2}^{n} w_{i,j,k-1/2}^{n}}{\Delta x} \frac{\Delta t}{\rho_{ijk}^{n}},$$

The formulas are actually a difference approximation for equations from equation [4.2], which are the laws of conservation of momentum and energy, under the assumption that the divergence terms are equal to zero. Thus, only the pressure forces at this stage are taken into account. Here, the following formulas are used to calculate the average values of the pressure and velocities at the relevant cell boundaries

$$p^n_{i+1/2,j,k} = \frac{p^n_{i+1,j,k} + p^n_{i,j,k}}{2}, \qquad p^n_{i-1/2,j,k} = \frac{p^n_{i,j,k} + p^n_{i-1,j,k}}{2},$$

$$u^n_{i+1/2,j,k} = \frac{u^n_{i+1,j,k} + u^n_{i,j,k}}{2}, \qquad u^n_{i-1/2,j,k} = \frac{u^n_{i,j,k} + u^n_{i-1,j,k}}{2}, \qquad [4.6]$$

$$v^n_{i+1/2,j,k} = \frac{v^n_{i+1,j,k} + v^n_{i,j,k}}{2}, \qquad v^n_{i-1/2,j,k} = \frac{v^n_{i,j,k} + v^n_{i-1,j,k}}{2},$$

and similar formulas are used to calculate the values $p^n_{i,j+1/2,k}$, $p^n_{i,j,k+1/2}$, ..., $w^n_{i,j-1/2,k}$, $w^n_{i,j,k-1/2}$, which determine the average value pressures and velocities from equation [4.3] on the corresponding boundaries of the cell with the number i,j,k.

At the Lagrangian stage, when passing from the time layer n to layer $n+1$, the intermediate values of velocity [4.4] are used to calculate the mass fluxes through the boundaries of the cell with the number i,j,k by the following formulas:

$$M^n_{i+1/2,j,k} = \rho^{n+1/2}_{i+1/2,j,k} u^{n+1/2}_{i+1/2,j,k} \Delta y \Delta z \Delta t,$$

$$M^n_{i-1/2,j,k} = \rho^{n+1/2}_{i-1/2,j,k} u^{n+1/2}_{i-1/2,j,k} \Delta y \Delta z \Delta t,$$

$$M^n_{i,j+1/2,k} = \rho^{n+1/2}_{i,j+1/2,k} v^{n+1/2}_{i,j+1/2,k} \Delta x \Delta z \Delta t,$$

$$M^n_{i,j-1/2,k} = \rho^{n+1/2}_{i,j-1/2,k} v^{n+1/2}_{i,j-1/2,k} \Delta x \Delta z \Delta t, \qquad [4.7]$$

$$M^n_{i,j,k+1/2} = \rho^{n+1/2}_{i,j,k+1/2} w^{n+1/2}_{i,j,k+1/2} \Delta x \Delta y \Delta t,$$

$$M^n_{i,j,k-1/2} = \rho^{n+1/2}_{i,j,k-1/2} w^{n+1/2}_{i,j,k-1/2} \Delta x \Delta y \Delta t,$$

where the average values $u^{n+1/2}_{i+1/2,j,k}$, $u^{n+1/2}_{i-1/2,j,k}$, ..., $w^{n+1/2}_{i,j,k-1/2}$ are calculated by formulas similar to equation [4.6] through the intermediate velocities from equation [4.4]. These values approximate the mass transfer through the boundaries of the relevant cell by the divergence term from the mass conservation law that is expressed by the first equation in equation [4.2]. Accordingly, the approximation of such divergence terms, which determine the transfer of moments and energy, is performed at the final stage. Here, the intermediate density fluxes are calculated by the following formulas:

$$\rho^{n+1/2}_{i+1/2,j,k} = \begin{cases} \rho^n_{i,j,k}, & \text{if } u^{n+1/2}_{i+1/2,j,k} \ge 0, \\ \rho^n_{i+1,j,k}, & \text{if } u^{n+1/2}_{i+1/2,j,k} < 0, \end{cases}$$

$$\rho^{n+1/2}_{i-1/2,j,k} = \begin{cases} \rho^n_{i-1,j,k}, & \text{if } u^{n+1/2}_{i-1/2,j,k} \ge 0, \\ \rho^n_{i,j,k}, & \text{if } u^{n+1/2}_{i-1/2,j,k} < 0, \end{cases} \qquad [4.8]$$

$$\rho^{n+1/2}_{i,j+1/2,k} = \begin{cases} \rho^n_{i,j,k}, & \text{if } v^{n+1/2}_{i,j+1/2,k} \ge 0, \\ \rho^n_{i,j+1,k}, & \text{if } v^{n+1/2}_{i,j+1/2,k} < 0. \end{cases}$$

Thus, in order to calculate intermediate flow density [4.8], which are used to calculate the mass fluxes [4.7] through the cell boundaries with the number i, j, k, the directions of the velocities $u^{n+1/2}_{i+1/2,j,k}$, $u^{n+1/2}_{i-1/2,j,k}$, $v^{n+1/2}_{i,j+1/2,k}$ are taken into account. The similar directions of the velocities $v^{n+1/2}_{i,j-1/2,k}$, $w^{n+1/2}_{i,j,k+1/2}$, $w^{n+1/2}_{i,j,k-1/2}$ are used to calculate the remaining intermediate density fluxes by the formulas

$$\rho^{n+1/2}_{i,j-1/2,k} = \begin{cases} \rho^n_{i,j-1,k}, & \text{if } v^{n+1/2}_{i,j-1/2,k} \ge 0, \\ \rho^n_{i,j,k}, & \text{if } v^{n+1/2}_{i,j-1/2,k} < 0, \end{cases}$$

$$\rho^{n+1/2}_{i,j,k+1/2} = \begin{cases} \rho^n_{i,j,k}, & \text{if } w^{n+1/2}_{i,j,k+1/2} \ge 0, \\ \rho^n_{i,j,k+1}, & \text{if } w^{n+1/2}_{i,j,k+1/2} < 0, \end{cases} \qquad [4.9]$$

$$\rho^{n+1/2}_{i,j,k-1/2} = \begin{cases} \rho^n_{i,j,k-1}, & \text{if } w^{n+1/2}_{i,j,k-1/2} \ge 0, \\ \rho^n_{i,j,k}, & \text{if } w^{n+1/2}_{i,j,k-1/2} < 0. \end{cases}$$

At the final stage, the mass flows [4.7] and speed and energy flows are used to calculate the final values of the parameters

$$\rho_{ijk}^{n+1},\quad u_{ijk}^{n+1},\quad v_{ijk}^{n+1},\quad w_{ijk}^{n+1},\quad p_{ijk}^{n+1},\quad E_{ijk}^{n+1}, \qquad [4.10]$$

which determine the fluid dynamics on the $n+1$ time layer for the cell with the number i,j,k. For example, the density is calculated by the formula

$$\rho_{ijk}^{n+1} = \rho_{ijk}^{n} - O_{ijk}^{-1}\Big(M_{i+1/2,j,k}^{n} - M_{i-1/2,j,k}^{n} + M_{i,j+1/2,k}^{n} - \qquad [4.11]$$

$$- M_{i,j-1/2,k}^{n} + M_{i,j,k+1/2}^{n} - M_{i,j,k-1/2}^{n}\Big),$$

where $O_{ijk} = \Delta x \Delta y \Delta z$. The velocities and energy are calculated using the formulas

$$u_{ijk}^{n+1} = (\rho_{ijk}^{n} / \rho_{ijk}^{n+1}) u_{ijk}^{n} -$$

$$-O_{ijk}^{-1}\Big(u_{i+1/2,j,k}^{n+1/2} M_{i+1/2,j,k}^{n} - u_{i-1/2,j,k}^{n+1/2} M_{i-1/2,j,k}^{n} + u_{i,j+1/2,k}^{n+1/2} M_{i,j+1/2,k}^{n} -$$

$$-u_{i,j-1/2,k}^{n+1/2} M_{i,j-1/2,k}^{n} + u_{i,j,k+1/2}^{n+1/2} M_{i,j,k+1/2}^{n} - u_{i,j,k-1/2}^{n+1/2} M_{i,j,k-1/2}^{n}\Big),$$

$$v_{ijk}^{n+1} = (\rho_{ijk}^{n} / \rho_{ijk}^{n+1}) v_{ijk}^{n} -$$

$$-O_{ijk}^{-1}\Big(v_{i+1/2,j,k}^{n+1/2} M_{i+1/2,j,k}^{n} - v_{i-1/2,j,k}^{n+1/2} M_{i-1/2,j,k}^{n} + v_{i,j+1/2,k}^{n+1/2} M_{i,j+1/2,k}^{n} -$$

$$-v_{i,j-1/2,k}^{n+1/2} M_{i,j-1/2,k}^{n} + v_{i,j,k+1/2}^{n+1/2} M_{i,j,k+1/2}^{n} - v_{i,j,k-1/2}^{n+1/2} M_{i,j,k-1/2}^{n}\Big),$$

$$w_{ijk}^{n+1} = (\rho_{ijk}^{n} / \rho_{ijk}^{n+1}) w_{ijk}^{n} - \qquad [4.12]$$

$$-O_{ijk}^{-1}\Big(w_{i+1/2,j,k}^{n+1/2} M_{i+1/2,j,k}^{n} - w_{i-1/2,j,k}^{n+1/2} M_{i-1/2,j,k}^{n} + w_{i,j+1/2,k}^{n+1/2} M_{i,j+1/2,k}^{n} -$$

$$-w_{i,j-1/2,k}^{n+1/2} M_{i,j-1/2,k}^{n} + w_{i,j,k+1/2}^{n+1/2} M_{i,j,k+1/2}^{n} - w_{i,j,k-1/2}^{n+1/2} M_{i,j,k-1/2}^{n}\Big),$$

$$E_{ijk}^{n+1} = (\rho_{ijk}^{n} / \rho_{ijk}^{n+1}) E_{ijk}^{n} -$$

$$-O_{ijk}^{-1}\left(E_{i+1/2,j,k}^{n+1/2}M_{i+1/2,j,k}^{n} - E_{i-1/2,j,k}^{n+1/2}M_{i-1/2,j,k}^{n} + E_{i,j+1/2,k}^{n+1/2}M_{i,j+1/2,k}^{n} -\right.$$

$$\left.-E_{i,j-1/2,k}^{n+1/2}M_{i,j-1/2,k}^{n} + E_{i,j,k+1/2}^{n+1/2}M_{i,j,k+1/2}^{n} - E_{i,j,k-1/2}^{n+1/2}M_{i,j,k-1/2}^{n}\right),$$

where the average intermediate (fluxes for) velocities and energy $u_{i+1/2,j,k}^{n+1/2}$, $u_{i-1/2,j,k}^{n+1/2}$, ..., $E_{i,j,k-1/2}^{n+1/2}$ are calculated by formulas similar to equations [4.8] and [4.9], where $\rho_{i,j,k}^{n}$ should be replaced by $u_{i,j,k}^{n}$, $v_{i,j,k}^{n}$, $w_{i,j,k}^{n}$, and $E_{i,j,k}^{n}$, respectively.

The specific types of the liquids or media under consideration are determined in this method through the definition of the form of the equation of state, which can be represented as $p = p(\rho,\varepsilon)$ or using the following formula:

$$p = p(\rho, j), \qquad [4.13]$$

where $j = \varepsilon - v^2 / 2$ denotes the internal energy. This formula allows us to calculate the value of p_{ijk}^{n+1} by values ρ_{ijk}^{n+1}, u_{ijk}^{n+1}, v_{ijk}^{n+1}, w_{ijk}^{n+1}, E_{ijk}^{n+1}, and complete the description of the transition from the time layer n to layer $n+1$.

To take into account the boundary conditions in the method, the region Ω occupied by the fluid is surrounded by a system of fictitious cells (Belotserkovskij and Davidov 1982), in which the values of parameters [4.3] are determined so that suitable boundary conditions were satisfied within the accepted accuracy on the boundary $\partial\Omega$. Contact boundaries can be taken into account in the same way.

Therefore, the discrete conservation laws are faithful within the accepted accuracy. For example, the total mass of the fluid under consideration is saved at every time step of such approximation. Moreover, the mass sources may be induced by boundary conditions and external forces that lead to corresponding modifications of conservation laws in equation [4.1] and also on the approximate level. The momentums and energy transports model in similar manners by averaging the intermediate velocities and energy on cell boundaries. Thus, large particle method is effective enough for numerical computing, since the approximations are rationale from the physical and mathematical point of view and conservation laws are correct on the discrete levels during the courses of corresponding numerical simulations.

4.3. The particle-in-cell method

The time approximation in the particle-in-cell method is also natural. Calculations are carried out step by step with a sufficiently small time interval Δt, starting from a given initial configuration of the liquid under consideration.

Spatial discretization in this method is more complex and takes into account the dynamics of motion of fluid particles. The domain Ω, occupied by the liquid, is divided into small cells, for example, with the steps $\Delta x, \Delta y, \Delta z$ in spatial variables as in the large particle method. Each such cell, as usual, is assigned the number i, j, k, and the values of density, velocities and energy refer to the center of the cell and are denoted, respectively, as

$$\rho_{ijk}^{n}, \quad u_{ijk}^{n}, \quad v_{ijk}^{n}, \quad w_{ijk}^{n}, \quad E_{ijk}^{n}, \qquad [4.14]$$

where the subscript n indicates the time step number as before.

The liquid filling each of these cells is considered as a collection of several particles or drops (which can also be considered as "molecules" or, more precisely, "computational molecules", since the size of such particles is much larger than the size of real liquid molecules). Each of these particles is assigned coordinates, mass, volume and full energy

$$x_s^n, \quad y_s^n, \quad z_s^n, \quad m_s^n, \quad o_s^n, \quad e_s^n, \qquad [4.15]$$

which are specified at an initial moment. The subscript s in these notations corresponds to the particle number and varies from one to S, where S denotes the number of all particles that fill the considered domain Ω. In addition, the velocities and total energy from equation [4.14] are also determined for each cell at the initial moment of time. Respectively, the density ρ_{ijk}^{n} for the cell is calculated by the formula

$$\rho_{ijk}^{n} = O_{ijk}^{-1} \sum_{s \in \{ijk\}} m_s^n, \qquad [4.16]$$

in particular, for $n = 0$, where the summation is carried out only over those particles that are in the cell i, j, k at time $t = n\Delta t$. Thus, it is possible to calculate the pressure p_{ijk}^{n} according to formula [4.13].

For the calculation by the particle-in-cell method for a homogeneous fluid, it is usually assumed that the mass m_s^n and the volume o_s^n of the considered particles do not depend on n and s. Thus, this method uses a discrete model of a continuous medium, and the space occupied by the fluid is divided by a grid of fixed cells. Inside such cells, a continuous medium is represented by particles, each of which carries a fixed mass of liquid according to Harlow (1964).

This approach takes advantage of the Lagrange and Euler concepts when considering fluid dynamics, since the calculations use a Lagrangian coordinate system, modeled by particles, superimposed on a fixed Euler grid. This is the reason for the high efficiency of the particle-in-cell method, and with such an approximation, the total mass of particles is preserved in the process of calculations at each time step, which follows from equality [4.16], which actually expresses the law of conservation of fluid mass during the development of dynamics.

The corresponding time step for modeling the motion of the considered homogeneous liquid by the particle-in-cell method, as in the large particles method, is divided into three stages (Euler, Lagrangian and final stages according to Belotserkovskij and Davidov (1982)) so that discrete conservation laws are fulfilled within the framework of the accepted accuracy of calculations determined by the quantities $\Delta x, \Delta y, \Delta z, \Delta t$ and the number of all particles S.

At the first stage, using the values from equations [4.14] and [4.16], the intermediate flow parameters defined in equation [4.4] are calculated by formulas [4.5] and [4.6]. Thus, at this stage, it is assumed that there is no particle mass flux through the boundaries of the Euler cells, which determines the Euler stage of the considered approximation, which corresponds to the transfer of moments and energy in each cell, defined by the pressure forces in accordance with the equations of moments and energy from equation [4.1].

At the second stage, the motion of the considered particles under the action of the calculated moments is taken into account, which corresponds to the Lagrangian stage of this approximation for the mass conservation equation. At this stage, the movement of particles for the time Δt is modeled (as shown in Figure 4.1) and the mass fluxes across the boundaries of the Euler cells are calculated.

The new particle coordinates with the number s are calculated by the formulas

$$x_s^{n+1} = x_s^n + u_s^n \Delta t, \quad y_s^{n+1} = y_s^n + v_s^n \Delta t, \quad z_s^{n+1} = z_s^n + w_s^n \Delta t, \qquad [4.17]$$

where the velocities u_s^n, v_s^n, w_s^n, of the particle under consideration are calculated from the intermediate values of the velocities from equation [4.4] using linear interpolation over cells lying near the cell with number i, j, k.

Using the comparison of the new coordinates of particles [4.17] with the coordinates from equation [4.15], it is determined whether a particular particle has remained in the previous cell or has passed into the neighboring one and into which one in particular (as illustrated in Figure 4.1). Based on such a comparison, the mass transfer from a specific cell to neighboring cells is modeled, and the mass fluxes through the corresponding grid boundaries are calculated by the following formulas:

$$M_{i+1/2,j,k}^n = (\Sigma_{in}\, m_s^n)_{i+1/2,j,k} - (\Sigma_{out}\, m_s^n)_{i+1/2,j,k},$$

$$M_{i-1/2,j,k}^n = (\Sigma_{in}\, m_s^n)_{i-1/2,j,k} - (\Sigma_{out}\, m_s^n)_{i-1/2,j,k},$$

$$M_{i,j+1/2,k}^n = (\Sigma_{in}\, m_s^n)_{i,j+1/2,k} - (\Sigma_{out}\, m_s^n)_{i,j+1/2,k},$$

$$M_{i,j-1/2,k}^n = (\Sigma_{in}\, m_s^n)_{i,j-1/2,k} - (\Sigma_{out}\, m_s^n)_{i,j-1/2,k}, \qquad [4.18]$$

$$M_{i,j,k+1/2}^n = (\Sigma_{in}\, m_s^n)_{i,j,k+1/2} - (\Sigma_{out}\, m_s^n)_{i,j,k+1/2},$$

$$M_{i,j,k-1/2}^n = (\Sigma_{in}\, m_s^n)_{i,j,k-1/2} - (\Sigma_{out}\, m_s^n)_{i,j,k-1/2},$$

where $\Sigma_{in}\, m_s^n$ denotes the summation over all particles that have fallen into the cell with number i, j, k during the time Δt through the boundaries $i + 1/2, j, k$, $i - 1/2, j, k, \ldots,$ $i, j, k - 1/2$ and $\Sigma_{out}\, m_s^n$ denotes summation over all particles that have left the considered cell across the appropriate boundaries during the time Δt.

At the third stage, the final transfer of moments and energy is calculated, which is the final stage of the considered time step. For example, the energy fluxes through the corresponding grid boundaries are calculated by the following formulas:

$$E_{i+1/2,j,k}^n = (\Sigma_{in}\, m_s^n e_s^n)_{i+1/2,j,k} - (\Sigma_{out}\, m_s^n e_s^n)_{i+1/2,j,k},$$

$$E_{i-1/2,j,k}^n = (\Sigma_{in}\, m_s^n e_s^n)_{i-1/2,j,k} - (\Sigma_{out}\, m_s^n e_s^n)_{i-1/2,j,k},$$

$$E^n_{i,j+1/2,k} = (\Sigma_{in}\, m^n_s e^n_s)_{i,j+1/2,k} - (\Sigma_{out}\, m^n_s e^n_s)_{i,j+1/2,k},$$

$$E^n_{i,j-1/2,k} = (\Sigma_{in}\, m^n_s e^n_s)_{i,j-1/2,k} - (\Sigma_{out}\, m^n_s e^n_s)_{i,j-1/2,k}, \qquad [4.19]$$

$$E^n_{i,j,k+1/2} = (\Sigma_{in}\, m^n_s e^n_s)_{i,j,k+1/2} - (\Sigma_{out}\, m^n_s e^n_s)_{i,j,k+1/2},$$

$$E^n_{i,j,k-1/2} = (\Sigma_{in}\, m^n_s e^n_s)_{i,j,k-1/2} - (\Sigma_{out}\, m^n_s e^n_s)_{i,j,k-1/2}.$$

At this stage, the equations of moments and energy from equation [4.2] are approximated without taking into account pressure forces, which is simulation for the transfer of moments and energy under the action of dynamic forces of inertia from a specific cell to neighboring cells. In this case, it is assumed that a particle passing from one cell to another transfers its values of mass, momentum and energy. Thus, based on conservation laws [4.2] by formulas [4.11] and [4.12] using values from equations [4.14], [4.18] and [4.19], the final values of flow parameters [4.10] on a new time layer are calculated. The considered approximation is rational from a physical and mathematical point of view, since the conservation laws are fulfilled at a discrete level during all stages of calculations. Therefore, the particle-in-cell method is quite effective and is often used to simulate fluid dynamics under the action of external forces and influences caused by boundary conditions.

A significant problem with this method is the need to use a sufficiently large number of particles for calculations. Indeed, the total number of cells under consideration should be large enough to guarantee an acceptable approximation accuracy for Euler cells, and the number of particles in each of these cells should also be large for a sufficient approximation accuracy for Lagrange particles. But, each of these particles is determined by mass, volume, energy and coordinates in space. Therefore, it is necessary to consider a sufficiently large amount of data and this data is recalculated at each time step. To solve this problem of computer processing of a large amount of data, the large particle method was developed. The large particle method is later introduced as the averaging of the particle-in-cell method over Euler cells in a sense. In the method, density, velocities and energy are specified for each cell at a given moment. But the volume of the particle now coincides with the volume of the cell. Therefore, the mass and energy of a particle are defined by the same quantities. Thus, the data is not that massive in the method.

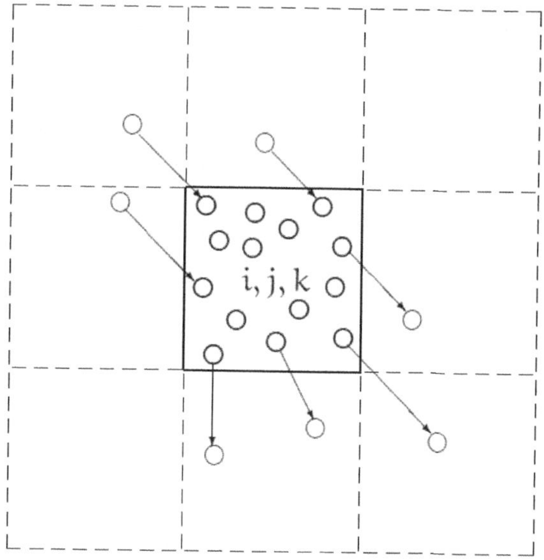

Figure 4.1. *Dynamics of particles in cells*

4.4. Modeling of heterogeneous fluid dynamics

Let us now consider a combination of the presented methods, which was developed for modeling the dynamics of a base fluid with inclusions of another fluid. The time approximation in such a combined method is determined as in the considered methods, and spatial sampling is defined as follows. The domain Ω, occupied by a heterogeneous fluid, is divided into small cells, for example, with the steps $\Delta x, \Delta y, \Delta z$ over spatial variables. Each such cell is assigned the number i, j, k as in the cases considered.

The base fluid filling each of these cells is considered as a "large" (or "big") particle, and inclusions are considered as a collection of "small" particles located in each such cell. The density, velocities, pressure and energy values assigned to large particles refer to the center of the cell and are denoted as

$$\rho_{ijk}^{bn}, \quad u_{ijk}^{bn}, \quad v_{ijk}^{bn}, \quad w_{ijk}^{bn}, \quad p_{ijk}^{bn}, \quad E_{ijk}^{bn}. \qquad [4.20]$$

These values are concretized at the initial time for $n = 0$. The system of inclusions is renumbered, for example, from one to S and is modeled by small particles. Small particles are assigned coordinates in space, mass, volume and energy from equation [4.15] and the pressure p_s^{sn}, which are also concretized at the initial time.

In addition, each cell is assigned the density, velocity and energy values related to the heterogeneous fluid and denoted as in equation [4.3]. It is assumed that

$$u_{ijk}^{n} = u_{ijk}^{bn}, \quad v_{ijk}^{n} = v_{ijk}^{bn}, \quad w_{ijk}^{n} = w_{ijk}^{bn}, \quad p_{ijk}^{n} = p_{ijk}^{bn} = p_s^{sn} \qquad [4.21]$$

for small particles with numbers s, which are in the cell with number i, j, k at the considered moment of time $n\Delta t$. Using equations [4.15] and [4.16], we can write

$$O_{ijk} = O_{ijk}^{bn} + O_{ijk}^{sn}, \qquad O_{ijk}^{sn} = \sum_{s \in \{ijk\}} o_s^{n}, \qquad [4.22]$$

which determines these quantities and expresses the law of conservation of volumes for large and small particles in a cell with a number $i, j, k.$ Similarly, we can represent the law of conservation of mass using the equalities

$$\rho_{ijk}^{n} O_{ijk} = \rho_{ijk}^{bn} O_{ijk}^{bn} + \rho_{ijk}^{sn} O_{ijk}^{sn}, \qquad \rho_{ijk}^{sn} O_{ijk}^{sn} = \sum_{s \in \{ijk\}} m_s^{n}, \qquad [4.23]$$

where the values of these quantities are determined by virtue of equations [4.3] and [4.20].

Moreover, for internal energies, it is assumed that

$$\rho_{ijk}^{n} J_{ijk}^{n} O_{ijk} = \rho_{ijk}^{bn} J_{ijk}^{bn} O_{ijk}^{bn} + \rho_{ijk}^{sn} J_{ijk}^{sn} O_{ijk}^{sn}, \qquad [4.24]$$

where J_{ijk}^{bn} and J_{ijk}^{sn} denote the internal energies of the base fluid and inclusions at the time moment, respectively. The last equality is not quite a conservation law, but rather precisely expresses the principle of internal energy additivity for the mixture by the mass of components, which is fulfilled for many hydrodynamic processes in accordance with Nigmatulin (1991).

The corresponding time step for modeling the motion of a heterogeneous fluid is divided into three stages, which are supplemented by a preliminary recalculation of parameters. At this preliminary stage, the internal energy of large and small particles in each cell is distributed between these particles so that the pressure in this cell is equal. Such a redistribution of energy in the cell is natural, since a large particle induces some pressure through its equation of state, small particles induce some pressure through their equations of state, and the coincidence of these pressures is characteristic of heterogeneous liquids. Furthermore, it is assumed here that the equations of state [4.13] of the considered liquids have the form of

$$p(\rho, j) = P(\rho) + \rho\Gamma(\rho) j,$$

where $P(\rho)$ and $\Gamma(\rho)$ are denoted the elastic component of the state equation and the Gruneisen coefficient for the relevant fluid or media (Nigmatulin 1991).

Therefore, the equality of pressures in large and small particles in the cell under consideration is written in the form of

$$P_b(\rho_{ijk}^{bn}) + \rho_{ijk}^{bn}\Gamma_b(\rho_{ijk}^{bn})J_{ijk}^{bn} = P_s(\rho_{ijk}^{sn}) + \rho_{ijk}^{sn}\Gamma_s(\rho_{ijk}^{sn})J_{ijk}^{sn}. \quad [4.25]$$

This equality, together with equality [4.24], can be viewed as a system of equations for J_{ijk}^{bn} and J_{ijk}^{sn}. The determinant of this system is given by the formula

$$\rho_{ijk}^{bn}O_{ijk}^{bn}\rho_{ijk}^{sn}\Gamma_s(\rho_{ijk}^{sn}) + \rho_{ijk}^{sn}O_{ijk}^{sn}\rho_{ijk}^{bn}\Gamma_b(\rho_{ijk}^{bn}).$$

Thus, under natural conditions on the Gruneisen coefficients, a solution to this system can be found. For example, it is possible to directly obtain the equality

$$\Lambda_{ijk}^{sn}J_{ijk}^{sn} = O_{ijk}\rho_{ijk}^{n}J_{ijk}^{n} + O_{ijk}^{bn}\Gamma_b^{-1}(\rho_{ijk}^{bn})(P_b(\rho_{ijk}^{bn}) - P_s(\rho_{ijk}^{sn})), \quad [4.26]$$

where $\Lambda_{ijk}^{sn} = \rho_{ijk}^{sn}O_{ijk}^{sn} + \rho_{ijk}^{sn}O_{ijk}^{bn}\Gamma_b^{-1}(\rho_{ijk}^{bn})\Gamma_s(\rho_{ijk}^{sn})$. And similarly, ones obtain that

$$\Lambda_{ijk}^{bn}J_{ijk}^{bn} = O_{ijk}\rho_{ijk}^{n}J_{ijk}^{n} - O_{ijk}^{sn}\Gamma_s^{-1}(\rho_{ijk}^{sn})(P_b(\rho_{ijk}^{bn}) - P_s(\rho_{ijk}^{sn})), \quad [4.27]$$

where $\Lambda_{ijk}^{bn} = \rho_{ijk}^{sn}O_{ijk}^{sn} + \rho_{ijk}^{bn}O_{ijk}^{sn}\Gamma_s^{-1}(\rho_{ijk}^{sn})\Gamma_s(\rho_{ijk}^{bn})$. It is interesting that the pressure difference for two fluids is included as the main equation in the filtration model of a two-phase immiscible liquid flow and is retained after homogenization of this model; further details can be found in Sandrakov (2000).

Another modified approach to the redistribution of internal energies in the cell with the i, j, k number is possible. Following Baum et al. (1975) and Nigmatulin (1991), we can distinguish the elastic and thermal components of internal energies

$$J_{ijk}^{bn} = J_b(\rho_{ijk}^{bn}) + c_b T_{ijk}^{bn}, \qquad J_{ijk}^{sn} = J_s(\rho_{ijk}^{sn}) + c_s T_{ijk}^{sn}, \qquad [4.28]$$

where T_{ijk}^{bn} and T_{ijk}^{sn} denote the temperatures of the base fluid and inclusions, respectively. In addition, c_b and c_s are the constants corresponding to the specific heat at a constant volume of large and small particles. Moreover, $J_b(\rho_{ijk}^{bn})$ and $J_s(\rho_{ijk}^{sn})$ are the elastic components for the relevant fluid or media (Nigmatulin 1991).

Therefore, the natural condition on the equality of temperatures in large and small particles in the cell under consideration can be written in the form of

$$c_b^{-1}(J_{ijk}^{bn} - J_b(\rho_{ijk}^{bn})) = c_s^{-1}(J_{ijk}^{sn} - J_s(\rho_{ijk}^{sn})). \qquad [4.29]$$

This equation and the five equations from equations [4.22]–[4.25] (without the second equation in equation [4.23], which is the law of the mass conservation of small particles in the cell under consideration) can be considered as a system of equations for the unknowns O_{ijk}^{bn}, O_{ijk}^{sn}, ρ_{ijk}^{bn}, ρ_{ijk}^{bn}, J_{ijk}^{bn}, J_{ijk}^{sn}. Some of these equations are linear, but not all of them. Such equations are usually solved iteratively. But, computer experiments have shown that the complicated modification does not introduce significant changes in comparison with the use of equations [4.26] and [4.27]. It is possible that this approach will be useful for modeling in porous media, which is discussed in the last section.

In any case, all values from equations [4.15], [4.20] and [4.21] are determined, where

$$p_{ijk}^{n} = P_b(\rho_{ijk}^{bn}) + \rho_{ijk}^{bn}\Gamma_b(\rho_{ijk}^{bn})J_{ijk}^{bn}$$

and, for example, $E_{ijk}^{bn} = J_{ijk}^{bn} + (u_{ijk}^{n})^2/2 + (v_{ijk}^{n})^2/2 + (w_{ijk}^{n})^2/2$ according to the representation for the total energy as the sum of the internal and kinetic energies.

In addition, at this stage, the internal energy is redistributed between the small particles in accordance with the following equality:

$$\rho_{ijk}^{sn} J_{ijk}^{sn} O_{ijk}^{sn} = \sum_{s \in \{ijk\}} m_s^n j_s^n,$$

where the summation is carried out only over those particles that are in the cell i, j, k at time $t = n\Delta t$ and the symbol $j_s^n = e_s^n - (u_s^n)^2 / 2 - (v_s^n)^2 / 2 - (w_s^n)^2 / 2$ denotes the internal energy, which is calculated according to equations [4.15] and [4.17].

Further calculations of this combinational method are carried out as follows. At the Euler stage, the values from equation [4.21] are used to determine intermediate values [4.4] by formulas [4.5] and [4.6]. At the Lagrangian stage, the motion of the considered particles under the action of the calculated moments is taken into account. At this stage, the movement of particles in time Δt is modeled by calculating new coordinates of the particle with the number s according to formulas [4.17]. Therefore, the fluxes of mass through the boundaries of the cells are calculated as the sum of fluxes of large and small particles calculated by formulas

$$\begin{aligned}
M_{i+1/2,j,k}^{n} &= M_{i+1/2,j,k}^{nb} + M_{i+1/2,j,k}^{ns}, \\
M_{i-1/2,j,k}^{n} &= M_{i-1/2,j,k}^{nb} + M_{i-1/2,j,k}^{ns}, \\
M_{i,j+1/2,k}^{n} &= M_{i,j+1/2,k}^{nb} + M_{i,j+1/2,k}^{ns}, \\
M_{i,j-1/2,k}^{n} &= M_{i,j-1/2,k}^{nb} + M_{i,j-1/2,k}^{ns}, \\
M_{i,j,k+1/2}^{n} &= M_{i,j,k+1/2}^{nb} + M_{i,j,k+1/2}^{ns}, \\
M_{i,j,k-1/2}^{n} &= M_{i,j,k-1/2}^{nb} + M_{i,j,k-1/2}^{ns},
\end{aligned} \qquad [4.30]$$

where fluxes for large particles are calculated in accordance with equations [4.7]–[4.9], and fluxes for small particles are calculated using the formulas in [4.18].

At the final stage, the moments and energy equations in equation [4.2] are approximated without taking into account pressure forces. For example, the energy fluxes through the corresponding grid boundaries are calculated by the following formulas:

$$E^n_{i+1/2,j,k} = E^{nb}_{i+1/2,j,k} + E^{ns}_{i+1/2,j,k},$$

$$E^n_{i-1/2,j,k} = E^{nb}_{i-1/2,j,k} + E^{ns}_{i-1/2,j,k},$$

$$E^n_{i,j+1/2,k} = E^{nb}_{i,j+1/2,k} + E^{ns}_{i,j+1/2,k},$$

$$E^n_{i,j-1/2,k} = E^{nb}_{i,j-1/2,k} + E^{ns}_{i,j-1/2,k}, \qquad [4.31]$$

$$E^n_{i,j,k+1/2} = E^{nb}_{i,j,k+1/2} + E^{ns}_{i,j,k+1/2},$$

$$E^n_{i,j,k-1/2} = E^{nb}_{i,j,k-1/2} + E^{ns}_{i,j,k-1/2}.$$

Here, the energy fluxes for small particles are calculated using the formulas in [4.19] and the fluxes for large particles are calculated in accordance with [4.7]–[4.9], where $\rho^n_{i,j,k}$ should be replaced by $E^n_{i,j,k}$. Similar formulas are used when calculating the average intermediate velocity fluxes.

Thus, it is simulation for the transfer of moments and energy under the dynamic force action for inertia from a specific cell to neighboring cells. Therefore, using formulas [4.11] and [4.12] through the values from equations [4.7], [4.18] and [4.19] according to equations [4.30] and [4.31], the final values of the flow parameters [4.10] at a new time instant are calculated; this corresponds to the fulfillment of conservation laws [4.1] at the discrete level in the process of all stages of calculations in modeling the dynamics of the considered heterogeneous fluid. This completes the description of all the steps. The method seems promising for modeling of diffusion and absorption processes in heterogeneous fluids and media.

4.5. Modeling of heterogeneous fluid dynamics with phase transitions

For the sake of concreteness, here we will consider a heterogeneous liquid with graphite particles, which can undergo phase transitions into diamond and, possibly, vice versa. To calculate the graphite–diamond phase transitions, it is customary to use a linear kinetic equation. According to Nigmatulin (1991), this equation has the form

$$\frac{dW}{dt} = -\frac{I_{12} - I_{21}}{\rho}, \qquad [4.32]$$

where W denotes the mass fraction of graphite in the considered medium, ρ is the medium density and $I = I_{12} - I_{21}$ characterizes the medium mass transfer per unit time per unit volume from the graphite phase to the diamond phase and vice versa (the intensity of phase transitions). Each of the components of the intensity I can only be nonnegative, and at least one of them is necessarily equal to zero. If there are no phase transitions, then $I_{12} = I_{21} = 0$. Thus, the value of I_{12} determines the rate of transition of graphite to diamond, and I_{21} is the rate of reverse transition.

The formulas for calculating such transfers, which are given in Nigmatulin (1991), depend on the pressure deviation (overcompression or overexpansion) from the saturation pressure $p_s = p_s(T)$ and have the following form:

$$I_{12} = \begin{cases} l_{12}(p - p_s), & \text{if} \quad \rho_g > 0 \quad \text{and} \quad p > p_s, \\ 0, & \text{if} \quad \rho_g = 0 \quad \text{or} \quad p_s \geq p, \end{cases}$$

$$I_{21} = \begin{cases} l_{21}(p_s - p), & \text{if} \quad \rho_d > 0 \quad \text{and} \quad p_s > p, \\ 0, & \text{if} \quad \rho_d = 0 \quad \text{or} \quad p \geq p_s, \end{cases}$$

where ρ_g and ρ_d denote the reduced densities of the graphite and diamond phases, respectively, and the equality $p = p_s(T)$ determines the pressure on the equilibrium line of the two phases (saturation pressure), which can be represented as

$$p_s = a + bT,$$

and the constants a, b, l_{12}, and l_{21} are in Baum et al. (1975) and Nigmatulin (1991).

As in the previous case, the domain Ω, occupied by a heterogeneous fluid, is divided into small cells, for example, with the steps $\Delta x, \Delta y, \Delta z$ over spatial variables. Each such cell is assigned the number i, j, k. The base fluid filling each of these cells is considered as a "big" particle, and inclusions are considered as a collection of "small" particles located in each such cell. The density, velocities, pressure and energy values assigned to big particles refer to the center of the cell and are denoted as in equation [4.20]. In addition, each cell is assigned the density, velocity and energy values related to the heterogeneous fluid and denoted as in equation [4.21]. These values are concretized also at the initial time for $n = 0$.

The system of inclusions is renumbered from one to S and is modeled by small particles. Small particles are assigned coordinates in space, mass, volume and energy from equation [4.15] and the pressure p_s^{sn}, which are also concretized at the initial time. Now, these values for small particles will be divided into two groups

$$x_s^n,\ y_s^n,\ z_s^n,\ m_{gs}^n,\ o_{gs}^n,\ e_{gs}^n,\ m_{ds}^n,\ o_{ds}^n,\ e_{ds}^n \qquad [4.33]$$

which refer to particles of graphite and diamond. It is assumed that equation [4.21] is satisfied and that at the initial moment of time all particles define graphite.

Using the notation from equation [4.22] and the division of particles into groups, the volume of particles in the cell with the number i, j, k can be represented as

$$O_{ijk}^{sn} = O_{ijk}^{gn} + O_{ijk}^{dn}, \qquad [4.34]$$

where $O_{ijk}^{gn} = \Sigma\, o_{gs}^n$ and $O_{ijk}^{dn} = \Sigma\, o_{ds}^n$ correspond to the volumes of graphite and diamond small particles. Here, the summation is carried out only over those graphite and diamond particles that are in the cell i, j, k at time $t = n\Delta t$. A similar representation can be written for particle masses.

Thus, in the notation from equation [4.24], the principle of internal energy additivity for the graphite and diamond particles can be written in the form of

$$\rho_{ijk}^{sn} J_{ijk}^{sn} O_{ijk}^{sn} = \rho_{ijk}^{gn} J_{ijk}^{gn} O_{ijk}^{gn} + \rho_{ijk}^{dn} J_{ijk}^{dn} O_{ijk}^{dn}, \qquad [4.35]$$

where J_{ijk}^{gn} and J_{ijk}^{dn} denote internal energies of relevant inclusions and J_{ijk}^{sn} is defined by equality [4.26] in the notation from equation [4.23].

The corresponding time step for modeling the motion of a heterogeneous fluid with phase transitions is divided into three stages, which are supplemented by a preliminary recalculation of parameters. At this preliminary stage, the internal energy of large and small particles in each cell is distributed between these particles so that the pressure in this cell is equal as in the previous case.

Moreover, the intermediate mass fraction of graphite W_{ijk}^{gn} in the considered cell with the number i,j,k is calculated via the Euler approximation for the kinetics equation, which can be represented in the following form:

$$W_{ijk}^{gn} = W_{ijk}^{gn-1} - (I_{12}^{n} - I_{21}^{n})\Delta t / \rho_{ijk}^{n}, \qquad [4.36]$$

where the notation from equation [4.32] is used. So ones have equalities for the masses

$$\rho_{ijk}^{sn} O_{ijk}^{sn} = \rho_{ijk}^{gn} O_{ijk}^{gn} + \rho_{ijk}^{dn} O_{ijk}^{dn}, \qquad \rho_{ijk}^{gn} O_{ijk}^{gn} = \rho_{ijk}^{sn} O_{ijk}^{sn} W_{ijk}^{gn} \qquad [4.37]$$

Furthermore, the equality of pressures of the relevant inclusions in the cell under consideration is written in the form of

$$P_g(\rho_{ijk}^{gn}) + \rho_{ijk}^{gn} \Gamma_g(\rho_{ijk}^{gn}) J_{ijk}^{gn} = P_d(\rho_{ijk}^{dn}) + \rho_{ijk}^{dn} \Gamma_d(\rho_{ijk}^{dn}) J_{ijk}^{dn}. \qquad [4.38]$$

where the notation similar to the notation from equation [4.25] is used. This equality, together with equality [4.35], can be viewed as a system of equations for J_{ijk}^{gn} and J_{ijk}^{dn}.

Therefore, as in the previous section, we can obtain equalities for internal energies that are analogous to equalities [4.26] and [4.27]. This approach assumes that the volume o_s^n of the considered particles does not depend on n and s. In addition, it is clear that equations [4.35] and [4.38] make sense only if diamond particles appeared in the relevant cell with number i,j,k.

Other modified approaches to the redistribution of volumes and internal energies in the cell with the i,j,k number are possible. For this, representations similar to [4.28] are used. Then, the natural condition on the equality of temperatures in relevant small particles in the cell under consideration can be written in the form of

$$c_g^{-1}(J_{ijk}^{gn} - J_g(\rho_{ijk}^{gn})) = c_d^{-1}(J_{ijk}^{dn} - J_d(\rho_{ijk}^{dn})).$$

These equations, [4.34] and [4.35], and the second equation in [4.36] can be considered as a system of equations for the unknowns O_{ijk}^{bn}, O_{ijk}^{sn}, J_{ijk}^{bn}, and J_{ijk}^{sn}.

Some of these equations are linear, but not all of them. Such equations are usually also solved iteratively.

Another modification is to consider the kinetics equation [4.32] for each small particle with coordinates from equation [4.33] and corresponding changes to the equations for volumes, masses and internal energies. However, computer experiments have shown that the complicated modifications do not introduce significant changes in comparison with the use equations from [4.34]–[4.38]. In fact, for a sufficiently fine Euler grid, all particles in a particular cell underwent phase transitions simultaneously. This explains the failure of these modifications and makes it possible to use the phase diagram to calculate phase transformations. This diagram is determined directly by the saturation pressure $p_s = p_s(T)$, which is used in equation [4.32].

Thus, all values from equations [4.20], [4.21] and [4.33] are defined, where the pressure is calculated according to formula [4.25]. Moreover, at this stage, the internal energy is redistributed between the small particles in accordance with the equalities

$$\rho_{ijk}^{gn} J_{ijk}^{gn} O_{ijk}^{gn} = \sum_{s \in \{ijk\}} m_g^n j_g^n, \qquad \rho_{ijk}^{dn} J_{ijk}^{dn} O_{ijk}^{dn} = \sum_{s \in \{ijk\}} m_d^n j_d^n,$$

where the summation is carried out only over those graphite and diamond particles that are in the cell i, j, k at time $t = n\Delta t$, and the symbols j_g^n and j_d^n denote the internal energies of relevant particles.

As in the previous section, further calculations of this combinational method are carried out as follows. At the Euler stage, the values from equation [4.21] are used to define intermediate values [4.4] by formulas [4.5] and [4.6]. At the Lagrangian stage, the motion of the considered particles under the action of the calculated moments is taken into account. At this stage, the movement of particles in time Δt is modeled by calculating new coordinates of the particle with the number s according to equation [4.17], which is the motion of particles along Lagrangian trajectories.

Thus, the fluxes of mass through the boundaries of the cells are calculated as the sum of fluxes of large, graphite and diamond particles calculated by formulas

$$M_{i+1/2,j,k}^{n} = M_{i+1/2,j,k}^{nb} + M_{i+1/2,j,k}^{ng} + M_{i+1/2,j,k}^{nd},$$

$$
\begin{aligned}
M^{n}_{i-1/2,j,k} &= M^{nb}_{i-1/2,j,k} + M^{gs}_{i-1/2,j,k} + M^{ds}_{i-1/2,j,k}, \\
M^{n}_{i,j+1/2,k} &= M^{nb}_{i,j+1/2,k} + M^{gs}_{i,j+1/2,k} + M^{ds}_{i,j+1/2,k}, \\
M^{n}_{i,j-1/2,k} &= M^{nb}_{i,j-1/2,k} + M^{gs}_{i,j-1/2,k} + M^{ds}_{i,j-1/2,k}, \\
M^{n}_{i,j,k+1/2} &= M^{nb}_{i,j,k+1/2} + M^{gs}_{i,j,k+1/2} + M^{ds}_{i,j,k+1/2}, \\
M^{n}_{i,j,k-1/2} &= M^{nb}_{i,j,k-1/2} + M^{gs}_{i,j,k-1/2} + M^{ds}_{i,j,k-1/2},
\end{aligned}
\tag{4.39}
$$

where fluxes for large particles are calculated in accordance with equations [4.7]–[4.9] and fluxes for graphite and diamond particles are calculated using the formulas [4.18].

At the final stage, the moments and energy equations in equation [4.2] are approximated without taking into account pressure forces. For example, the energy fluxes through the corresponding grid boundaries are calculated by the following formulas:

$$
\begin{aligned}
E^{n}_{i+1/2,j,k} &= E^{nb}_{i+1/2,j,k} + E^{gs}_{i+1/2,j,k} + E^{ds}_{i+1/2,j,k}, \\
E^{n}_{i-1/2,j,k} &= E^{nb}_{i-1/2,j,k} + E^{gs}_{i-1/2,j,k} + E^{ds}_{i-1/2,j,k}, \\
E^{n}_{i,j+1/2,k} &= E^{nb}_{i,j+1/2,k} + E^{gs}_{i,j+1/2,k} + E^{ds}_{i,j+1/2,k}, \\
E^{n}_{i,j-1/2,k} &= E^{nb}_{i,j-1/2,k} + E^{gs}_{i,j-1/2,k} + E^{ds}_{i,j-1/2,k}, \\
E^{n}_{i,j,k+1/2} &= E^{nb}_{i,j,k+1/2} + E^{gs}_{i,j,k+1/2} + E^{ds}_{i,j,k+1/2}, \\
E^{n}_{i,j,k-1/2} &= E^{nb}_{i,j,k-1/2} + E^{gs}_{i,j,k-1/2} + E^{ds}_{i,j,k-1/2}.
\end{aligned}
\tag{4.40}
$$

Here, the energy fluxes for graphite and diamond particles are calculated using the formulas [4.19], and the fluxes for large particles are calculated in accordance with equations [4.7]–[4.9], where $\rho^n_{i,j,k}$ should be replaced by $E^n_{i,j,k}$. Similar formulas are used when calculating the average intermediate velocity fluxes.

Thus, formulas [4.39] and [4.40] are used to simulate the transfer of moments and energy under the dynamic force action for inertia from a specific cell to neighboring cells. Therefore, using formulas [4.11] and [4.12] according to equations [4.39] and [4.40], the final values of the flow parameters [4.10] at a new time instant are calculated; this corresponds to the fulfillment of conservation laws [4.1] at the discrete level in the process of all stages of calculations in modeling the dynamics of the considered heterogeneous fluid in which phase transitions are possible. This completes the description of all the steps. It is essential to emphasize that, for example, formulas [4.5] use the division by the density of the heterogeneous medium. Whereas in multiphase models, it may be necessary to divide, for example, by the density of diamond that can be very small in the simulation.

The presented method is designed to the numerical modeling of the following physical processes. Let us consider graphite drops distributing uniformly in some fluid. More exactly, there is heterogeneous medium with graphite particles and the medium may be considered under high pressure as "fluid" with the corresponding state equation. For example, we consider a cylinder of the medium that consists of copper with graphite particles. Let the cylinder be in an outside explosive tube device. By inducing detonation shock waves in the outside explosive tube device, we can observe dynamics of such shock waves in computer experiments by the method. The results of such experiments were in strong agreement with the results of physical experiments and depended significantly on the distribution density of graphite particles and the intensity of shock waves. Such physical experiments are very fast-paced, and computer simulation allows us to have a fairly complete picture of the pressures, energies and velocities at various points in the medium at suitable times.

4.6. Modeling of viscous fluid dynamics and porous media

The combinational method for modeling of heterogeneous fluid dynamics with phase transitions was developed for numerical simulation of fast physical processes. Such processes take place within several tens of microseconds. Therefore, the effects of viscosity were not significant enough to simulate such processes. However, the presented method is also relevant for modeling processes in which the effects of viscosity are significant. Such processes are usually not fast-paced.

In addition, when modeling and fast-flowing processes, it is customary to introduce "artificial" viscosity to improve the stability of computational algorithms

(Belotserkovskij and Davidov 1982), which can also be considered as a simulation of fluid dynamics taking into account the effects of viscosity. Viscosity effects are also relevant when simulating fluid dynamics in porous media. Let us emphasize here that when freezing particles in the presented method, we receive a model of the motion of a liquid in a porous medium with a suitable choice of particle sizes.

The laws of conservation of mass, moments and energy for viscous fluids can be represented (Lions 1998) in the following differential form:

$$\frac{\partial \rho}{\partial t} + \operatorname{div}(\rho \mathrm{v}) = 0,$$

$$\frac{\partial \rho \mathrm{v}}{\partial t} + \operatorname{div}(\rho \mathrm{v} \otimes \mathrm{v}) + \nabla p = \qquad [4.41]$$

$$= \operatorname{div}(\mu \nabla \mathrm{v}) + \nabla((\mu + \lambda)\operatorname{div} \mathrm{v}),$$

$$\frac{\partial \rho \varepsilon}{\partial t} + \operatorname{div}(\rho \mathrm{v} \varepsilon) + \operatorname{div}(p \mathrm{v}) = \operatorname{div}(\tau \cdot \mathrm{v}) + \operatorname{div}(k \nabla T),$$

where notations from equation [4.2] are used and T is the notation for the temperature which is proportional to the internal energy up to the elastic component in accordance with equalities [4.28]. Moreover, k is the coefficient of thermal conductivity, μ and λ are the Lame coefficients of dynamic and bulk viscosities, respectively, which are determined by the type of fluid and satisfy the inequalities

$$k > 0, \quad \mu > 0, \quad 2\mu + 3\lambda \geq 0.$$

It is usually assumed (Lions 1998) that the equality $3\lambda = -2\mu$ and the Stokes law of state are satisfied, according to which

$$\tau = 2\mu D - ((2/3)\mu \operatorname{div} \mathrm{v})\mathrm{I},$$

where I is the unit tensor and the strain rate tensor D are defined as the half-sum of the tensor $\nabla \mathrm{v}$ and the transposed tensor $(\nabla \mathrm{v})^*$. The laws of conservation of mass, moments and energy for viscous fluids are given in the integral form in Lions (1998).

The system of equations [4.41] differs from the system of equations [4.2] only by the presence of additional terms due to viscous and thermal effects. Thus, all of the basic relationships of the presented method can be preserved (taking into account additional terms at the Euler stage), which will only lead to the addition of suitable difference relations to the difference approximations [4.5]. For example, the first two relations from equation [4.5] can be replaced by taking into account the viscosity from the following difference explicit–implicit approximation:

$$u_{ijk}^{n+1/2} - \frac{u_{i+1jk}^{n+1/2} - 2u_{ijk}^{n+1/2} + u_{i-1jk}^{n+1/2}}{\rho_{ijk}^{n} 3(\Delta x)^2} 4\mu\Delta t - \frac{u_{ij+1k}^{n+1/2} - 2u_{ijk}^{n+1/2} + u_{ij-1k}^{n+1/2}}{\rho_{ijk}^{n} (\Delta y)^2} \mu\Delta t -$$

$$- \frac{u_{ijk+1}^{n+1/2} - 2u_{ijk}^{n+1/2} + u_{ijk-1}^{n+1/2}}{\rho_{ijk}^{n} (\Delta z)^2} \mu\Delta t = u_{ijk}^{n} - \frac{p_{i+1jk}^{n} - p_{i-1jk}^{n}}{\rho_{ijk}^{n} 2\Delta x} \Delta t +$$

$$+ \frac{v_{i+1j+1k}^{n} - v_{i+1j-1k}^{n} - v_{i-1j+1k}^{n} + v_{i-1j-1k}^{n}}{\rho_{ijk}^{n} 3\Delta x \Delta y} \mu\Delta t +$$

$$+ \frac{w_{i+1jk+1}^{n} - w_{i+1jk-1}^{n} - w_{i-1jk+1}^{n} + w_{i-1jk-1}^{n}}{\rho_{ijk}^{n} 3\Delta x \Delta z} \mu\Delta t,$$

$$v_{ijk}^{n+1/2} - \frac{v_{i+1jk}^{n+1/2} - v_{ijk}^{n+1/2} + v_{i-1jk}^{n+1/2}}{\rho_{ijk}^{n} (\Delta x)^2} \mu\Delta t - \frac{v_{ij+1k}^{n+1/2} - 2v_{ijk}^{n+1/2} + v_{ij-1k}^{n+1/2}}{\rho_{ijk}^{n} 3(\Delta y)^2} 4\mu\Delta t -$$

$$- \frac{v_{ijk+1}^{n+1/2} - 2v_{ijk}^{n+1/2} + v_{ijk-1}^{n+1/2}}{\rho_{ijk}^{n} (\Delta z)^2} \mu\Delta t = v_{ijk}^{n} - \frac{p_{ij+1k}^{n} - p_{ij-1k}^{n}}{\rho_{ijk}^{n} 2\Delta y} \Delta t +$$

$$+ \frac{u_{i+1j+1k}^{n} - u_{i-1j+1k}^{n} - u_{i+1j-1k}^{n} + u_{i-1j-1k}^{n}}{\rho_{ijk}^{n} 3\Delta x \Delta y} \mu\Delta t +$$

$$+ \frac{w_{ij+1k+1}^{n} - w_{ij+1k-1}^{n} - w_{ij-1k+1}^{n} + w_{ij-1k-1}^{n}}{\rho_{ijk}^{n} 3\Delta x \Delta z} \mu\Delta t.$$

The use of an explicit–implicit scheme in the case under consideration is natural, since the stability conditions for explicit schemes are rather stringent according to, for example, Grishchenko et al. (1997) and Grigoriev et al. (2002). It is also natural that the third equation from equation [4.5] can be replaced taking into account the viscosity by the following difference explicit–implicit approximation:

$$w_{ijk}^{n+1/2} - \frac{w_{i+1jk}^{n+1/2} - 2w_{ijk}^{n+1/2} + w_{i-1jk}^{n+1/2}}{\rho_{ijk}^{n}(\Delta x)^2}\mu\Delta t - \frac{w_{ij+1k}^{n+1/2} - 2w_{ijk}^{n+1/2} + w_{ij-1k}^{n+1/2}}{\rho_{ijk}^{n}(\Delta y)^2}\mu\Delta t -$$

$$-\frac{w_{ijk+1}^{n+1/2} - 2w_{ijk}^{n+1/2} + w_{ijk-1}^{n+1/2}}{\rho_{ijk}^{n}3(\Delta z)^2}4\mu\Delta t = w_{ijk}^{n} - \frac{p_{ijk+1}^{n} - p_{ijk-1}^{n}}{\rho_{ijk}^{n}2\Delta z}\Delta t +$$

$$+\frac{u_{i+1jk+1}^{n} - u_{i-1jk+1}^{n} - u_{i+1jk-1}^{n} + u_{i-1jk-1}^{n}}{\rho_{ijk}^{n}3\Delta x\Delta y}\mu\Delta t +$$

$$+\frac{v_{ij+1k+1}^{n} - v_{ij-1k+1}^{n} - v_{ij+1k-1}^{n} + v_{ij-1k-1}^{n}}{\rho_{ijk}^{n}3\Delta x\Delta z}\mu\Delta t.$$

Here, it is assumed that the coefficient μ is constant. If the coefficient is not constant, a more complex approximation should be used.

Similar changes should be made to the remaining ratio for $E_{ijk}^{n+1/2}$ from equation [4.5] to take into account the viscous and thermal effects. Of course, it is more convenient to represent the last equation in equation [4.41] in the form of

$$\frac{\partial \rho cT}{\partial t} - \mathrm{div}(k\nabla T) = \mathrm{div}(\tau\cdot \mathrm{v}) - \mathrm{div}(\rho \mathrm{v}\varepsilon) - \mathrm{div}(p\mathrm{v}) - (\kappa + j(\rho))_t^{'},$$

since the energy can be represented in the form $\varepsilon = \kappa + j(\rho) + cT$ in accordance with equation [4.28], where c is the specific heat capacity at a constant volume and κ is the kinetic energy in the following form: $\kappa = u^2/2 + v^2/2 + w^2/2$. Using such a representation, it is easy to write an explicit–implicit approximation for the last equation of equation [4.41], which should be taken into account in equation [4.5].

Therefore, practically without changing the stages of the presented method, with the exception of the Eulerian one, it is possible to take into account the viscous and thermal effects when modeling the processes under consideration. At the Euler stage, the above relations lead to an explicit–implicit scheme and the need to invert the difference operator for this approximation. Well-known splitting methods can be used to invert such an operator (e.g. Grigoriev et al. 2002; Lyashko et al. 2020).

In addition, to simulate the dynamics of fluids in porous media, it is necessary to freeze particles in the cells that have suitable sizes setting $u_s^n = v_s^n = w_s^n = 0$ in the formula [4.17] for calculating the Lagrangian coordinates. It is also useful to take into account equations [4.22]–[4.25] for a more accurate calculation of thermal effects, since the porous medium can also be heated and cooled. In turn, such effects can significantly affect the dynamics of the fluid under consideration.

Thus, a new method of modeling for heterogeneous fluid dynamics with phase transitions is presented. The method is based on the relevant approximation of conservation laws for masses, momentums and energies in integral and differential forms. The approximation is natural, and numerical simulations are realized as direct computer experiments. The method seems to be much more adequate to the physical and mathematical essence of the dynamics because conservation laws are fulfilled on the discrete level with a suitable degree of approximation. The calculation algorithm of this method is implemented as a computer simulation of the dynamics of a multiphase carrier fluid containing particles that can undergo graphite–diamond phase transitions. Such transitions are modeled on the basis of the theory of phase transformations and the laws of thermodynamics. A modification of this method has also been developed to take into account the effects of viscosity when simulating the dynamics of a multiphase fluid in porous media. A model of the motion of such a liquid in a porous medium is obtained by freezing the motion of particles of the corresponding size in the presented method.

4.7. References

Bacchini, F., Olshevsky, V., Poedts, S., Lapenta G. (2017). A new particle-in-cell method for modeling magnetized fluids. *Comput. Phys. Commun.*, 210, 79–91. doi: 10.1016/j.cpc.2016.10.001.

Baum, F.A., Orlenko, L.P., Stanyukovich, K.P., Chelyshev, V.P., Shekhter, B.I. (1975). *Physics of an Explosion* (in Russian). Nauka, Moscow.

Belotserkovskij, O.M. and Davidov, Yu.M. (1982). *The Large Particles Method in Gas Dynamics* (in Russian). Nauka, Moscow.

Boyko, S.B. and Sandrakov, G.V. (2012). Mathematical modeling of pase transitions graphite–diamond dynamics (in Russian). *J. Num. Appl. Math.*, 2(108), 88–109.

Boyko, S.B. and Sandrakov, G.V. (2013). Parameter computing of hydrodynamics processes with phase transitions. *Bull. Taras Shevchenko National Univ. of Kyiv*, Special Vol., 11–16.

Boyko, S.B., Mischenko, V.V., Sandrakov, G.V. (2007). The numerical investigation method for evaporated plasma. *J. Num. Appl. Math.*, 2(95), 3–12.

Dikalyuk, A.S. and Kuratov, S.E. (2018). Numerical modeling of plasma devices by the particle-in-cell method on unstructured grids. *Math. Models and Computer Simulations*, 10, 198–208. doi: 10.1134/S2070048218020059.

Grigoriev, Yu.N., Vshivkov, V.A., Fedoruk, M.P. (2002). *Numerical Particle-In-Cell Methods: Theory and Applications*. De Gruyter, Utrecht.

Grishchenko, O.Yu., Lyashko, S.I., Molodtsov, O.I. (1997). *Numerical Modelling of Relaxation Processes Gas Dynamics* (in Ukrainian). Kyiv University, Kyiv.

Harlow, F. (1964). Numerical particles-in-cells method. In *Fundamental Methods in Hydrodynamics*, Alder, B. (ed.). Academic Press, New York.

Kormann, K. and Sonnendrücker, E. (2021). Energy-conserving time propagation for a structure-preserving particle-in-cell Vlasov–Maxwell solver. *J. Comput. Physics*, 425, 109890. doi: 10.1016/j.jcp.2020.109890.

Koukouvinis, P., Kyriazis, N., Gavaises, M. (2018). Smoothed particle hydrodynamics simulation of a laser pulse impact onto a liquid metal droplet. *PLOS ONE*, 13(9), 1–13. doi: 10.1371/journal.pone.0204125.

Li, S. and Liu, W.K. (2004). *A Meshfree Particle Method.* Springer-Verlag, Berlin.

Lions, P.-L. (1998). *Mathematical Topics in Fluid Mechanics.* Clarendon Press, Oxford.

Liu, G.R. and Liu, M.B. (2003). *Smoothed Particle Hydrodynamics. A Meshfree Particle Method.* World Scientific Publishing, New Jersey.

Liu, M.B. and Liu, G.R. (2010). Smoothed particle hydrodynamics (SPH): An overview and recent developments. *Arch. Comput. Method Eng.*, 17, 25–76. doi: 10.1007/s11831-010-9040-7.

Lyashko, S.I., Sandrakov, G.V., Semenov, V.V., Klyushin, D.A., Vedel, Y.I., Grigorenko, V.A., Denisov, S.V. (2020). *Mathematical Modelling and Computational Mathematics* (in Ukrainian). Kyiv University, Kyiv.

Nigmatulin, R.I. (1991). *Dynamics of Multiphase Media.* Hemisphere, New York.

Olshevsky, V., Bacchini, F., Poedts, S., Lapenta, G. (2019). Slurm: Fluid particle-in-cell code for plasma modeling. *Comput. Phys. Commun.*, 235, 16–24. doi: 10.1016/j.cpc.2018.06.014.

Parreiras, E.A., Vieira, M.B., Machado, A.G., Renhe, M.C., Giraldi, G.A. (2021). A particle-in-cell method for anisotropic fluid simulation. *Comput. Graph.* doi: 10.1016/j.cag.2021.08.010.

Sandrakov, G.V. (1997). The homogenization of nonstationary equations with contrast coefficients. *Dokl. Math.*, 56(1), 586–589.

Sandrakov, G.V. (1998). Homogenization of dynamic problems of the elasticity theory with very different coefficients. *Dokl. Akad. Nauk*, 362, 449–452.

Sandrakov, G.V. (2000). Homogenization of the filtration of a two-phase immiscible liquid flow. *Dokl. Math.*, 62(2), 186–189.

Sandrakov, G.V. (2004a). Multiphase diffusion models in homogenization. *Dokl. Akad. Nauk*, 394(2), 171–174.

Sandrakov, G.V. (2004b). Multiphase models of nonstationary diffusion arising from homogenization. *Comp. Math. Math. Phys.*, 44(10), 1741–1756.

Sandrakov, G.V. (2006). Homogenization of nonlinear equations and variational inequalities with obstacles. *Dokl. Math.*, 73(2), 178–181. doi: 10.1134/S1064562406020062.

Sandrakov, G.V. (2007). Multiphase homogenized diffusion models for problems with several parameters. *Izvest. Mathe.*, 71(6), 1193–1252. doi: 10.1070/IM2007v071n06ABEH002387.

Sandrakov, G. (2019). Modeling of heterogeneous hydrodynamics processes with phase transition. *Model. Control Inf. Technol.*, 3, 67–68. doi: 10.31713/MCIT.2019.18.

Sandrakov, G. (2020). A modified method for modeling of heterogeneous hydrodynamics processes. *Model., Control Inf. Technol.*, 4, 63–66. doi: 10.31713/MCIT.2020.06.

Sandrakov, G.V. and Boyko, S.B. (2011). Mathematical modeling of complex heterogeneous fluid dynamics (in Russian). *J. Num. Appl. Math.*, 1(104), 109–120.

Wang, Z.B., Chen, R., Wang, H., Liao, Q., Zhu, X., Li, S.Z. (2016). An overview of smoothed particle hydrodynamics for simulating multiphase flow. *Appl. Math. Model.*, 40(23–24), 9625–9655. doi: 10.1016/j.apm.2016.06.030.

Wang, Z., Qin, H., Sturdevant, B., Chang, C.S. (2021). Geometric electrostatic particle-in-cell algorithm on unstructured meshes. *J. Plasma Physics*, 87(4), 905870406. doi: 10.1017/S0022377821000702.

Yudovich, V.I. (2003). Eleven great problems of mathematical hydrodynamics. *Mosc. Math. J.*, 3(2), 711–737. doi: 10.17323/1609-4514-2003-3-2-711-737.

5

Mathematical Models and Control of Functionally Stable Technological Process

Volodymyr PICHKUR, Valentyn SOBCHUK and Dmytro CHERNIY
Taras Shevchenko National University of Kyiv, Ukraine

This chapter offers an approach to modeling industrial enterprises providing production according to the set standard with observance of admissible tolerances by standards. The mathematical model has the form of a discrete control system. To design control, we use the properties of generalized inverse matrices. We present an algorithm of control of production processes providing release of production according to the set standard, whilst following the requirements of the corresponding standards. This approach makes it possible to model technological processes in the enterprises of metallurgy, energy and chemical industry. This is important given the operating conditions of such enterprises under the constant influence of internal and external destabilizing factors.

5.1. Introduction

Currently, improving the efficiency of production enterprise management is closely related to improving the system of operational production planning for enterprises. The main purpose of operational production planning is to ensure a

Computational Methods and Mathematical Modeling in Cyberphysics and Engineering Applications 1,
coordinated by Dmitri KOROLIOUK, Sergiy LYASHKO and Nikolaos LIMNIOS.

coordinated, comprehensive, rhythmic course of production and production with the fullest and most uniform use of all production resources, along with a timely fulfillment of obligations in order to produce finished products on the basis of a uniform and complete course of production at each production chain, as well as achieving a minimum duration of the production cycle and, accordingly, reducing the volume of work in progress. Achieving the above goal is ensured by solving a rather complex set of functions at the operational production planning stage, which, as a rule, includes: calculations of capacity utilization of the enterprise as a whole and production units in particular; calculations of calendar-planning standards of organization of the production process, including the duration of the production cycle for manufacturing finished products and their elements, definition of the size of unfinished production, calculations of parties of start of products, definition of terms of advanced start of products in production on phases and stages of processing of details with production of products; development of calendar schedules of launch-release of products; establishment of operational production tasks to production divisions of the enterprise (shops, sections, workplaces); operational regulation, accounting and control of planned tasks; network planning tasks; scheduling tasks; ensuring the functional stability of systems, etc.

As a result of the accumulated experience in this field, a large number of methods of operational production planning and management of an industrial enterprise have been developed. The most well-known of these are the following regulatory documents: MRP II (Manufacturing Resource Planning), ERP (Enterprise Resource Planning), APS systems (Advanced Planning & Scheduling) and MES (Manufacturing Execution Systems) (Zagidullin 2011). These system solutions have been tested over the last 15–20 years at various manufacturing plants in many industries.

A common feature of the above systems is that they implement the principles characteristic of integrated automated enterprise management systems, when automating a wide range of management functions, covering tasks of strategic, production, financial planning, operational management of supply, procurement, inventory combined with design automation, technological preparation of production, etc.

Today, the greatest attention in the study of automation of enterprise management is paid to ERP systems. Their concept is based on automation and informatization of business processes of sales planning, volume production and financial planning, supply chain management, accounting tasks, etc. The advantages of ERP systems include the implementation of a process approach to industrial enterprise management, which is the basis of international ISO standards and, in fact, defined the representation of ERP systems as a common international practice of creating industrial enterprise management systems.

However, ERP systems have significant drawbacks. In particular, in most ERP systems, planning is based on the application of the MRPII standard. This standard provides only volume planning, usually at the level of the monthly plan. This is due to the fact that the planning of works, technological operations for machines and other technological equipment in ERP systems is carried out without taking into account the current load of production units and the state of the work in progress. In a rather primitive form, the duration of the production process is determined by simply summing up the duration of all its operations, although in reality the duration of the production process depends on the load of the equipment at the time of the order. This is the reason why any detailed ERP plan will be almost impossible to implement. It is usually not possible to adjust such plans quickly.

In these conditions, one of the effective options is to create an enterprise-class production planning system (an APS system) or, with a high level of production automation, MES's, which provide more accurate production planning. These systems appeared on the market in the late 1990s and have the main purpose of compiling a detailed (in comparison with ERP systems) shop and inter-shop work schedule for the entire fleet of equipment on all long-term order portfolios, taking into account the actual state of work in progress.

APS systems make up the work schedule of the entire enterprise, all its production sites and work centers. But the tasks are not issued for individual jobs (this is the task of MES's), but for individual shops. As a result, each shop receives its own schedule for a long period of work. In this case, in comparison with MRPII algorithms, APS algorithms, when scheduling simultaneously, take into account both the needs for materials and the capacity of the enterprise, taking into account their current and planned loading. APS algorithms take into account readjustments and some other parameters of the technological environment. APS systems also allow us to solve planning tasks based on the configuration. Thus, when it comes to planning tasks, under tasks of a complete set of production processes, we will be able to understand the procedure responsible for manufacturing this or that unit of planning from a portfolio of orders ensuring that all necessary materials, technological and auxiliary resources, accessories, equipment, tools, rules and documentation are available. If all this is available, then the manufacture of this planning unit can be planned in time.

A modern manufacturing enterprise is complex, consisting of a large number of processes that are combined into a single set of management for all its resources – a single coordinated mechanism. Managing such complexes requires high-quality modeling for each production process. It is necessary to obtain constructive conditions for the functional stability (Maksymuk et al. 2020) of the respective processes – that is, the conditions that guarantee the effective loading of production centers and the performance of each of the functions assigned to them, for the stable production of quality competitive products. In fact, the problem of approaches to the organization of production process planning, the conditions of existence of functionally stable production processes of industrial enterprises and the algorithm for controlling such a process are investigated below.

5.2. Analysis of production process planning procedure

We will consider this planning procedure in more detail. Suppose that there are planning units e_i, represented by the technological process in the form of a set of operations $\{e_{ij}, j = 1, \ldots, p_i\}$. For each operation, the resources necessary for performance are known: resources, the equipment, the tool, equipment, accessories, documentation, etc.

In essence, this is nothing more than an iterative process of obtaining an acceptable schedule, taking into account the new restrictions imposed on the new iteration. In some cases, the planning process is further simplified. First, plan one detail, then another, until the whole set of details is planned. Thus, by simplifying the algorithm for constructing the production schedule, APS systems provide the ability within the available computing power to obtain acceptable production schedules and to more accurately predict delivery times. At the same time, APS systems do not set themselves complex tasks, such as minimizing the time of reconfiguration, transport operations, reducing the number of equipment involved, etc., as taking into account these requirements will inevitably complicate algorithms and the impossibility of obtaining schedules for large dimensions. In this regard, APS systems have a limited set of planning criteria.

Next, we note that APS systems should, if possible, work closely with MES's, which are defined as operational systems, aimed at forming work schedules not only for machines but also for vehicles, warehousing control systems, teams of adjusters, etc. Thus, the main functions of MES's are not only the planning but also the scheduling of production processes in shop.

Typically, MES's include advanced tools to support the technological training of a particular type of production. Otherwise, they may have the means to integrate with process automation systems. Such information automated control systems are called PLM (Product Lifecycle Management) systems and are characterized by the fact that they erase the functional and information gap between the contours of operational management and the technical preparation of production.

In fact, there is evidence that the creation of new generation control systems is closely linked to the development of a methodology for creating a single information space of the enterprise and integrated models for automated information management systems based on a systems approach that contains formalized descriptions of production and business processes; material objects and resources involved in the operational management of production and its design and technological training. It is obvious that this methodology for creating a single information space should be supported by appropriate basic information technologies that provide: re-engineering of business processes for operational management and technical preparation of production in order to ensure the transition from function-oriented to process organization of enterprise management; description and visual representation of the electronic structure of the product, giving it the status of an enterprise standard; management of business processes of operational planning and technical preparation of production with the use of the electronic system of a product, focused on support for functional tasks of an APS, MES and PLM systems of information automated control system.

It should be noted that the priority component in the above list of components of the information automated control system, from our point of view, is a system of medium-term and operational planning. The emphasis on the importance of this system is due to the fact that effective production planning is the cheapest way to increase the capacity of production systems. And an important point here is the choice of the appropriate method of production planning.

Analysis of known methods for solving problems of production planning shows that the latter, for all its diversity, can be divided into two classes: network planning methods and methods related to the general theory of schedules.

The advantages of network planning methods include their relative simplicity, focus on the process representation of the general scheme of work on the production program orders, the natural perception of management network schedules and binding of works (technological operations), which are included in time, etc. The disadvantages of these methods include the fact that the planning of production operations in them is carried out using simplified schemes of distribution of production capacity for technological units, without taking into account the state of work in progress, etc.

The advantages of the general theory of schedules, which uses the methods of simulation modeling, in the formation of production programs, include the ability to take into account a large number of technological and organizational constraints, the current load of production units and the state of work in progress. Thus, the general theory of schedules only makes it possible to solve planning problems with a small dimension; at the same time, the narrow orientation of a method is directed on an increase in loading of the production equipment without accurate management of terms of the performance of the production program in the absence of process representation.

A combined approach is often used to solve the problems of production planning in practice, presenting the problems of network planning and the distribution of available production resources of the work network as an extension of the general problem of building production schedules.

We will assume that the production system has a set of planning units $\{z\}$, $z = 1,\dots,Z$ from the portfolio of orders represented by technological processes from a set of operations $\{o_{zir}\}$, $z = 1,\dots,Z$; $i = 1,\dots,I$, $r = 1,\dots,R$. The set of operations $\{o_{zir}\}$, $z = 1,\dots,Z$; $i = 1,\dots,I$, $r = 1,\dots,R$ determines the set of works on the network $\mathcal{S} = \{s\}$, where $s = o_{zir}$ corresponds to the i-th operation of the z-th order using the resource r-th type. We will also assume that the production system has many types of resources $\mathfrak{R} = \{r\}$, $r = 1,\dots,R$, which can be production sites, work centers, etc. The set $\mathfrak{R}$ is considered as a service system for performance $\{o_{zir}\}$.

The order is established by the technological implementation of works in the network, which can be represented in the form of a network graph, i.e. in the form of a contourless digraph $\mathcal{S} = \langle I,*,**,V \rangle$ with one initial vertex " $*$ ", final vertex " $**$ ", set works $\mathcal{I}$ and set arcs $\mathcal{V}$, determining the technological order of works.

According to the technology, each work i is performed using a certain type of resource of the r-th type, i.e. it is performed on the equipment of the r-th production unit, which makes it possible to identify any of the works by the triple $\langle zir \rangle$ – in this case, as the initial characteristic of the work $\langle zir \rangle$ is its volume w_{zir} (complexity of execution) and the intensity of execution N_{zir}. Here, the intensity of work $\langle zir \rangle$ means the number of units of the r-th resource allocated for its implementation. Taking into account the introduced performance characteristics, its duration t_{zir} is defined as some function of the form $t_{zir} = f(w_{zir}, N_{zir})$, in particular, $t_{zir} = {}^{w_{zir}}/_{N_{zir}}$.

Then, after determining $\tau_{zir}^{\text{begin}}$ as the time of the beginning of the performance of work iz, and $\tau_{zir}^{\text{compl}}$ as the time of its completion, the production schedule of the process of execution of an order can formally be represented as a set of pairs of the form

$$\langle\tau_{zir}^{\text{begin}}, \tau_{zir}^{\text{compl}}\rangle, z = 1, \ldots, Z, i = 1, \ldots, I, r = 1, \ldots, R.$$

The entered notations make it possible to allocate the general scheme of formation of the production schedule on the execution of orders as the procedure consisting of steps that are cyclically repeated: to select from the list of unconsidered orders Z the next order $\{z\}$ to include in the schedule and set a temporary reference point equal to the day number of the order; to select from the network schedule of the order $\{Z\}$ the maximum front of the works claiming inclusion in the schedule, i.e. such in which all the works following technology are already included in the schedule; to select from the formed front of works operation with the greatest complexity w_{zir} (this heuristic provides primary inclusion in the schedule of works of a critical direction); to include in the schedule the allocated work, having checked conditions of availability for its performance of all accessories, necessary materials, technological and auxiliary resources, equipment, the necessary tool, documentation; and to appoint for its performance a "minimum sufficient" number of units released after the performance of the next production operation of a resource, taking into account the execution requirements in terms of the order, preventing the increase in the duration of the critical direction of the order and ensuring the uniformity of the technological process. Then, the moments of time of the beginning and end of the allocated work are defined.

At each cycle of this procedure, one work of the selected order is built into the schedule. Therefore, after a finite number of repetitions of steps 2–4, equal to the power of the set of operations $\{o_{zir}\}$, the process of building a schedule for the selected order will be completed. In this case, in case of a shortage of resources on a given data, to perform any work on the set $\{o_{zir}\}$ all work on this set is excluded from the schedule and the order is re-considered subject to the allocation of the last work of its critical direction greater (e.g. per unit) than the resource.

Note that the described methodology of scheduling makes it possible, on the one hand, to carry out production planning, taking into account the current technological and organizational constraints, and on the other hand, to implement the principles of "pull" planning (pull scheduling), ensuring the execution of orders on the principle of just in time with minimization of work in progress.

The proposed methodology is recursive and makes it possible to build production schedules with verification of compliance with such integrated criteria of schedule efficiency as the criterion of the minimum calendar duration of the schedule, the criterion of the minimum cost of schedules, etc.

Thus, the formed production schedule that is focused on the process representation of the general scheme of the organization of works in a production system makes it possible to carry out a transition from production planning on time intervals and shops to out-of-order-oriented management. It also makes it possible, on its basis of process-oriented budget planning systems, to create focused control and cost management.

Consider below the actual criterion that will ensure the continuity of the modern enterprise. We will consider that everything described above is executed; the enterprise works in the conditions of resource maintenance with accurately planned schedules of release of finished goods. Let us actually explore how this process works at the atomic level. Let us explore how it works at the level of a separate production process at a certain production site (work center), which produces a specific type of product.

Without the automation of the process of control of parameters of production processes at modern enterprises, it is impossible to organize the serial release of qualitative production. In fact, to solve such problems in industrial enterprises, to ensure the stability of production processes through real-time control of key production parameters, a mathematical model is proposed. This model can be integrated into an automated enterprise management system.

5.3. Mathematical model of the production process management system of an industrial enterprise

Considering the problem of designing the control that ensures the execution of the production process in accordance with the established standards, the authors have reviewed the mathematical model of the production control system of an industrial enterprise. Having analyzed the problems of the automation of the enterprise management systems, the authors intend to suggest an algorithm for the automation of the production process using the technique of generalized inversion (Sobchuk et al. 2020).

The production process of a modern enterprise consists of a set of measures to produce finished, semi-finished or other products. One of the main tasks of industrial development is the introduction of the new products, machine and equipment designs, automation tools, the latest technologies, etc. Each product industry has its own specifics depending on the type of production, purpose, size and accuracy of the machines, as well as the level of production and technical equipment. In general, automation of the production is a stage of machine production, characterized by the release of the human factor from the direct performance of the management functions in the production processes and the delegation of these functions to the information and computing systems (Barabash et al. 2020). Control is a purposeful action on the object to ensure its operation in the optimal or specified mode within acceptable tolerances.

Automation of the production processes does not exclude a person completely from the value chain. Rather, automation means the most rational distribution of computing and production load at each production center. The proportions of such distributions depend on the specific enterprise and the goals of automation. The enterprise automation processes are subject to certain requirements, without which they become inefficient and difficult to implement.

First, a process management model is a must. At present, a significant number of the enterprises operate on the basis of a system-functional approach, which preceded the process approach. The complexity of the transition to the process management model depends on the scale and specifics of the enterprise.

Second, compliance of the current model of the enterprise processes with the technical criteria used in their automation is a very important requirement.

Nowadays, at modern enterprises, it is impossible to organize a serial production of the quality products without automating the process of control over the parameters of production processes. We suggest a mathematical model for solving the similar problems of the manufacturing plants, to ensure the stability of production processes with real-time control of key production parameters. This model can be integrated into an automated enterprise management system. The property of the functional stability of complex technical systems must be realized. This means that the technological process must perform its main technological tasks as intended under the influence of external and internal destabilizing factors (Maksymuk et al. 2020).

The automation stage allows re-engineering of the processes. The purpose of re-engineering is to find and overcome the bottlenecks in the enterprise. It is necessary to monitor the production potential, to identify the opportunities for expansion of the production system of the enterprise, the resources that are not used rationally and so on. It is important to prepare the companies for the re-engineering process. It is necessary to bring the structures of its processes in the most efficient configuration and in the most efficient form. It is necessary to ensure strict compliance with the requirements for the technological process at each production center in accordance with the standards, with deviations only within the permissible tolerance standards.

In practice, the principal characteristics of integrated automated enterprise management systems are widely implemented for this purpose. These systems automate a wide range of management functions, including the tasks of strategic, production and financial planning, operational management of supply, procurement and inventory. In addition, they automate the tasks of design, technological and technical preparation of production, etc.

Automation of the modern technological processes leads to the creation of complex dynamic models. Many processes are described by nonlinear dynamical systems of complex structure (Lyashko and Nomirovskii 2003; Dovgiy et al. 2017; Lyashko et al. 2019; Pichkur et al. 2020). The behavior of such models has a fractal structure (Barabash et al. 2019). At the same time, problems of control arise in such systems (Garashchenko and Pichkur 2000; Matvienko 2007; Pichkur and Sasonkina 2012).

Producing usually consists of a number of stages, at each of which there are certain requirements for the parameters and characteristics of the raw materials, semi-finished and finished products. We denote by $x(i)$ the vector of parameters at i-th stage, $i = 1,2, \dots N$. At each stage, there is an external influence $u(i)$ on the production process to obtain the desired parameters (work effect, energy effect, chemical or other technological influences at each stage). It is clear that the final quality of the product, as well as the intermediate quantity at each stage, depends on the strict adherence to the technology and one must ensure the endurance of the necessary parameters at each previous step. We assume that this requirement holds.

Also we denote by $A(i)$ a matrix of dependence of product quality indicators at $i + 1$-st stage on the indicators at i-th stage, and by $C(i)$ a matrix that determines the structure of influence on the production process $u(i)$. Then, the mathematical model of the technological process can be written as follows:

$$x(t+1) = A(t)x(t) + C(t)u(t), t = 0,1, \dots, N-1, \qquad [5.1]$$

$$x(t) \in \mathbb{R}^n,\ A(t) \in \mathbb{R}^{n\times n},\ \ C(t) \in \mathbb{R}^{n\times m},\ \ u(t) \in \mathbb{R}^m.$$

Here, we denote by $\mathbb{R}^n$ the n-dimensional Euclidean space with the Euclidean norm $\|\cdot\|$ in it, $x = (x_1, x_2, \dots, x_n)^T$ is a state vector of system [5.1], $u = (u_1, \dots, u_m)^T$ is a control vector, $A(t)$ is an $n \times n$ matrix and $C(t)$ is an $n \times m$

matrix, $t = 0,1,\dots,N-1$. Let $I_N = \{0,1,\dots,N\}$, $x(t, x_0, u)$ be a solution of system [5.1], $t \in I_N$ under the control $u(t)$, $t \in I_{N-1}$.

We admit that there is an accurately defined set of certain works and a number of criteria to fulfill during the realization of this process. This means that we know characteristics of the process at the initial stage, requirements for products at the end of the process and intermediate characteristics of products at control points at stages of this process. At the same time, in the automation of such processes, in practice, it is necessary to set control tasks describing the design conditions for the control function u providing controlled purposeful execution of the process. In addition, it is advisable to provide conditions for practical stability for these processes (Bashnyakov et al. 2001; Pichkur 2019; Pichkur and Linder 2021; Pichkur et al. 2021).

5.4. Control design

The main problem that we analyze consists of finding the control function providing execution of the process, so that the result of the process ensures ultimately in $x(N)$ products that meet all the quality characteristics required by current standards for it.

The purpose of designing a control function u is to ensure that the process is performed in such a way that we end up with a product that meets all the characteristics required by the standards. If at the end of the process the product has deviations from the specified standard parameters, then such deviations are guaranteed to fall into the set of permissible tolerances, which are defined by current standards for such products (Pichkur and Sobchuk 2021). This means that there is a desired final state $x_N \in \mathbb{R}^n$ and a positive parameter $\varepsilon > 0$ such that

$$\|x(N) - x_N\| < \varepsilon.$$

Let us define a set of admissible controls. To do this, we consider space $\ell_2^{(m)}$ of sequences of vectors from $\mathbb{R}^m$ such that if $\in \ell_2^{(m)}$, then $\sum_{t=0}^{\infty} \| u(t) \|^2 < \infty$, $u(t) \in \mathbb{R}^m$, $t = 0,1,2,\dots$. $\ell_2^{(m)}$ is a real Hilbert space with the inner product

$$\langle u, v \rangle_{\ell_2} = \sum_{t=0}^{\infty} \langle u(t), v(t) \rangle, \quad u, v \in \ell_2^{(m)}$$

and the norm

$$\| u \|_{\ell_2} = \sqrt{\sum_{t=0}^{\infty} \| u(t) \|^2} < \infty.$$

We assume that a control function u is admissible if $u \in \ell_2^{(m)}$ and $u(t) = 0$, $t = N+1, N+2, \ldots$.

Denote $\Theta(t) = A_{t-1}A_{t-2} \ldots A_1 A_0$, $\Theta(t,s) = A_{t-1}A_{t-2} \ldots A_s$, $\Theta(t,t) = E$, where E is the identity matrix, $t, s \in I_N$. The problem is finding a control that moves the system [5.1] from the initial state $x(0) = x_0$ to the nearest point $x(N)$ to a given state x_N (Matvienko 2007). Hence, we obtain the problem of minimization

$$I(u) = \| x(N, x_0, u) - x_N \| \tag{5.2}$$

on the solutions of system [5.1] with the initial condition $x(0) = x_0$. Here, $x(N, x_0, u)$ denotes the value of the solution of system [5.1] with the initial condition $x(0) = x_0$ for an admissible control $u \in \ell_2^{(m)}$ at the moment $t = N$. Since

$$x(N, x_0, u) = \Theta(N)x_0 + \sum_{k=0}^{N-1} \Theta(N,k)C(k)u(k),$$

then the substitution of the last equality in [5.2] gives

$$I(u) = \| x(N, x_0, u) - x_N \| = \| \Theta(N)x_0 + \sum_{k=0}^{N-1} \Theta(N,k)C(k)u(k) - x_N \| =$$

$$= \| \sum_{k=0}^{N-1} W(k)u(k) - c \| \tag{5.3}$$

where $c = x_N - \Theta(N)x_0$,

$$W(t) = \Theta(N,t)C(t),\ W^T(t) = (w_1(t)\ w_2(t)\ \ldots\ w_n(t)),$$

$w_j(t) \in \mathbb{R}^m$, $t \in I_{N-1}$, $j = 1,2,\ldots,n$ are the vectors describing the matrix $W(t)$ rows, $t \in I_{N-1}$. At the same time, we can observe that

$$H = \{w_1(\cdot), w_2(\cdot), \ldots, w_n(\cdot)\} \subset \ell_2^{(m)}$$

and $w_j(t) = 0, t = N, N+1, ..., j = 1,2, ..., n$.

We define a linear manifold $L = Lin\ H$. Since $\ell_2^{(m)}$ is the Hilbert space, then $\ell_2^{(m)}$ decomposes into a direct sum $\ell_2^{(m)} = L \oplus L^\perp$, where $L^\perp$ is an orthogonal complement to L. Any control $u \in \ell_2^{(m)}$ can be represented as follows:

$$u(t) = u_0(t) + v(t),\ t = 0,1, ... \tag{[5.4]}$$

Here, $u_0 \in L$, $v \in L^\perp$. Therefore, $\langle u_0, v\rangle_{\ell_2} = 0$ for $u_0 \in L$. Since $u_0 \in Lin\ H$, there exists a vector $\lambda = (\lambda_1, \lambda_2, ..., \lambda_n)^T$ such that

$$u_0(t) = \lambda_1 w_1(t) + \lambda_2 w_2(t) + \lambda_n w_n(t) = W^T(t) \quad \lambda \in Im(W^T(t)).$$

Here, $Im(\cdot)$ denotes an image of a linear operator. From [5.4], we have

$$u(t) = W^T(t)\lambda + v(t), \quad t = 0,1, ... \tag{[5.5]}$$

Observe that $\sum_{k=0}^{N-1} W(k)v(k) = 0$ since $v \in L^\perp$. Substituting [5.5] in [5.3] gives

$$J(\lambda) = \| \sum_{k=0}^{N-1} W(k)W^T(k)\lambda - c \| \to \min_{\lambda \in \mathbb{R}^n}, \tag{[5.6]}$$

$c = x_N - \Theta(N)x_0$. We see that $v \in L^\perp$ does not affect the solution [5.2], it plays the role of an invariant and it can be considered zero. Using the properties of generalized inversion (Albert 1972), we obtain the solution of problem [5.6]

$$\hat{\lambda} = \Phi^+(N)c + z, \tag{[5.7]}$$

where $\Phi(N) = \sum_{k=0}^{N-1} W(k)W^T(k)$, $\Phi^+(N)$ is a generalized inverse matrix to the matrix $\Phi(N)$, $z \in Ker(\Phi(N))$ is an arbitrary vector (Albert 1972).

Since $Ker(\Phi(N)) = Z_N\mathbb{R}^n$, $Z_N = Z(\Phi(N)) = E - \Phi^+(N)\Phi(N)$ is the projection operator onto $Ker(\Phi(N))$, then [5.7] has the following representation:

$$\hat{\lambda} = \Phi^+(N)c + Z_N p, \tag{[5.8]}$$

where $p \in \mathbb{R}^n$, $c = x_N - \Theta(N)x_0$ (Albert 1972). Formula [5.8] describes the set of all solutions of problem [5.6]. Note that among the vectors that solve the problem [5.6], the vector $\hat{\lambda} = \Phi^+(N)c$ has the smallest norm. This follows from the properties of generalized inverse matrices. Substituting [5.8] in [5.5] at $v(t) = 0$, $t \in I_{N-1}$ gives

$$u(t) = W^T(t)\Phi^+(N)(x_N - \Theta(N)x_0) + W^T(t)Z_N p, \qquad [5.9]$$

$$p \in \mathbb{R}^n, t \in I_{N-1}.$$

Formula [5.9] solves problem [5.2]. If $p = 0$, then

$$u(t) = W^T(t)\Phi^+(N)(x_N - \Theta(N)x_0), \qquad t \in I_{N-1}.$$

Substituting [5.9] in [5.3], we obtain

$$I(u) = J(\hat{\lambda}) = \| \Phi(N)\hat{\lambda} - c \| = \| \Phi(N)\Phi^+(N)c + \Phi(N)Z_N p - c \|.$$

Since $Z_N p \in Ker(\Phi(N)$, then $\Phi(N)Z_N p = 0$. Therefore

$$I(u) = \| \Phi(N)\Phi^+(N)c - c \| = \| (I - \Phi(N)\Phi^+(N))c \|.$$

Since $Y(\Phi(N)) = \Phi(N)\Phi^+(N)$ is a projector onto the image $Im(\Phi(N))$ of the matrix $\Phi(N)$, $Z(\Phi^T(N)) = E - \Phi(N)\Phi^+(N)$ is a projector onto the kernel $Ker(\Phi^T(N))$ of the matrix $\Phi^T(N)$ (Albert 1972), then

$$I(u) = \| Y(\Phi(N))c - c \| = \| Z(\Phi^T(N))c \|. \qquad [5.10]$$

Formula [5.10] shows how accurately we can move system [5.10] from the point $x(0) = x_0$ to the state $x(N) = x_N$. From [5.10], it follows that $I(u) = 0$ if and only if $(\Phi(N))c = c$ or $Z(\Phi^T(N)) = 0$. Thus, the following statement is true.

THEOREM 5.1. – The control function

$$u(t) = W^T(t)\Phi^+(N)(x_N - \Theta(N)x_0) + W^T(t)Z_N p \qquad [5.11]$$

moves system [5.1] from the initial state $x(0) = x_0$ to the nearest point $x(N)$ to a given state x_N. Here, $p \in \mathbb{R}^n$, $t \in I_{N-1}$, $Z_N = Z(\Phi(N)) = E - \Phi^+(N)\Phi(N)$. Moreover

$$\| x(N) - x_N \| = \| Y(\Phi(N))c - c \| = \| Z(\Phi^T(N))c \|$$

describes Euclidean distance from $x(N)$ to x_N, where $c = x_N - \Theta(N)x_0$. If $Y(\Phi(N))c = c$ so that $c = x_N - \Theta(N)x_0$ belongs to the image $Im(\Phi(N))$ of the matrix $\Phi(N)$, then [5.11] moves system [5.1] from $x(0) = x_0$ to the point $x(N) = x_N$.

Note that control function [5.11] solves the problem for arbitrary x_0, x_N if and only if $Z(\Phi^T(N)) = 0$ or in equivalent form $\Phi(N)\Phi^+(N) = E$. Since the matrix $\Phi(N)$ is symmetric of $n \times n$, this means that $\Phi^+(N) = \Phi^{-1}(N)$.

THEOREM 5.2.– Among the control functions that move system [5.1] from $x(0) = x_0$ to the nearest state $x(N)$ to the point x_N, the function

$$u_*(t) = W^T(t)\Phi^+(N)(x_N - \Theta(N)x_0),\ t \in I_{N-1} \qquad [5.12]$$

has the smallest norm in $\ell_2^{(m)}$.

Proof. From the proof of Theorem 5.1, it follows that the admissible control u moving system [5.1] from $x(0) = x_0$ to the nearest point to x_N satisfies [5.4], where $u_0 = u_* + z_0 \in L$, u_* is determined by [5.12], $z_0(t) = W^T(t)Z_N p$, $p \in \mathbb{R}^n$, $t \in I_{N-1}$, $v \in L^\perp$. Then

$$\| u \|_{\ell_2}^2 = \langle u, u\rangle_{\ell_2} = \langle u_0 + v_0, u_0 + v_0\rangle_{\ell_2} = \langle u_0, u_0\rangle_{\ell_2} + \langle v_0, v_0\rangle_{\ell_2} \geq$$

$$\geq \langle u_0, u_0\rangle_{\ell_2} = \langle u_* + z_0, u_* + z_0\rangle_{\ell_2} = \langle u_*, u_*\rangle_{\ell_2} + \langle z_0, z_0\rangle_{\ell_2} + 2\langle u_*, z_0\rangle_{\ell_2}.$$

Since $Z_N p \in Ker\Phi(N)$ we have $\Phi(N)Z_N p = 0$ and

$$\langle u_*, z_0\rangle_{\ell_2} = \sum_{t=0}^{N-1} \langle W^T(t)\Phi^+(N)(x_N - \Theta(N)x_0), W^T(t)Z_N p\rangle =$$

$$= \langle \Phi^+(N)(x_N - \Theta(N)x_0), \Phi(N)Z_N p\rangle = 0.$$

Finally, we obtain

$$\| u \|_{\ell_2}^2 \geq \langle u_*, u_*\rangle_{\ell_2} + \langle z_0, z_0\rangle_{\ell_2} \geq \langle u_*, u_*\rangle_{\ell_2} = \| u_* \|_{\ell_2}^2.$$

The last inequality proves the theorem.

5.5. Algorithm of control of production process

We offer an algorithm of control design for the production process, which ensures the production according to a standard in compliance with the permissible standards of tolerances at a production plant:

1) given the initial state $x(0) = x_0$ and the final state $x(N) = x_N$, the parameter $\varepsilon > 0$ determining the set of possible deviations (tolerances) for the product from the requirements of the standard, the matrices $A(t)$, $C(t)$, $t = 0,1, \ldots N-1$;

2) find the matrices $\Theta(N) = A_{N-1}A_{N-2} \ldots A_1A_0$, $\Theta(N,t) = A_{N-1}A_{N-2} \ldots A_t$, $W(t) = \Theta(N,t)C(t)$, $t = 0,1, \ldots N-1$;

3) find the matrix $\Phi(N) = \sum_{k=0}^{N-1} W(k)W^T(k)$;

4) find the generalized inverse matrix $\Phi^+(N)$;

5) find the control function $u(t) = K(t)(x_N - \Theta(N)x_0)$, where $K(t) = W^T(t)\Phi^+(N)$ for all $t = 0,1, \ldots N-1$;

6) find the matrix $Z_N = E - \Phi^+(N)\Phi(N)$. If the condition

$$\|Z_N(x_N - \Theta(N)x_0)\| < \varepsilon$$

is true, then the control $u(t)$ solves the problem with specified tolerances

$$\|x(N) - x_N\| < \varepsilon.$$

Otherwise the control $u(t)$ ensuring the production of products with given tolerances does not exist. Here is the end of the algorithm description.

5.6. Conclusion

A modern enterprise is a set of a large number of atomic processes that are combined into a single set of management of all its resources as a single coordinated mechanism. Fulfilling the requirements for the organization of processes in a modern enterprise, in the vast majority of cases we will deal with the matrix topology of the organization of production processes. This topology of the organization of processes provides a functionally stable organization of the production enterprise. Its information space thoroughly combines the operation of ERP systems and MES's with feedback from each atomic production process.

In a modern enterprise, many parallel business processes operate at the same time, and quite often not only linear ones. It is important to note that to obtain high quality, the level of service, flexibility, low cost, etc. processes must be simple and the following conditions must be met: implementation of integrated processes; production centers (units) make decisions, i.e. perform the part of the work that was previously assigned to the management function; stages of the process are performed naturally; there are different versions of the production process; the work is performed where it is appropriate to perform, which makes it possible to achieve optimal solutions; reducing the share of inspection and control work; minimization of approvals; single coordination center and combination of centralized and decentralized operations.

This requires the organization of a topology of production processes that meet these requirements – in particular, the organization of production with symmetrical directions of production, which increase the efficiency of production processes while increasing productivity. The rapid dynamics of development of modern enterprises requires the evolution of approaches for high-quality modeling of modern production processes, the development of effective algorithms for managing production processes and the constant modernization of automated information management systems.

In this chapter, we have analyzed the modern approaches to the automation of production process management in the industrial enterprises, and we have solved the problem of designing a control function that ensures the implementation of the production process. As a result of the process, we get a finished product that meets all the characteristics required by the current standards. We propose an algorithm for automating the atomic process of production.

The research results are important for the design, modernization and integration of the enterprise information systems into one generalized enterprise information system. This will ensure their high efficiency in operation. The lack of such solutions in our country and abroad makes research results a priority.

We see the prospects for further research in the design and improvement of the models and methods for constructing functionally stable technological processes that are integrated into the information system of the enterprise. This approach ensures the efficiency of the information infrastructure during the time required to perform the technological processes and sustainable operation of the enterprise as a whole. In doing so, we will take into account the specific needs of the enterprises operating in the sectors that have a continuous production cycle, such as metallurgy, energy, the chemical industry and so on.

5.7. Acknowledgments

This work was partially supported by the Ministry of Education and Science of Ukraine: Grant of the Ministry of Education and Science of Ukraine for perspective development of a scientific direction "Mathematical sciences and natural sciences" at Taras Shevchenko National University of Kyiv.

5.8. References

Albert, A. (1972). *Regression and the Moore-Penrose Pseudoinverse*. Elsevier, Burlington.

Barabash, O.V., Kopiika, O.V., Zamrii, I.V., Sobchuk, V.V., Musienko, A.P. (2019). Fractal and differential properties of the inversor of digits of Q_s-representation of real number: Modern mathematics and mechanics. In *Modern Mathematics and Mechanics: Fundamentals, Problems and Challenges*, Sadovnichiy, V.A. and Zgurovsky, M.Z. (eds). Springer, Cham.

Barabash, O.V., Tverdenko, H.V., Sobchuk, V.V., Musienko, A.P., Lukova-Chuiko, N.V. (2020). The assessment of the quality of functional stability of the automated control system with hierarchic structure. *IEEE 2nd International Conference on System Analysis Intelligent Computing (SAIC)*, Kyiv, 158–161.

Bashnyakov, A.N., Pichkur, V.V., Hitko, I.V. (2001). On maximal initial data set in problems of practical stability of discrete system. *Journal of Automation and Information Sciences*, 43(3), 1–8.

Dovgiy, S.O., Lyashko, S.I., Cherniy, D.I. (2017). Algorithms of the discrete singularity method of computational technologies. *Cybernetics and System Analysis*, 53(6), 147–159.

Garashchenko, F.G. and Pichkur, V.V. (2000). Structural optimization of dynamic systems by use of generalized Bellman's principle. *Journal of Automation and Information Sciences*, 32(3), 1–6.

Lyashko, S.I. and Nomirovskii, D.A. (2003). Generalized solutions and optimal controls in systems describing the dynamics of a viscous stratified fluid. *Differential Equations*, 39(1), 90–98.

Lyashko, S.I., Klyushin, D.A., Timoshenko, A.A., Lyashko, N.I., Bondar, E.S. (2019). Optimal control of intensity of water point sources in unsaturated porous medium. *Journal of Automation and Information Sciences*, 51(7), 24–33.

Maksymuk, O.V., Sobchuk, V.V., Salanda, I.P., Sachuk, Yu.V. (2020). A system of indicators and criteria for evaluation of the level of functional stability of information heterogenic networks. *Mathematical Modeling and Computing*, 7(2), 285–292.

Matvienko, V.T. (2007). Control of trajectories set by linear dynamic systems with discrete argument. *Journal of Automation and Information Sciences*, 39(11), 4–10.

Pichkur, V.V. (2019). Maximum sets of initial conditions in practical stability and stabilization of differential inclusions. In *Modern Mathematics and Mechanics: Fundamentals, Problems and Challenges*, Sadovnichiy, V.A. and Zgurovsky, M. (eds). Springer, Berlin.

Pichkur, V.V. and Linder, Y.M. (2021). Practical stability of discrete systems: Maximum sets of initial conditions concept. In *Contemporary Approaches and Methods in Fundamental Mathematics and Mechanics. Understanding Complex Systems*, Sadovnichiy, V.A. and Zgurovsky, M. (eds). Springer, Berlin.

Pichkur, V.V. and Sasonkina, M.S. (2012). Practical stabilization of discrete control systems. *International Journal of Pure and Applied Mathematics*, 81(6), 877–884.

Pichkur, V.V. and Sobchuk, V.V. (2021). Mathematical model and control design of a functionally stable technological process. *Journal of Optimization, Differential Equations and their Applications (JODEA)*, 29(1), 32–41.

Pichkur, V.V., Kapustyan, O.V., Sobchuk, V.V. (2020). *Theory of Dynamical Systems*. Vezha, Lutsk.

Pichkur, V.V., Linder, Y.M., Tairova, M.S. (2021). On the practical stability of discrete inclusions with spatial components. *Journal of Mathematical Sciences*, 254(2), 280–286.

Sobchuk, V.V., Pichkur, V.V., Barabash, O.V., Laptiev, O.A., Kovalchuk, I.R., Zidan, A.A. (2020). Algorithm of control of functionally stable manufacturing processes of enterprises. *2nd IEEE International Conference on Advanced Trends in Information Theory*, 206–210.

Zagidullin, R.R. (2011). *Management of Machine-Building Production by Means of MES, APS, ERP Systems*. TNT, Stary Oskol [in Russian].

6

Alternative Direction Multiblock Method with Nesterov Acceleration and Its Applications

Vladislav HRYHORENKO[1], Nataliia LYASHKO[2], Sergiy LYASHKO[1] and Dmytro KLYUSHIN[1]

[1] *Taras Shevchenko National University of Kyiv, Ukraine*

[2] *V.M. Glushkov Institute of Cybernetics, National Academy of Sciences of Ukraine, Kyiv, Ukraine*

6.1. Introduction

Proximal algorithms in machine learning are useful for solving the problem of optimizing a complex function. By proximal algorithm, we mean an algorithm where, when finding the optimal function, we compute the proximal operator. Algorithms such as the proximal gradient method, proximal Newton method, ISTA, ADMM, forward-backward and Douglas–Ratchford splitting and Nesterov acceleration are considered. We also consider the problem of optimizing a complex function, where the objective function contains a non-smooth or non-convex function. Many optimization problems in statistics have the form: $\arg\min_{x \in X} F(x) = l(x) + \phi(x)$, where $l(x)$ depends on certain observed data and $\phi(x)$ is

Computational Methods and Mathematical Modeling in Cyberphysics and Engineering Applications 1, coordinated by Dmitri KOROLIOUK, Sergiy LYASHKO and Nikolaos LIMNIOS. © ISTE Ltd 2024.

a regularization expression. As a rule, $l(x)$ is a smooth convex function and $\phi(x)$ is a convex lower semi-continuous but non-smooth function (Parikh and Boyd 2013). This allows us to find the solution of the optimization problem for a class of objective functions that are not differentiable. Another useful feature of proximal methods is the ability to apply approaches that significantly increase computational efficiency (Nesterov acceleration). One of the most effective proximal methods is the ADMM method. We will consider its already known modifications that improve the rate of convergence and propose a new approach based on a combination of multiblock ADMM and the calculation of Nesterov acceleration at each iteration.

6.2. Proximal operators

Let $f(x)$ be a lower semi-continuous function and $\gamma > 0$ a scalar. Then, semi-continuous below Moreau envelopes f and a proximal operator $prox_{\gamma f}(x)$ with parameter γ are defined as

$$\underset{\gamma f}{prox}(x) = \arg\min_{z} \left\{ f(z) + \frac{1}{2\gamma} \|z - x\|_2^2 \right\}$$

Intuitively, a Moreau envelope is a regularized version of f. It approximates $\min_{x \in C} f(x)$ and has the same set of minimum values. A proximal operator defines a solution of a minimization problem for a Moreau envelope. It allows us to find a minimum value of $\min_{x \in C} f(x)$ and a solution in a neighborhood of ρ, and via γ, we control an error.

A proximal operator is an analogue of a gradient step. To justify this statement, let us consider a Moreau envelope $f^{\gamma}(x)$ approximating f. The derivative of a Moreau envelope is

$$\partial f^{\gamma}(x) = \partial \inf_{z} \left\{ f(z) + \frac{1}{2\gamma} \|z - x\|_2^2 \right\} = \frac{1}{\gamma} [x - \hat{z}(x)]$$

where $\hat{z}(x) = prox_{\gamma f}(x)$ is a minimum point. Hence

$$\underset{\gamma f}{prox}(x) = x - \gamma \partial f^{\gamma}(x).$$

Thus, the calculation of the proximal operator can be considered equivalent to a gradient step for a regularized initial function with a parameter γ.

A proximal operator is a generalization of an Euclidean projection (Patrinos and Bemporad 2013). Consider a case where $f(x)=I_C(x)$ is a set with indicator functions of some convex set *C*. Then, $prox_f(x)=\arg\min_{z\in C}\|x-z\|_2^2$ is a common Euclidean projection on *C*. For other functions, the proximal operator can be considered as a generalized projection. Bounded optimization problem $\min_{x\in C} f(x)$ has a solution equivalent to a problem for an unbounded proximal operator. Proximal approaches are directly related to convex relaxation and quadratic majorization, due to the addition of an expression like $\frac{\rho}{2}\|x-v\|^2$ to an objective function, where ρ is a constant that limits the operator or Hessian function. We can choose which variables this expression contains and the order in which to perform optimization steps.

There is a close connection between the proximal operator and fixed point theory: $prox_{\gamma f}(x^*)=x^*$ iff x^* minimizes $f(x)$. To make sure of this, consider the proximal minimization algorithm, where we start at a point and recompute the proximal operator

$$x^{t+1}=\underset{\gamma f}{prox}(x^t)=x^t-\gamma\nabla f^\gamma(x^t)$$

If iterations converge, we will reach a minimum point x^* of the Moreau envelope, and hence, the minimum points of the initial function. At this point, we have

$$\nabla f^\gamma(x^*)=0,$$

then

$$prox_{\gamma f}(x^*)=x^*.$$

Another important property of the proximal operator is Moreau decomposition for the proximal operator f^* conjugated to f

$$x = \underset{\lambda f}{prox}(x) + \lambda \underset{f^*/\lambda}{prox}(\lambda x),$$

$$I - \underset{\lambda f}{prox}(x) = \lambda \underset{f^*/\lambda}{prox}(\lambda x).$$

This allows us to easily change the step and perform the same calculations in conjugate space.

One of the simplest proximal methods is the proximal gradient method, which is an important starting point for the study of more complex methods.

Suppose that the objective function has the form $F(x)=l(x)+\varphi(x)$, where $l(x)$ is a differentiable function and $\varphi(x)$ is the non-differentiable function. A typical example of such a problem is a generalized linear model with a non-differentiable penalty function to increase the sparseness. The proximal gradient method is well suited for this task. It consists of only two basic steps that are repeated at each iteration to convergence.

1) Gradient step

Let us define an intermediate value v^t computing the gradient in the direction $\nabla l(x)$: $v^t = x^t - \gamma \nabla(x^t)$

2) Computing the proximal operator

Calculate the value of the proximal operator of the differential function at the intermediate point v^t (Patrinos and Bemporad 2013):

$$x^{t+1} = \underset{\gamma\varphi}{prox}(v^t) = \underset{\gamma\varphi}{prox}\{x^t - \gamma \nabla(x^t)\}$$

The convergence can be justified in at least two ways.

1) As an MM algorithm

Suppose that $l(x)$ has a Lipschitz continuous gradient with module λ_l. This allows us to find a majorant function: for any $\gamma \in (0, 1/\lambda_l]$

$$l(x) + \varphi(x) \le l(x_0) + (x - x_0)^T \nabla l(x_0) + \frac{1}{2\gamma} \|x - x_0\|_2^2 + \varphi(x).$$

At the point $x = x_0$, the inequality converts into equality. The optimal value of the expression on the right side of the inequality has the form

$$\hat{x} = \arg\min_{x} \left\{ \varphi(x) + \frac{1}{2\gamma} \|x - u\|_2^2 \right\},$$

where $u = x_0 - \gamma \nabla l(x_0)$.

From the fact that we can consider this method as an MM algorithm, the following consequences for convergence follow. Suppose that $l(x)$ is a convex function, and $\nabla l(x)$ is a Lipschitz continuous gradient with module λ_l, i.e. $\|\nabla l(x) - \nabla l(z)\|_2 \leq \lambda_l \|x - z\|_2$, $\forall x, z$, φ is a convex and closed function. This fact proves that $prox_{\gamma\varphi}$ is meaningful. The optimal value is finite and attained in x^*. If these conditions are met, the proximal gradient method converges with the rate $1/t$ with a fixed step $\gamma = 1/\lambda_l$.

2) As a fixed point of an "forward-backward" operator

The proximal gradient method can also be used as a tool to find a fixed point of a forward-backward operator obtained from standard optimality conditions from subdifferential calculus. A necessary and sufficient condition for the fact that x^* minimizes $l(x)$ is

$$0 \in \partial \left\{ l(x^*) + \varphi(x^*) \right\} = \nabla l(x^*) + \partial \varphi(x^*).$$

Let us use this fact to prove that x^* is a fixed point of the operator

$$x^* = \underset{\gamma\varphi}{prox} \left\{ x^* - \gamma \nabla l(x^*) \right\}.$$

Let I be the identity operator. Note that the optimal conditions are equivalent to the following:

$$0 \in \gamma \nabla l(x^*) - x^* + x^* + \gamma \partial \varphi(x^*)$$
$$(I - \gamma \nabla l) x^* \in (I + \gamma \partial \varphi) x^*,$$
$$x^* = (I + \gamma \partial \varphi)^{-1} (I - \gamma \nabla l) x^* = \underset{\gamma\varphi}{prox}(x^* - \gamma \nabla l(x^*)).$$

The latter statement follows from the fact that the proximal operator is a resolvent of the subdifferential operator: $prox_{\gamma\varphi}(x)=(I+\gamma\partial\varphi)^{-1}(x)$. Then, to find the solution, we recalculate the operator for which x^* is a fixed point:

$$x^{t+1}=\underset{\gamma\varphi}{prox}\{x^t-\gamma^t\nabla l(x^t)\}.$$

Thus, we have a proximal gradient method. We now show that the proximal operator is a resolvent of the subdifferential operator. By definition, if $z\in(I+\gamma\partial l)^{-1}x$, then $x\in(I+\gamma\partial l)z$, $x\in z+\gamma\partial l(z)$, $0\in\frac{1}{\gamma}(z-x)+\partial l(x)$, $0\in\partial_z\left\{\frac{1}{2\gamma}\|z-x\|_2^2+l(x)\right\}$.

For 0 to belong to the subdifferential of the function on the right-hand side, it is necessary and sufficient that the equality

$$z=\arg\min_u\left\{\frac{1}{2\gamma}\|u-x\|_2^2+l(x)\right\}=\underset{\gamma l}{prox}(x)$$

Thus, $z=prox_{\gamma l}(x)$ iff $z\in(I+\gamma\partial l)^{-1}x$. It should be noted that $(I+\gamma\partial l)^{-1}$ is single-valued; therefore, it is a function despite the fact that ∂l is a set (Polson et al. 2015).

Consider the application of the proximal gradient method to the quadratic form of the likelihood function as a problem of weighted least squares, with a penalty function $\varphi(x)$. Then, $\nabla l(x)=A^T\Omega Ax-A^T\Omega y$ and the proximal gradient method takes the form

$$x^{t+1}=\underset{\gamma\phi}{prox}\left\{x^t-\gamma^t A^T\Omega(Ax^t-y)\right\}.$$

This algorithm is widely known as IST, or iterative shrinkage thresholding (Tibshirani 1996). The cost of calculations on each iteration: (1) multiplying the value on the current iteration by A and (2) multiplying the residual error Ax^t-y by $A^T\Omega$. Usually, the proximal operator for ϕ is easy to calculate, as, for example, in the case when the penalty function is L_1 or L_2-norms, and the impact of these calculations on the overall complexity of the algorithm is negligible.

The proximal gradient method is a generalization of the classical gradient approach. It requires calculations of first-order derivatives, and its rate of convergence can be improved by calculating second-order derivatives, where the resulting algorithm is based on a quasi-Newtonian procedure.

The quadratic estimate obtained by the definition of the proximal operator uses a linear approximation $l(x)$, but it is natural to use a higher order of approximation (Patrinos and Bemporad 2013). Let us put

$$F_H(x,z)=l(z)+\nabla,$$

$$l(z)^T(x-z)+\frac{1}{2}(x-z)^T H_z(x-z).$$

Then, compute the proximal operator

$$\underset{F_h}{prox}(z)=z-(\gamma^{-1}I+H_z)^{-1}\nabla l(z).$$

Instead of calculating Hessian directly, we can use the quasi-Newtonian approximation. Newton's proximal method can even be applied to some non-convex problems.

One of the advantages of proximal algorithms is that we can accelerate the execution of the algorithm steps by a step that adds a perturbation parameter for the intermediate variable z before computing the forward and inverse steps:

$$z^{t+1}=x^t+\theta_{t+1}(\theta_t^{-1}-1)(x^t-x^{t-1}),$$

$$x^{t+1}=\underset{\gamma^{-1}\varphi}{prox}(z^{t+1}-\gamma^{-1}\nabla l(z^{t+1})),$$

3 $\theta_t=2/(t+1)$, $\theta_{t+1}(\theta_t^{-1}-1)=(t-1)/(t+2)$.

If the function φ is convex, then the proximal problem is strictly convex, and more complex techniques can be used to speed up calculations.

6.3. ADMM (alternating direction method of multipliers)

The initial problem $\min_x l(z)+\varphi(z)$ is equivalent to the problem $\min_x l(z)+\varphi(x)$ subject to $x-z=0$. Here, we introduced an additional variable z and rewrote the initial restrictions in the form $x-z=0$. Entering other additional parameters is also possible. For example, consider an exponential model with a forecast and a generating function $\psi(z)$: $p(y)=p_0(y)\exp\{yz-\psi(z)\}$.

In the generalized linear model, the parameter z for the prediction is a linear regression, where $z_i=a_i^T x$. In this case, we may represent $l(x)$ in the form: $l(x)=\sum_{i=1}^{N} l_i(x)$, where $l_i(x)=\psi(a_i^T x)-y_i(a_i^T x)$.

Substituting $z_i=a_i^T x$, we obtain the following problem:

$$\min_{x,z} \sum_{i=1}^{N} \{\psi(z_i)-y_i z_i\}+\varphi(x) \text{ subject to } Ax-z=0$$

A similar optimization problem can be obtained when we consider convex variational forms. For example, we can obtain such an equality for densities $p(x),q(x)$ and constants μ,k:

$$-\log p(x)=-\sup_{z>0}\log(p_N(x;\mu+\kappa/z,z^{-1})q(z)=\inf_{z>0}\left\{\frac{z}{2}(x-\mu-\kappa/z)^2-\log(\sqrt{z}q(z))\right\},$$

where $p_N(x;\mu,\sigma^2)$ is a density of a Gaussian distribution with a median μ and a variance σ^2.

The advantage of this splitting approach is that the penalty function is no longer related to the objective function of the initial problem. The standard approach to prove this is to set and solve a problem that is dual to the initial optimization (Polson et al. 2015).

An additional variable allows us to consider the problem $\min_x F(x)$ as an extended problem $\min_{x,z} F(x,z)$, where an extended problem $F(\cdot,\cdot)$ may be easily minimized. Also, the proximal operator allows us to change the minimization step for a non-smooth penalty function. Moreover, solving complex problems of large dimension can be reduced to a group of smaller subproblems, and a general solution can be obtained from the solution of subproblems.

Let us start with a simple problem

$$\min_x l(x)$$
$$\textit{subject to } Ax=y.$$

Consider the Lagrangian of this problem

$$L(x,z)=l(x)+z^T(Ax-y)=l(x)+(A^T z)^T x-z^T y.$$

A dual function is $g(z)=\inf_x L(x,z)=-l^*(-A^T z)-y^T z$, and a dual problem has the form $\max_z g(z)$.

Let p^* and d^* be optimal values of the initial and the dual problem, respectively. These optimal values are the same. Also, the following conditions are satisfied:

$$x^*=\arg\min_x L(x,z^*)\Leftrightarrow 0\in\partial_x L(x^*,z^*)$$

The idea of a dual descent is to solve a dual problem with a gradient descent method

$$\nabla g(z)=\nabla_z L(\hat{x}_z,z), \text{ where } \hat{x}_z=\arg\min_x L(x,z).$$

The expression on the right-hand side is a residual for the constraints of the objective function

$$\hat{x}_z=\arg\min_x L(x,z).$$

$$\nabla_z L(x,z)=Ax-y.$$

Therefore, for each iteration, a dual gradient descent method consists of two steps:

$$x^{k+1} = \arg\min_x L(x, z^k)$$

$$z^{k+1} = z^k + \alpha_k (Ax^{k+1} - y),$$ where α_k is a step length.

Consider the same problem as mentioned previously:

$\min_x l(x)$ subject to $Ax = y$ with the Lagrangian $L(x,z) = l(x) + z^T (Ax - y)$

The augmented Lagrangian approach is to stabilize the error in the intermediate step of the iteration by adding to the Lagrangian expression similar to that used in the ridge regression

$$L_\gamma(x,z) = l(x) + z^T (Ax - y) + \frac{\gamma}{2} \|Ax - y\|_2^2.$$

One way to show this is to consider a standard Lagrangian for an equivalent problem

$$\min_x l(x) + \frac{\gamma}{2} \|Ax - y\|_2^2$$

subject to $Ax = y$,

For any x, from an attainable set of the initial problem, the object function did not change. The dual function $g_\gamma(z) = \inf_x L_\gamma(x,z)$ is differentiable and strictly convex.

Now, we can apply a dual gradient descent method to the modified problem

$$x^{k+1} = \arg\min_x \left\{ l(x) + z^T (Ax^k - y) + \frac{\gamma}{2} \|Ax - y\|_2^2 \right\},$$

$$z^{k+1} = z^k + \alpha_k (Ax^{k+1} - y).$$

Therefore, the update on each iteration of the independent variable does not change compared to the standard dual gradient descent. But the formula by which we recompute the independent variable x contains a regularization expression, i.e. the value of which depends on the correction parameter γ.

Let us introduce the substitution $u=\gamma^{-1}z$ and rewrite the extended Lagrangian with $r=Ax-y$ as:

$$L\gamma(x,u)=l(x)+\gamma u^{T}(Ax-y)+\frac{\gamma}{2}\|Ax-y\|_2^2=l(x)+\frac{\gamma}{2}\|r+u\|_2^2-\frac{\gamma}{2}\|u\|_2^2$$

Hence

$$x^{k+1}=\arg\min_{x}\left\{l(x)+\frac{\gamma}{2}\left\|Ax-y+u^k\right\|_2^2\right\},$$

$$u^{k+1}=u^k+(Ax^{k+1}-y)$$

6.4. Bregman iteration

The method of extended Lagrangian for the LASSO problem is named the Bregman iteration (Tibshirani 1996) and has the form

$$\min_{x}\|x\|_1 \text{ subject to } Ax=y,$$

where y is the observation, x is an unknown signal and the matrix A contains fewer observations than the number of coordinates in the vector x. The extended Lagrangian corresponding to this problem has the form

$$L_{\gamma}(x,u)=\|x\|_1+\frac{\gamma}{2}\|Ax-y+u\|_2^2-\frac{\gamma}{2}\|u\|_2^2$$

with the step

$$x^{k+1}=\arg\min_{x}\left\{\|x\|_1+\frac{\gamma}{2}\|Ax-z_k\|_2^2\right\},$$

$$z^{k+1}=y+z^k-Ax^{k+1},$$

where $z^k=y-u^k$

Thus, each intermediate step of the Bregman iteration is similar to the Lasso regression. Combining the idea of separating variables with extended Lagrangian, we obtain the ADMM (alternating-direction method of multipliers). Advanced Lagrangian has the form

$$L_\gamma(x,z,u)=l(z)+\varphi(x)+\frac{\gamma}{2}\|x-z+u\|_2^2+\frac{\gamma}{2}\|u\|_2^2$$

ADMM is similar to the dual gradient method for this problem, except that we optimize the Lagrangian in x and z separately. The ADMM iteration is as follows:

$$z^{k+1}=\arg\min_z\left\{l(z^k)+\frac{\gamma}{2}\left\|x^k-z^k+u^k\right\|_2^2\right\},$$

$$x^{k+1}=\arg\min_x\left\{\varphi(x^k)+\frac{\gamma}{2}\left\|x^k-z^{k+1}+u_k\right\|_2^2\right\},$$

$$u^{k+1}=u^k+x^{k+1}-z^{k+1}.$$

The first two formulas are the proximal operators for the functions $l(x)$ and $\varphi(x)$, respectively.

One way to interpret the ADMM method is to use a proximal gradient method for a dual problem. Nesterov acceleration can also be applied, but under additional conditions, in particular the strict convexity of the function $l(x)$.

6.5. Forward-backward envelope (FBE)

Suppose that we minimize $F=l+\varphi$, where l is a strictly convex function that has a Lipschitz continuous gradient with module λ_l, i.e. $\left|\nabla^2 l(x)\right|\leq\lambda_l$. The penalty function φ is lower semi-continuous and convex. Define FBE

$$F_\gamma^{FB}(x):=\min_v\left\{l(x)+\nabla l(x)^T(v-x)+\varphi(v)+\frac{1}{2\gamma}\|v-x\|^2\right\}$$

$$=l(x)-\frac{\gamma}{2}\|\nabla l(x)\|^2+\varphi^\gamma(x-\gamma\nabla l(x))$$

If $\gamma \in (0, \lambda_l^{-1})$, then $I - \gamma \nabla^2 l(x)$ is a symmetrical and positively defined matrix. The fixed point of $F_\gamma^{FB}(x)$ is a solution x^* of the equation $x = prox_{\gamma\varphi}(x - \gamma \nabla l(x))$, since

$$\nabla F_\gamma^{FB}(x) = (I - \gamma \nabla^2 l(x)) G_\gamma(x),$$
$$G_\gamma(x) = \gamma^{-1}(x - P_\gamma(x)), P_\gamma(x) = prox_{\gamma\varphi}(x - \gamma \nabla l(x)).$$

Therefore

$$F_\gamma^{FB}(x) \le F(x) - \frac{\gamma}{2} \|G_\gamma(x)\|^2,$$
$$F(P_\gamma(x)) \le F_\gamma^{FB}(x) - \frac{\gamma}{2}(1 - \gamma\lambda_l) \|G_\gamma(x)\|^2.$$

Thus, if $\gamma \in (0, \lambda_l^{-1})$, then FBE is monotonically decreasing, and its fixed point exists.

6.6. Douglas–Rachford envelope (DRE)

The Douglas–Rachford envelope has the form

$$F_\gamma^{DR}(x) = l^\gamma(x) - \frac{\gamma}{2} \left\| \nabla l^\gamma(x) \right\|_2^2 + \varphi^\gamma(x - 2\gamma \nabla l^\gamma(x)) =$$

$$= \min_z \left\{ l(x^*) + \nabla l(x^*)^T (z - x^*) + \varphi(z) + \frac{1}{2\gamma} \left\| z - x^* \right\|^2 \right\},$$

where l^γ is a Moreau envelope for l and $x^* = prox_{\gamma l}(x)$. We may interpret this fact as a case of FBE for a proximal γl

$$F_\gamma^{DR}(x) = F_\gamma^{FB}(\underset{\gamma l}{prox}(x))$$

Computing the gradient of this envelope leads to the following proximal algorithm that coincides with the solution $\min_x \{ l(x) + \varphi(x) \}$:

$$w^{t+1} = \underset{\gamma l}{prox}(x^t),$$

$$z^{t+1} = \underset{\gamma\varphi}{prox}(2w^t - x^t),$$

$$x^{t+1} = x^t + (z^t - w^t).$$

There are many ways to reduce the DR algorithm to a more convenient form. For example, we can enter an additional variable $v = w - x$, and the algorithm will have the form

$$w^{t+1} = \underset{\gamma l}{prox}(x^t - v^t),$$

$$x^{t+1} = \underset{\gamma\varphi}{prox}(w^t + v^t),$$

$$v^{t+1} = v^t + (w^t - x^t).$$

6.7. Proximal algorithms for complex functions

Now, consider the problem of optimizing a complex objective function given as follows:

$$\min_x F(x) := l(x) + \varphi(Bx).$$

The composition of maps $\varphi(Bx)$ arises in multidimensional statistical models that take into account structural models and correlations. Therefore, any practical approach to estimating statistical models must be able to take into account to some extent the impact of these maps. The approaches described in this chapter use partitions, proximal operators and Moreau envelopes.

Combinations of these methods can easily be used together to optimize a wide variety of object functions. Solving an optimization problem, we need to determine the exact steps to solve the subproblems within it. In some cases, the solution in analytical form may not exist or be too difficult to calculate. On the contrary, the exact solution of such a problem, which has similar critical points, can be easily calculated. Such problems can take the form of an envelope, and as a result, the solution may be a proximal operator. In some cases, especially if the optimization is performed on several variables, the use of proximal operators and their properties can greatly simplify the iterative formulas for finding the optimal value and allow the transition from basic to dual space and vice versa.

Note that the quadratic term of the proximal operator can be considered as a quadratic penalty for the linear Lagrangian constraint. In cases where the separation of the objective function is used, the use of the proximal operator leads to results similar to the extended Lagrangian. In particular, the addition of a quadratic term in the problem leads to the use of the ADMM technique of the method with which each iteration is optimized for variables x and z. Then, the extended Lagrangian for ADMM has the form

$$\varphi(w)+l(x)+z^T(Bx-w)+\frac{\rho}{2}\|Ax-z\|^2$$

$$F_{SP}(x,w,z)+\frac{\rho}{2}\|Ax-z\|^2$$

We can consider a direct proximal solution for some values of the variables of the objective function of this problem, for example

$$z^*=prox_{F_{SP}(x,w,z)/\rho}(Ax)$$

The objective function has the form

$$F_{SP}(x,w,z^*)+\frac{\rho}{2}\left\|Ax-z^*\right\|^2$$

If z^* is a linear function of w, then we may consider another proximal step $w^*=prox_{F_{SP}(x,w,z^*)/\rho}(\ldots)$. If the exact solution for a complex proximal operator cannot be found, linearization can be applied to $\frac{\rho}{2}\|Ax-z\|^2$:

$$F_{SP}(x,w,z)+\frac{\rho}{2}\|Ax-z\|^2\le F_{SP}(x,w,z)+\frac{\rho}{2\lambda_A}\|x-z\|^2$$

To demonstrate the above statement, we consider an example of how proximal operators, their properties and approaches can be used to solve the problem of optimizing a complex function.

Let $l(x), \varphi(x)$ be the convex function with Lipschitz continuous derivatives. Consider the following problem:

$$\max_z \inf_x \left\{ l(x) + z^T (Bx) - \varphi^*(z) \right\}$$

and note that the minimum of $l(x) + z^T (Bx)$ is attained if

$$x^* = \underset{\lambda_l (l(x)+z^T Bx)}{prox} (x^*) .$$

Applying the property of a proximal operator

$$\underset{g(z)+u^T z}{prox} (q) = \underset{g}{prox}(q - u)$$

to g(z), q, z and u, we have

$$x^* = \underset{\lambda_l (l(x)+z^T Bx)}{prox} (x^*) = \underset{\lambda_l l}{prox}(x^* - \lambda_l B^T z)$$

Now, we must solve the subproblem with z:

$$\max_z \left\{ l(x^*) + z^T (Bx^*) - \varphi^*(z) \right\} = -\min_z \left\{ \varphi^*(z) - z^T (Bx^*) - l(x^*) \right\}$$

We can take another step to minimize the problem $\varphi^*(z) - z^T (Bx^*)$ with a variable with the constant λ_ϕ.

Hence

$$z^* = \underset{\lambda_\varphi \phi^*}{prox}(z^* + \lambda_\phi Bx^*)$$

Despite the fact that, in some cases, we fail to calculate a proximal solution for ϕ^*, we can still find x. Using a Moreau envelope, we can easily move on to an alternative problem:

$$\underset{\lambda_\phi \phi^*}{prox}(z^* + \lambda_\phi Bx^*) = \frac{1}{\lambda_\phi}\left(I - \underset{\phi/\lambda_\phi}{prox}\right)\circ(\lambda_\phi(z^* + Bx^*))$$

From the above, we obtain an iterative algorithm

$$x^* = \underset{\lambda_l l}{prox}(x^* - \lambda_l B^T z^*)$$

$$z^* = \frac{1}{\lambda_\phi}\left(I - \underset{\phi/\lambda_\phi}{prox}\right)\circ(\lambda_\phi(z^* + Bx^*))$$

Dividing the last step in two and introducing the substitution $\lambda_l = \lambda_\phi = 1$, we have

$$x^* = \underset{l}{prox}(x^* - B^T u^*),$$

$$w^* = \underset{\varphi}{prox}(u^* + Bx^*),$$

$$u^* = u^* - (w^* - Bx^*).$$

Thus, we have obtained an iterative procedure similar to the ADMM method, which shows us the wide application of the proximal operator and its properties in solving optimization problems.

6.8. Fast alternative directions methods

The standard ADMM method is used to solve the optimization problem

$$\begin{aligned} &\textit{minimize} && f_1(x_1) + f_2(x_2) \\ &\textit{subject to} && A_1 x_1 + A_2 x_2 = b \end{aligned}$$

where f_i are closed and strictly convex functions, $i = 1,2$. As a rule, $f_1(x)$ is a smooth function, $f_2(x)$ is a non-smooth function (e.g. a penalty function in the LASSO problem), and l_1, l_2 are regularization parameters, where l_1 is a regularization parameter in the least-square method with a penalty function. Such problems are typical in deep learning of neural networks (Scardapane et al. 2017), bioinformatics (Guo et al. 2019), computational statistics (Song 2018), etc.

The main idea of ADMM is to divide the initial optimization problem into smaller subproblems, each of which is much easier to solve (Boyd et al. 2011). The global linear and sublinear convergence of the multiblock ADMM method has been studied in many papers. At each iteration of the basic algorithm, we minimize the extended Lagrangian separately for each variable x_i

$$L_\rho(x_1,x_2,\lambda) = \sum_{i=1}^{N} f_i(x_i) - \lambda^T\left(\sum_{i=1}^{N} A_i x_i - b\right) + \frac{\rho}{2}\left\|\sum_{i=1}^{N} A_i x_i - b\right\|_2^2$$

The ADMM iterative scheme is as follows:

$$\begin{cases} x_1^{k+1} = \arg\min_{x_1} L(x_1, x_2^k, \lambda^k, \rho), \\ x_2^{k+1} = \arg\min_{x_2} L(x_1^{k+1}, x_2, \lambda^k, \rho), \\ \lambda^{k+1} = \lambda^k - \rho(A_1 x_1^{k+1} + A_2 x_2^{k+1} - b) \end{cases}$$

The aim of this study is to find a modification of ADMM with a better rate of convergence than existing methods. To achieve this goal, we considered a combination of multiblock ADMM and fast ADMM (Nesterov acceleration).

The Nesterov method is often used to accelerate the gradient method. Nesterov's iterative scheme has the order of convergence $O\left(1/k^2\right)$. It is the optimal rate of convergence for the class of Lipschitz functions (Nesterov 1983). The acceleration is achieved by calculating the relaxation step at each iteration. Nesterov's iterative acceleration scheme for the gradient objective function minimization method $F(x)$, $x \in \mathbb{R}^n$, $F : R^n \rightarrow R$ is defined as follows:

$$\begin{cases} x_k = y_k - \tau \nabla F(y_k), \\ \alpha_{k+1} = (1 + \sqrt{4\alpha_k^2 + 1}) / 2, \\ y_{k+1} = x_k + (\alpha_k - 1)(x_k - x_{k-1}) / \alpha_{k+1}. \end{cases}$$

Parameters satisfy the conditions $\alpha_0 = 1$, $x_0 = y_1 \in \mathbb{R}^n$, $\tau < 1/K$, where K is a Lipschitz constant for ∇F.

This approach has been widely used to accelerate proximal methods (Nesterov 2013). Since then, Nesterov's approach has been successfully applied to partitioned iterative schemes (Beck and Teboulle 2009). In Goldstein et al. (2014), this approach was generalized for the ADMM method, where it was proposed to add the correction parameter proposed in Nesterov (1983) to the usual ADMM iterative scheme at each iteration step. The method converges if both terms of the objective function are convex.

The iterative scheme for the accelerated ADMM with extended Lagrangian is as follows:

$$\begin{cases} x_1^{k+1} = \arg\min_{x_1} L(x_1, x_2^k, \lambda^k, \rho), \\ x_2^{k+1} = \arg\min_{x_2} L(x_1^{k+1}, x_2, \lambda^k, \rho), \\ \lambda^{k+1} = \lambda^k - \rho(A_1 x_1^{k+1} + A_2 x_2^{k+1} - c), \\ \alpha_{k+1} = \dfrac{1 + \sqrt{1 + 4\alpha_k^2}}{2}, \\ \hat{x}_{k+1}^{(i)} = x_k^{(i)} + \dfrac{\alpha_k - 1}{\alpha_{k+1}}\left(x_k^{(i)} - x_{k-1}^{(i)}\right), \\ \hat{\lambda}_{k+1} = \lambda_k + \dfrac{\alpha_k - 1}{\alpha_{k+1}}(\lambda_k - \lambda_{k-1}) \end{cases}$$

where $x_2^0 \in \mathbb{R}^{m_2}$, $\lambda^0 \in \mathbb{R}^n$, $\tau > 0$, $\alpha_0 = 1$.

The rate of convergence $O\left(1/k^2\right)$ of this method for smooth convex functions was proved in Goldstein et al. (2014). It is more effective than the classical ADMM.

The article (Wang et al. 2014) considered the solution of a convex multiblock optimization problem using ADMM and proposed its modification, which extends this approach by splitting the existing problem into subproblems and solving them separately at each iteration. The global convergence of the multiblock ADMM method has been studied in many papers (Lin et al. 2015a, 2015b, 2016; Sun 2015, Tao and Yuan 2018, etc.). The modified method has better performance than the classical two-block approach.

The purpose of this chapter is to develop and study a new modification of the ADMM method with a higher rate of convergence compared to existing methods. To achieve this task, we propose to combine multiblock ADMM and accelerated ADMM (Nesterov acceleration).

Consider an optimization problem with blocks, where the objective function is the sum of N convex functions:

$$\begin{aligned} &\mathit{minimize} \sum_{i=1}^{N} f_i(x_i) \\ &\mathit{subject\ to} \sum_{i=1}^{N} A_i x_i = b. \end{aligned}$$

The multiblock convex optimization problem has many practical applications. For example, the RPCA (Robust Principal Component Analysis) problem (Candès et al. 2011) occurs when it is necessary to restore the sparse component of the matrix L_0 using strongly disturbed measurements $M = L_0 + S_0$.

The classic iterative scheme for multiblock ADMM is (Wang et al. 2014)

$$\begin{cases} x_i^{k+1} = \arg\min\limits_{x_i \in X_i} f_i(x_i) + \dfrac{\beta}{2}\left\| \sum\limits_{j=1}^{i-1} A_j x_j^{k+1} + A_i x_i + \sum\limits_{j=i+1}^{m} A_j x_j^k - b - \dfrac{\lambda^k}{\beta} \right\|^2, \\ \lambda^{k+1} = \lambda^k - \beta\left(\sum\limits_{i=1}^{m} A_i x_i^{k+1} - b \right), \end{cases}$$

where $x_i^0 \in \mathbb{R}^{m_i}$, $\lambda^0 \in \mathbb{R}^n$, $\beta > 0$.

The convergence of ADMM has been proven for the case of two variables. For multiblock ADMM, convergence has been proven in Chen et al. (2013, 2016) and Hong and Luo (2017). A method is convergent if three conditions are met: (1) for each i, the matrix A_i has full rank and they are orthogonal, (2) the parameter β is sufficiently small and (3) for each subproblem, the proximal step is added without restrictions posed on A_i (Hong and Luo 2017).

We propose a new modification of the multiblock ADMM based on the addition of a Nesterov acceleration similar to the modification of the two-block ADMM described in Goldstein et al. (2014). At each iteration, we calculate the Nesterov step and use it to correct the results:

$$\alpha_{k+1} = \frac{1+\sqrt{1+4\alpha_k^2}}{2}, \alpha_0 = 1$$

$$\hat{x}_i^{(k+1)} = x_i^{(k)} + \frac{\alpha_k - 1}{\alpha_{k+1}}\left(x_i^{(k)} - x_i^{(k-1)}\right),$$

$$\hat{\lambda}_{k+1} = \lambda_k + \frac{\alpha_k - 1}{\alpha_{k+1}}(\lambda_k - \lambda_{k-1})$$

where $x_i^0 \in \mathbb{R}^{m_i}$, $\lambda^0 \in \mathbb{R}^n$, $\beta > 0$.

This allows us to increase the rate of convergence. The proposed iterative scheme is as follows:

$$\begin{cases} x_i^{(k)} = \arg\min\limits_{x_i \in X_i} f_i(x_i) + \dfrac{\beta}{2}\left\| \sum\limits_{j=1}^{i-1} A_j \hat{x}_j^{(k)} + A_i x_i + \sum\limits_{j=i+1}^{m} A_j x_j^{(k-1)} - b - \dfrac{\lambda_k}{\beta} \right\|^2, \\ \lambda^k = \hat{\lambda}^k - \beta\left(\sum\limits_{i=1}^{m} A_i x_i^{(k)} - b \right), \\ \alpha_{k+1} = \dfrac{1+\sqrt{1+4\alpha_k^2}}{2}, \\ \hat{x}_i^{(k+1)} = x_i^{(k)} + \dfrac{\alpha_k - 1}{\alpha_{k+1}}\left(x_i^{(k)} - x_i^{(k-1)}\right), \\ \hat{\lambda}_{k+1} = \lambda_k + \dfrac{\alpha_k - 1}{\alpha_{k+1}}(\lambda_k - \lambda_{k-1}). \end{cases}$$

where $x_i^0 \in \mathbb{R}^{m_i}$, $\lambda^0 \in \mathbb{R}^n$, $\beta > 0$, $\alpha_0 = 1$.

6.9. Numerical experiments

In this section, we compare the performance of the modified Nesterov-accelerated multiblock ADMM with conventional multiblock and classical two-block ADMM methods. To do this, consider the exchange problem, the basis pursuit problem and the constrained LASSO problem.

Similar problems often arise when studying the optimal control of classical and non-classical equations of mathematical physics (Lyashko et al. 2019; Tymoshenko et al. 2019; Klyushin et al. 2020; Hryhorenko et al. 2021, etc.).

6.9.1. *Exchange problem*

Consider a network with N agents that exchange by n products. Each agent i has a cost function $f_i : \mathbb{R}^n \to \mathbb{R}$. The exchange problem has the form

$$\min_{\{x_i\}} \sum_{i=1}^{N} f_i(x_i),$$

$$\textit{subject to} \sum_{i=1}^{N} x_i = 0.$$

Consider quadratic cost functions $f_i(x_i)=\frac{1}{2}\|C_i x_i - d_i\|^2$, where $C_i \in \mathbb{R}^{p \times n}, d_i \in \mathbb{R}^p$. At each iteration step, we have

$$x_i^{k+1} = \arg\min_{x_i} \frac{1}{2}\|C_i x_i\|^2 + \frac{\rho}{2}\left\|x_i - b_i^k\right\|^2$$

where b_i^k is a vector that does not depend on x.

For the experiment, we will generate a random vector x_i^* obeying the standard Gaussian distribution, where C_i are random Gaussian matrices and $d_i = C_i x_i^*$. The dimension of the test problem was set as $n = 100, N = 100, p = 80$. To test the operation of the algorithm, 400 iterations were calculated, and plots of the dependence of the objective function and the error on the number of iterations are provided in Table 6.1.

Iteration	Error		
	Multiblock ADMM with Nesterov acceleration	Multiblock ADMM	ADMM
1	1.1×10^{3}	2.5×10^{2}	7.2×10^{6}
100	1.39×10^{-22}	5.21×10^{-8}	1.4×10^{4}
200	1.36×10^{-22}	3.17×10^{-21}	8.73×10^{1}
300	1.35×10^{-22}	2.83×10^{-21}	5.29×10^{-1}
400	1.34×10^{-22}	1.08×10^{-21}	3.89×10^{-3}

Table 6.1. *Dependence of the errors from the iteration numbers in the exchange problem*

6.9.2. *Basis pursuit problem*

The basis pursuit problem has the following form:

$$\begin{aligned}&minimize\ \|x\|_1\\&subject\ to\ Ax = b,\end{aligned}$$

where $x \in \mathbb{R}^m$, $A \in \mathbb{R}^{n\times m}$ and $b \in \mathbb{R}^n$. This problem usually arises when we have an undefined system of linear equations, and we need to restore the x vector from a small number of observations x b ($n \ll m$) (Wang et al. 2014). Split data into m blocks: $x = (x_1, x_2, ..., x_m)$ and $A = (A_1, A_2, ..., A_m)$, and apply the iteration scheme of the multiblock ADMM.

Each iteration may be computed as

$$x_i^{(k)} = \arg\min_{x_i \in X_i} |x_i|_1 + \frac{\beta}{2}\left\| \sum_{j=1}^{i-1} A_j \hat{x}_j^{(k)} + A_i x_i + \sum_{j=i+1}^{m} A_j x_j^{(k-1)} - b - \frac{\lambda_k}{\beta} \right\|^2 =$$

$$= shrinkage\left(\frac{\beta A_i^T}{\sqrt{a_i^T a_i}} \left[b + \frac{\lambda^k}{\beta} - \sum_{j=1}^{i-1} A_j \hat{x}_j^{(k)} - \sum_{j=i+1}^{m} A_j x_j^{(k-1)} \right], \frac{1}{\beta(a_i^T a_i)} \right).$$

For testing, we will generate a random vector x^* with a k ($k \ll n$) non-zero element obeying the standard Gaussian distribution. The matrix A is also generated according to the standard normal distribution, and it is divided into m blocks. The generated dataset has n = 500 rows and $m = n \times 2.5$ columns. More than 300 iterations of the algorithm were calculated, and the dependence of the average error on the number of iterations for 100 runs of the algorithm was detected. For multiblock ADMM and multiblock ADMM with Nesterov acceleration, the objective function was presented as the sum of five functions. The dependence of objective function values on the number of iterations is shown in Table 6.2.

Iteration	Error		
	Multiblock ADMM with Nesterov acceleration	Multiblock ADMM	ADMM
1	5.48×10^{0}	4.30×10^{0}	2.85×10^{-3}
100	2.98×10^{-8}	7.73×10^{-14}	9.71×10^{-5}
200	5.14×10^{-16}	5.95×10^{-14}	8.13×10^{-5}
300	5.02×10^{-16}	3.56×10^{-14}	1.37×10^{-5}

Table 6.2. *Dependence of the errors from the iteration numbers in the basis pursuit problem*

6.9.3. *Constrained LASSO problem*

The formulation of this task is as follows:

$$\begin{aligned}&\textit{minimize}\ \frac{1}{2}\|Ax-b\|_2^2+\lambda\|x\|_1\\&\textit{subject to}\ Cx=d, x\geq 0,\end{aligned}$$

where $x \in \mathbb{R}^m$, $A \in \mathbb{R}^{n\times m}$ and $b \in \mathbb{R}^n$. The problem often arises in statistics, machine learning and pattern recognition. For testing, we will generate a random vector x^* using the standard Gaussian distribution. The matrices A and C are also generated using the standard Gaussian distribution. They are divided into m blocks. The generated dataset contains n = 1,000 rows and 500 columns. A total of 200 iterations of the algorithm were calculated. The experiment again showed that the Nesterov-accelerated multiblock ADMM converges faster (see Table 6.3).

Iteration	Error		
	Multiblock ADMM with Nesterov acceleration	Multiblock ADMM	ADMM
1	8.84×10^{0}	8.91×10^{0}	1.50×10^{0}
50	7.31×10^{-3}	5.51×10^{-2}	7.75×10^{-1}
100	3.93×10^{-7}	6.32×10^{-7}	8.68×10^{-2}
150	1.39×10^{-11}	5.81×10^{-9}	9.79×10^{-3}
200	1.01×10^{-12}	4.57×10^{-11}	3.72×10^{-4}

Table 6.3. *Dependence of the errors from the iteration numbers in the constrained LASSO problem*

6.10. Conclusion

Multiblock ADMM is much more efficient than standard two-block ADMM, and applying Nesterov's acceleration at each iteration of multiblock ADMM significantly increases the rate of convergence of the method. ADMM is a very handy tool for solving a large number of optimization problems in statistics, machine learning and pattern recognition, and is very well suited for distributed computing. This chapter discusses several existing methods and techniques, which made it possible to increase the effectiveness of ADMM. A new approach is proposed, which is based on a combination of multiblock ADMM and calculation of the Nesterov acceleration. For the proposed method, a theoretical proof of convergence is given. To demonstrate the effectiveness in comparison with existing methods, several optimization problems were solved, in particular, the exchange problem, the basis pursuit problem and the constrained LASSO problem.

6.11. References

Beck, A. and Teboulle, M. (2009). A fast iterative shrinkage-thresholding algorithm for linear inverse problems. *SIAM J. Imaging Sci.*, 2, 183–202.

Boyd, S., Parikh, N., Chu, E., Peleato, B., Eckstein, J. (2011). Distributed optimization and statistical learning via the alternating direction method of multipliers. *Found. Trends Mach. Learn.*, 3(1), 1–122.

Candès, E., Li, X., Ma, Y., Wright, J. (2011). Robust principal component analysis? *J. ACM*, 58(1), 1–37.

Chen, C., Shen, Y., You, Y. (2013). On the convergence analysis of the alternating direction method of multipliers with three blocks. *Abstr. Appl. Anal.*, 183961 [Online]. Available at: https://www.hindawi.com/journals/aaa/2013/183961/ [Accessed 4 February 2022].

Chen, C., He, B., Ye, Y. (2016). The direct extension of ADMM for multi-block convex minimization problems is not necessarily convergent. *Math. Program.*, 155(1), 57–79.

Deng, W., Lai, M.-J., Peng, Z., Yin, W. (2016). Parallel multiblock ADMM with o(1/*k*) convergence. *J. Sci. Comput.*, 71, 712–736.

Goldstein, T. and Setzer, S. (2010). High-order methods for basis pursuit. UCLA CAM Report, 10–41.

Goldstein, T., O'Donoghue, B., Setzer, S. (2014). Fast alternating direction optimization methods. *SIAM J. Imaging Sci.*, 7(3), 1588–1623.

Guo, Y., Wu, C., Guo, M., Zou, Q., Liu, X., Keinan, A. (2019). Combining sparse group lasso and linear mixed model improves power to detect genetic variants underlying quantitative traits. *Front. Genet.*, 10(271) [Online]. Available at: https://www.frontiersin.org/articles/10.3389/fgene.2019.00271/full [Accessed 4 February 2022].

Hong, M. and Luo, Z.-Q. (2017). On the linear convergence of the alternating direction method of multipliers. *Math. Progr.*, 162(1–2), 165–199.

Hryhorenko, V.A., Klyushin, D.A., Lyashko, S.I. (2021). Multiblock ADMM with Nesterov acceleration. *Probl. Control Inform.*, 4, 5–18 [in Russian].

Klyushin, D.A., Lyashko, S.I., Lyashko, N.I., Bondar, O.S., Tymoshenko, A.A. (2020). Generalized optimization of processes of drug transport in tumors. *Cybern. Syst. Anal.*, 56(5), 758–765.

Lin, T.Y., Ma, S.Q., Zhang, S.Z. (2015a). On the sublinear convergence rate of multi-block ADMM. *J. Oper. Res. Soc. China*, 3(3), 251–274.

Lin, T.Y., Ma, S.Q., Zhang, S.Z. (2015b). On the global linear convergence of the ADMM with multiblock variables. *SIAM J. Optim.*, 25(3), 1478–1497.

Lin, T.Y., Ma, S.Q., Zhang, S.Z. (2016). Iteration complexity analysis of multi-block ADMM for a family of convex minimization without strong convexity. *J. Sci. Comput.*, 69(1), 52–81.

Lyashko, S.I., Klyushin, D.A., Timoshenko, A.A., Lyashko, N.I., Bondar, E.S. (2019). Optimal control of intensity of water point sources in unsaturated porous medium. *J. Automat. Inf. Scien.*, 51(7), 24–33.

Nesterov, Y. (1983). A method of solving a convex programming problem with convergence rate $O(1/k^2)$. *So. Math. Dokl.*, 27, 372–376.

Nesterov, Y. (2013). Gradient methods for minimizing composite functions. *Math. Program.*, 140(1), 125–161.

Parikh, N. and Boyd, S. (2013). Proximal algorithms. *Found. Trends Optim.*, 1(3), 123–231.

Patrinos, P. and Bemporad, A. (2013). Proximal Newton methods for convex composite optimization. *52nd IEEE Conference on Decision and Control*, 2358–2363.

Polson, N., Scott, J., Willard, B. (2015). Proximal algorithms in statistics and machine learning. *Stat. Sci.*, 30(4), 559–581.

Scardapane, S., Comminiello, D., Hussain, A., Uncini, A. (2017). Group sparse regularization for deep neural networks. *Neurocomputing*, 241, 81–89.

Song, Q. (2018). An overview of reciprocal L1-regularization for high dimensional regression data. *Wiley Interdiscip. Rev. Comput. Stat.*, 10, e1416 [Online]. Available at: https://wires.onlinelibrary.wiley.com/doi/abs/10.1002/wics.1416 [Accessed 4 February 2022].

Sun, D.F., Toh, K.C., Yang, L.Q. (2015). A convergent 3-block semiproximal alternating direction method of multipliers for conic programming with 4-type constraints. *SIAM J. Optim.*, 25(2), 882–915.

Tao, M. and Yuan, X. (2018). Convergence analysis of the direct extension of ADMM for multiple-block separable convex minimization. *Adv. Comput. Math.*, 44(3), 773–813.

Tibshirani, R. (1996). Regression shrinkage and selection via the lasso. *J. R. Stat. Soc. Series B Stat. Methodol.*, 58(1), 267–288.

Tymoshenko, A., Klyushin D., Lyashko, S. (2019). Optimal control of point sources in Richards–Klute equation. In *Advances in Intelligent Systems and Computing*, Hu, Z., Petoukhov, S., Dychka, I., He, M. (eds). Springer, Cham.

Wang, X., Hong, M., Ma, S. (2014). Solving multiple-block separable convex minimization problems using two-block alternating direction method of multipliers. *Pac. J. Optim.*, 11(4), 645–667.

7

Modified Extragradient Algorithms for Variational Inequalities

Vladimir V. SEMENOV and Sergey V. DENISOV
Taras Shevchenko National University of Kyiv, Ukraine

7.1. Introduction

In this chapter, different variants of the modified extragradient algorithm for solving variational inequalities with monotone operators acting in the Hilbert spaces are considered. Algorithms use the variable step sizes which are updated over each iteration by some cheap computations. These step sizes are found without the prior knowledge of the Lipschitz constant of an operator. The main focus of the chapter is on proving the convergence of algorithms. The main materials of the chapter were first published in articles (Lyashko et al. 2011; Denisov et al. 2015; Verlan et al. 2015).

7.2. Preliminaries

Let H be a real Hilbert space with scalar product $(\cdot,\cdot)$ and induced norm $\|\cdot\|$. We will denote strong convergence as $\to$, and the weak one as $\overset{w}{\to}$.

Computational Methods and Mathematical Modeling in Cyberphysics and Engineering Applications 1,
coordinated by Dmitri KOROLIOUK, Sergiy LYASHKO and Nikolaos LIMNIOS.

Let C be a non-empty closed convex subset of H, and let us consider a mapping $A: H \to H$.

DEFINITION 7.1.– *The next problem is called a variational inequality problem:*

$$\text{find } x \in C : (Ax, y - x) \geq 0 \quad \forall y \in C. \tag{7.1}$$

Let us denote the solution set of [7.1] as $VI(A, C)$.

DEFINITION 7.2.– *Mapping $A: H \to H$ is called monotone, if*

$$(Ax - Ay, x - y) \geq 0 \quad \forall x, y \in H.$$

Everywhere later, we assume that the next conditions hold:

– **(A1)** $VI(A, C) \neq \varnothing$;

– **(A2)** Mapping $A: H \to H$ is monotone and uniformly continuous on bounded sets.

REMARK 7.1.– If space H is finite dimensional, then it is enough for A to be monotone and continuous.

REMARK 7.2.– Variational inequalities are one of the central objects in applied nonlinear analysis. The study of variational inequalities began in the 1960s in works by Fichera (1964), Stampacchia (1965), Hartman and Stampacchia (1966) and Lions and Stampacchia (1967). These works were attempts to solve the problems of variational calculus, optimal control and the theory of boundary value problems with one-sided conditions. A good overview of the first applications of variational inequalities is given by books by Lions (1969), Baiocchi and Capello (1984) and Kinderlehrer and Stampacchia (2000). Also note the thorough monographs of Soviet mathematicians Vainberg (1974) and Ivanenko and Melnik (1988). The relationship between convex analysis, the theory of maximal monotone operators and variational inequalities was established in the works of Moreau (1967) and Brezis (1972). Very quickly, variational irregularities became a powerful tool for solving problems of optimization, optimal control, economics and mathematical physics (Lions 1968; Bensoussan 1974; Ekeland and Temam 1976; Glowinski et al. 1981; Dafermos 1982; Aubin and Ekeland 1984; Harker and Pang 1990; Nagurney 1999; Patriksson 1999). In optimization and optimal control, variational inequalities is a convenient way of writing optimality conditions. The famous monograph by Lions (1968) is a systematic exposition of the theory of optimal control of systems with distributed parameters from the point of view of variational inequalities. In an article by

Bensoussan (1974), the author formulated the problem of finding the Nash equilibrium in the form of variation inequality. The problem of finding an equilibrium flow distribution in the transport network was presented in the form of a variational inequality in Dafermos (1982). Starting from these works, active research began on various problems of finding equilibrium using the methods of the theory of variational inequalities (Harker and Pang 1990). The remarkable monograph by Nagurney (1999) is devoted to actual equilibrium problems, such as price equilibria or transport equilibria corresponding to the behavioral principles of Wardrop. In mathematical physics models, variational inequalities are applied in the theory of elasticity, the theory of elastoplastic materials, the theory of porous media and the theory of non-Newtonian fluids (Lions 1969; Duveau and Lions 1980).

Let us recall several well-known facts, important for the main results.

Let P_C be the metric projection operator to a closed convex set $C \subseteq H$, so $P_C x$ is a unique element of C, which has the property

$$\|P_C x - x\| = \min_{z \in C} \|z - x\|.$$

Element $P_C x$ can be characterized as (Kinderlehrer and Stampacchia 2000):

$$y = P_C x \Leftrightarrow y \in C \text{ and } (y - x, z - y) \geq 0 \ \forall x \in C, \qquad [7.2]$$

$$y = P_C x \Leftrightarrow y \in C \text{ and } \|y - z\|^2 \leq \|x - z\|^2 - \|y - x\|^2 \ \forall x \in C. \qquad [7.3]$$

It follows from inequality [7.2] that $x \in VI(A, C)$ if and only if

$$x = P_C(x - \lambda Ax),$$

where $\lambda > 0$ (Kinderlehrer and Stampacchia 2000).

The following Lemma holds:

LEMMA 7.1.– *Let mapping $A : H \to H$ be monotone and continuous, and set $C \subseteq H$ is a convex and closed. Then, $x \in VI(A, C)$ if and only if $x \in C$ and*

$$(Ay, y - x) \geq 0 \quad \forall y \in C.$$

In particular, the set $VI(A,C)$ is convex and closed.

We will use the well-known Opial lemma to prove the weak convergence of sequences of Hilbert space elements.

LEMMA 7.2.– (Opial 1967). *Let a sequence* x_n *of elements from Hilbert space* H *converge weakly to an element* $x \in H$. *Then,* $\forall\, y \in H \setminus \{x\}$, *we have*

$$\liminf_{n\to\infty} \|x_n - x\| < \liminf_{n\to\infty} \|x_n - y\|.$$

PROOF.– To prove the lemma, it is enough to note that in equality

$$\|x_n - y\|^2 = \|x_n - x\|^2 + \|x - y\|^2 + 2(x_n - x, x - y)$$

the last term tends to zero with $n \to \infty$.

The next facts also play an important role in proving the main results.

LEMMA 7.3.– *Let* (ξ_n) *be a sequence of non-negative numbers, which satisfy recurrent inequality*

$$\xi_{n+1} \le (1-\alpha_n)\xi_n + \alpha_n\beta_n,$$

where sequences (α_n) and (β_n) have the next properties:

1) $\alpha_n \in [0,1)$ and $\sum_{n=1}^{\infty} \alpha_n = +\infty$;

2) $\limsup_{n\to\infty} \beta_n \le 0$.

Then, $\lim_{x\to\infty} \xi_n = 0$.

PROOF.– For any $\epsilon > 0$, there exists such $N \ge 1$ that $\beta_n < \epsilon$ for all $n \ge N$.

For $n > N$ from recurrent inequality, we get

$$\xi_{n+1} \leq \left(\prod_{k=N}^{n} (1-\alpha_k) \right) \xi_N + \left(1 - \prod_{k=N}^{n} (1-\alpha_k) \right) \epsilon .$$

Taking into account that $\sum_{n=1}^{\infty} \alpha_n = +\infty$ is equivalent to $\prod_{n=1}^{\infty} (1-\alpha_n) = 0$, we get

$$\limsup_{n \to \infty} \xi_n \leq \epsilon$$

So, $\lim_{x \to \infty} \xi_n = 0$.

LEMMA 7.4.– (Mainge 2008). *Let numeric sequence* (a_n) *have subsequence* (a_{n_k}) *with the next property*

$$a_{n_k} < a_{n_k+1} \quad \forall k \geq 1.$$

Then, there exists a non-decreasing sequence of natural numbers (m_k) such that

$$m_k \to +\infty \text{ and } a_{m_k} \leq a_{m_k+1}, a_k \leq a_{m_k+1} \; \forall k \geq n_1.$$

PROOF.– Let us denote $m_k = \max\{i \leq k : a_i < a_{i+1}\}$ for any $k \geq n_1$. The sequence m_k is correctly defined, non-decreasing and $m_k \to +\infty$. Inequality $a_{m_k} \leq a_{m_k+1}$ obviously holds. We need to prove that for all $k \geq n_1$ also holds the next inequality:

$$a_k \leq a_{m_k+1} \qquad [7.4]$$

Let us consider three possible cases:

– (i) $m_k = k$;

– (ii) $m_k = k-1$;

– (iii) $m_k < k-1$.

In case (i), inequality [7.4] matches with $a_{m_k} \le a_{m_k+1}$. In case (ii), equation [7.4] is obvious. And in the last case for all $i \in \{m_k+1, m_k+2, ..., k-1\}$ we have $a_i \ge a_{i+1}$, or, the same, $a_{m_k+1} \ge a_{m_k+2} \ge ... \ge a_{k-1} \ge a_k$.

REMARK 7.3.– Lemma 7.4 is an effective tool for studying the convergence of iterative processes that do not have the Fejer property with respect to the set of solutions (Mainge 2008; Denisov and Semenov 2011; Semenov 2014a, 2014b).

Let us show some useful properties for the next function:

$$t \mapsto \|x - P_C(x - tAx)\|, \quad t \ge 0.$$

LEMMA 7.5.– *For an element $x \in H$ and numbers $\alpha \ge \beta > 0$, the next inequalities hold:*

$$\frac{\|x - P_C(x - \alpha Ax)\|}{\alpha} \le \frac{\|x - P_C(x - \beta Ax)\|}{\beta},$$

$$\|x - P_C(x - \beta Ax)\| \le \|x - P_C(x - \alpha Ax)\|.$$

PROOF.– Let

$$x_\alpha = P_C(x - \alpha Ax), \quad x_\beta = P_C(x - \beta Ax).$$

From the inequality in equation [7.2] follows

$$\left(\frac{x_\alpha - x + \alpha Ax}{\alpha}, x_\beta - x_\alpha\right) \ge 0, \quad \left(\frac{x_\beta - x + \beta Ax}{\beta}, x_\alpha - x_\beta\right) \ge 0.$$

Adding these two inequalities, we get

$$0 \le \left(\frac{x - x_\alpha}{\alpha} - \frac{x - x_\beta}{\beta}, x_\alpha - x_\beta\right) = \left(\frac{x - x_\alpha}{\alpha} - \frac{x - x_\beta}{\beta}, (x - x_\beta) - (x - x_\alpha)\right).$$

And this leads to

$$0 \le -\|x-x_\alpha\|^2 - \frac{\alpha}{\beta}\|x-x_\beta\|^2 + \|x-x_\alpha\|\|x-x_\beta\| + \frac{\alpha}{\beta}\|x-x_\alpha\|\|x-x_\beta\|.$$

So,

$$0 \ge \left(\|x-x_\alpha\| - \frac{\alpha}{\beta}\|x-x_\beta\| \right)\left(\|x-x_\alpha\| - \|x-x_\beta\| \right).$$

This allows getting the results we have had to prove:

$$\|x-x_\alpha\| - \frac{\alpha}{\beta}\|x-x_\beta\| \le 0, \quad \|x-x_\beta\| \le \|x-x_\alpha\|.$$

DEFINITION 7.3.– *Mapping $T: H \to H$ is called non-expansive, if*

$$\|Tx-Ty\| \le \|x-y\| \quad \forall x, y \in H.$$

Projection operator P_C is non-expansive.

DEFINITION 7.4.– *Mapping $T: H \to H$ is called quasi-non-expansive (mapping with Fejer property), if $F(T) = \{x \in H : Tx = x\} \ne \varnothing$ and*

$$\|Tx-y\| \le \|x-y\| \quad \forall x \in H, \ \forall y \in F(T).$$

It is known that the fixed point set of quasi-non-expansive mapping $F(T)$ is closed and convex (Vasin and Eremin 2005; Bauschke and Combettes 2011).

REMARK 7.4.– Quasi-non-expansive (Fejer) operators and the iterative processes generated by them are of great importance in optimization algorithms (see Eremin and Mazurov (1979) and Nurminski (2008)). But in many other areas of computational mathematics, these successful constructions are not known well enough. The book by Vasin and Eremin (2005) is one of the best sources on iterative algorithms with the Fejer property for solving equations, inequalities and optimization problems.

DEFINITION 7.5.– *Mapping* $S: C \to H$ *is called demiclosed at the point* $y \in H$ *if for a sequence* (x_n), $x_n \in C$ *from* $x_n \overset{w}{\to} x$ *and* $x_n \to y$ *follows* $Sx = y$.

It is known that for a non-expansive mapping $T: C \to H$, the mapping $I - T$ is demiclosed in zero (Bauschke and Combettes 2011).

7.3. Overview of the main algorithms for solving variational inequalities and approximations of fixed points

Among many methods for solving [7.1], the projective ones are conceptually and practically the simplest. Their theoretical background is the equivalence of the variational inequality [7.1] and the problem of finding a fixed point of the operator $P_C(I-\lambda A)$, where $\lambda > 0$ and P_C is the metric projection operator on the set C (Kinderlehrer and Stampacchia 2000). Projective methods for variational inequalities originate from the works of Goldstein (1964) and Levitin and Polyak (1966), where they independently proposed a gradient projection method for a conditional optimization problem. For equation [7.1], an analogue of this method generates a sequence (X_n) according to the rule

$$x_{n+1} = P_C(x_n - \lambda A x_n), \tag{7.5}$$

where $\lambda > 0$. For the convergence of (X_n) to the solution of equation [7.1], it is necessary to impose stronger conditions on A than just monotonicity. In particular, if A is strongly monotone or inversely strongly monotone (co-coercive), then the sequence (X_n) converges to solution [7.1] (strongly in the first case, weakly in the second). In the first case, the value of step size λ is chosen from interval $(0, 2\mu / L^2)$, where $\mu > 0$, $L > 0$ are the strong monotonicity and Lipschitz constants of the operator A. In the second case, it is chosen from $(0, 2\alpha)$, where $\alpha > 0$ is the co-coercivity constant of A. Recall that an operator $A: H \to H$ is called co-coercive (inversely strongly monotone) if there exists a positive constant α such that

$$(Ax - Ay, x - y) \geq \alpha \|Ax - Ay\|^2 \quad \forall x, y \in H.$$

If the operator A is only monotonic, the simplest scheme [7.5] does not guarantee convergence. Korpelevich in 1976 proposed the so-called extragradient algorithm (Korpelevich 1976) for solving [7.1] in Euclidean space (see also Antipin 1976):

$$\begin{cases} y_n = P_C\left(x_n - \lambda A x_n\right), \\ x_{n+1} = P_C\left(x_n - \lambda A y_n\right), \end{cases} \qquad [7.6]$$

where $\lambda \in (0,1/L)$. She proved the convergence of the sequences (X_n) and (Y_n) to a solution of inequality [7.1]. A large number of publications have been devoted to the generalization and study of this algorithm. For example, Nadezhkina and Takahashi (2006b) proved its weak convergence in an infinite-dimensional Hilbert space. In 2010, Russian researchers Zykina and Melenchuk (2010) proposed the following two-step extragradient algorithm ($\lambda>0$):

$$\begin{cases} y_n = P_C\left(x_n - \lambda A x_n\right), \\ z_n = P_C\left(y_n - \lambda A y_n\right), \\ x_{n+1} = P_C\left(x_n - \lambda A z_n\right). \end{cases}$$

Censor et al. (2011b) proposed a modification of the Korpelevich algorithm with one metric projection onto a feasible set C – the so-called subgradient extragradient algorithm, which has the form:

$$\begin{cases} y_n = P_C\left(x_n - \lambda A x_n\right), \\ T_n = \left\{z \in H : \left(x_n - \lambda A x_n - y_n, z - y_n\right) \le 0\right\}, \\ x_{n+1} = P_{T_n}\left(x_n - \lambda A y_n\right), \end{cases} \qquad [7.7]$$

where $\lambda \in (0,1/L)$. The second projection in [7.7] has an explicit form – so this method is attractive when projecting onto a feasible set C, which is a difficult problem. Note that algorithm [7.7] was independently proposed by Lyashko et al. (2011) in the context of equilibrium programming problems. In Censor et al. (2011b) and Lyashko et al. (2011), the weak convergence of sequences (x_n) and (y_n), generated by this algorithm, to some solution of the variational inequality [7.1] was proven.

An alternative to the extragradient method is the following scheme, proposed by Tseng (2000):

$$\begin{cases} y_n = P_C\left(x_n - \lambda A x_n\right), \\ x_{n+1} = y_n + \lambda\left(A x_n - A y_n\right), \end{cases} \qquad [7.8]$$

where $\lambda \in (0,1/L)$. The sequences (x_n) and (y_n) converge weakly to some solution of inequality [7.1]. Both algorithms [7.7] and [7.8] have the same iterative step complexity – single projection on C and two calculations of the operator A.

Popov (1980) proposed an interesting method similar to the extragradient one but with only one operator A calculation on the iterative step:

$$\begin{cases} x_{n+1} = P_C\left(x_n - \lambda A y_n\right), \\ y_{n+1} = P_C\left(x_{n+1} - \lambda A y_n\right), \end{cases} \qquad [7.9]$$

where $\lambda \in (0,1/3L)$. Convergence of [7.9] for a finite-dimensional space is proved in Popov (1980). In 2014, Malitsky and Semenov (2014) proved a weak convergence of [7.9] for variational inequalities in an infinite-dimensional Hilbert space and proposed a modification of Popov's algorithm with one metric projection onto a feasible set:

$$\begin{cases} T_n = \left\{ z \in H : \left(x_n - \lambda A y_{n-1} - y_n, z - y_n\right) \le 0 \right\}, \\ x_{n+1} = P_{T_n}\left(x_n - \lambda A y_n\right), \\ y_{n+1} = P_C\left(x_{n+1} - \lambda A y_n\right), \end{cases}$$

where $\lambda \in (0,1/3L)$.

Malitsky (2015) published and proved weak convergence of the so-called projective reflecting algorithm:

$$\begin{cases} x_{n+1} = P_C\left(x_n - \lambda A y_n\right), \\ y_{n+1} = 2x_{n+1} - x_n, \end{cases}$$

where $\lambda \in \left(0, (\sqrt{2}-1)L^{-1}\right)$.

In an infinite dimensional Hilbert space, the above algorithms do not guarantee strong convergence. Therefore, the issue of constructing algorithms that would guarantee strong convergence is important. To construct strongly convergent algorithms, the iterative regularization technique (Bakushinskii and Goncharskii 1989) is applied to the basic method or the basic method is supplemented with one of the strongly convergent schemes for approximating fixed points (Halpern 1967; Nakajo and Takahashi 2003; Takahashi et al. 2008). For example, Nadezhkina and Takahashi (2006a) proposed a method combining the Korpelevich extragradient method [7.6] with a hybrid method (Nakajo and Takahashi 2003). The history of the hybrid method begins in 1968 with the work of Haugazeau (1968). Later, his idea was successfully generalized and improved (regardless of the pioneering work (Haugazeau 1968)) in Combettes (1999), Solodov and Svaiter (1999a, 2000), Nakajo and Takahashi (2003) and Burachik et al. (2005). Nadezhkina and Takahashi (2006a) proved that the sequence (X_n) generated by

$$\begin{cases} y_n = P_C(x_n - \lambda A x_n), \\ z_n = P_C(x_n - \lambda A y_n), \\ C_n = \{z \in H : \|z - z_n\| \le \|z - x_n\|\}, \\ Q_n = \{z \in H : (x_n - z, x_1 - x_n) \ge 0\}, \\ x_{n+1} = P_{C_n \cap Q_n} x_1, \end{cases}$$

where $\lambda \in (0, 1/L)$, $x_1 \in C$ strongly converges to the solution of [7.1]. Also, the next strongly convergent method was obtained based on the scheme (Takahashi et al. 2008):

$$\begin{cases} x_1 \in C, C_1 = C, \\ y_n = P_C(x_n - \lambda A x_n), \\ z_n = P_C(x_n - \lambda A y_n), \\ C_{n+1} = \{z \in C_n : \|z - z_n\| \le \|z - x_n\|\}, \\ x_{n+1} = P_{C_{n+1}} x_1, \end{cases} \qquad [7.10]$$

where $\lambda \in (0, 1/L)$. The main drawback of [7.10] is the increasing complexity of convex sets C_n on which the point x_1 is projected. It is desirable to modify the scheme [7.10] to get rid of that behavior.

Having adopted the ideas of Censor et al. (2011b) and Nadezhkina and Takahashi (2006a), Censor et al. (2011a) proposed a strongly convergent method:

$$\begin{cases} x_1 \in H, \lambda \in (0,1/L), \\ y_n = P_C(x_n - \lambda A x_n), \\ T_n = \{z \in H : (x_n - \lambda A x_n - y_n, z - y_n) \le 0\}, \\ z_n = P_{T_n}(x_n - \lambda A y_n), \\ C_n = \{z \in H : \|z - z_n\| \le \|z - x_n\|\}, \\ Q_n = \{z \in H : (x_n - z, x_1 - x_n) \ge 0\}, \\ x_{n+1} = P_{C_n \cap Q_n} x_1. \end{cases}$$

Malitsky and Semenov (2015) proposed a hybrid algorithm based on similar principles:

$$\begin{cases} x_1 = y_1 \in H, C_1 = Q_1 = H, \\ y_{n+1} = P_C(x_n - \lambda A y_n), \\ C_n = \{z \in H : \| y_{n+1} - z \|^2 \le \| x_n - z \|^2 + \ell_n\}, \\ \ell_n = k \| x_{n-1} - x_n \|^2 - (1 - k^{-1} - \lambda L) \| y_{n+1} - y_n \|^2 + \lambda L \| y_{n-1} - y_n \|^2, \\ Q_n = \{z \in H : (x_n - z, x_1 - x_n) \ge 0\}, \\ x_{n+1} = P_{C_n \cap Q_n} x_1, \end{cases}$$

where $\lambda \in (0,1/2L)$, $k > (1-2\lambda L)^{-1}$.

All the above-mentioned algorithms require $\lambda > 0$ to be less than or equal to some constant, which is proportional to $1/L$. The Lipschitz constant is usually hard or even impossible to determine. Often, we can estimate its upper bounds, but that leads to lowered estimation of the step size and increases the iteration number. Khobotov (1987) had modified [7.6], avoiding the use of the Lipschitz constant. The main idea is to recalculate $y_n = P_C(x_n - \lambda_n A x_n)$ on every step of the extragradient method, until

$$\|Ay_n - Ax_n\| \le \frac{\theta}{\lambda_n} \|y_n - x_n\|,$$

where $\theta \in (0,1)$. And then, the next approximation is calculated as

$$x_{n+1} = P_C(x_n - \lambda_n A y_n).$$

The paper by Marcotte (1991) shows that such an approach can be more successful than the classic extragradient method.

Russian mathematician Konnov (1993) proposed another idea, which is based on separation of the point x_n from the solution set [7.1]. This idea was widely adopted and later improved and generalized (Iusem and Svaiter 1997; Solodov and Svaiter 1999b; Solodov 2001; Iusem and Nasri 2011). For example, the next algorithm was proposed in 1999 by Solodov and Svaiter (1999b):

1) For $x_n \in C$, we calculate

$$y_n = P_C(x_n - Ax_n).$$

2) Calculate

$$j_n = \min\left\{j \geq 0: \ (A(x_n + \alpha^j (y_n - x_n)), x_n - y_n) \geq \sigma \| x_n - y_n \|^2\right\},$$

where $\alpha, \sigma \in (0,1)$.

3) Let $z_n = x_n + \alpha^{j_n}(y_n - x_n)$ and

$$T_n = \{z \in H: \ (Az_n, z - z_n) \leq 0\}.$$

4) Calculate x_{n+1}:

$$x_{n+1} = P_{C \cap T_n} x_n.$$

It can be shown that hyperplanes $H_n = \{z \in H: (Az_n, z - z_n) = 0\}$ separate point x_n from the solution set of variational inequality [7.1].

The theory of fixed points is a powerful tool for studying nonlinear models. At the same time, the classical fixed point existence theorems by Brouwer, Schauder and Kakutani are non-constructive. It is natural to identify classes of operators for which it is possible to constructively find fixed points. The most famous example of such a class is contraction mapping. In the classical proof of the Banach–Picard–Caciopolli theorem, a sequence converging to a fixed point is clearly constructed.

For the reader interested in a broader class of non-expansive operators, we note that the theory has actively been developing since 1965. A good introduction to this theory and description of the main fixed point approximation algorithms can be found in the book by Bauschke and Combettes (2011).

Browder (1965a) proves a theorem on the existence of a fixed point of a non-expansive mapping acting in a convex closed and bounded subset of a Hilbert space, and in Browder (1965b), he proves a similar theorem for uniformly convex Banach spaces. These results simultaneously were obtained by Göhde (1965) and Kirk (1965).

The main schemes providing the approximation of fixed points of non-expansive operators (assuming their existence) are the algorithms by Krasnoselsky–Mann (Mann 1953; Krasnoselsky 1955), Halpern (Halpern 1967; Lions 1977; Wittmann 1992; Bauschke 1996; Xu 2002a, 2002b; Suzuki 2007; Semenov 2013), the Nakajo–Takahashi hybrid algorithm (Nakajo and Takahashi 2003) and the algorithm of external approximations (shrinking algorithm) by Takahashi–Takeuchi–Kubota (Takahashi et al. 2008).

Let $C \subseteq H$ be a non-empty convex and closed subset of Hilbert space H, and let $T: C \to C$ be a non-expansive operator for which $F(T) = \{x \in C : x = Tx\} \neq \emptyset$. The Krasnoselsky–Mann method generates a sequence x_n with the next scheme:

$$x_{n+1} = \alpha x_n + (1-\alpha) T x_n,$$

where $\alpha \in (0,1)$. It is known that the sequence (x_n) weakly converges to a point from $F(T)$ (Opial 1967).

Halpern (1967) investigated the convergence of the next algorithm:

$$x_{n+1} = \alpha_n y + (1-\alpha_n) T x_n, \qquad [7.11]$$

where $y \in C$, and (α_n) is a sequence of numbers from (0,1). He proved that conditions

$$\lim_{n\to\infty} \alpha_n = 0, \ \sum_{n=1}^{\infty} \alpha_n = +\infty, \qquad [7.12]$$

are necessary for a strong convergence of sequence (x_n) to a fixed point of non-expansive mapping T and established sufficient conditions for this convergence. Between others, he proved that algorithm [7.11] converges if $\alpha_n = \frac{1}{(n+1)^q}$, where $q \in (0,1)$.

Lions (1977), who later obtained his Fields Medal, improved Halpern's result and proved strong convergence of [7.11] with conditions [7.12] and

$$\lim_{n\to\infty} (\alpha_{n+1} - \alpha_n) / \alpha_{n+1}^2 = 0.$$

Wittmann (1992) proved a strong convergence of [7.11] to the fixed point of mapping T using [7.12] and the next condition:

$$\sum_{n=0}^{\infty} \left|\alpha_{n+1} - \alpha_n\right| < +\infty.$$

And finally, Xu (2002a, 2002b) proved the theorem about the strong convergence of algorithm [7.11] using conditions [7.12] and

$$\lim_{n\to\infty} (\alpha_{n+1} - \alpha_n) / \alpha_{n+1} = 0.$$

The next question is still open (Suzuki 2007): Is [7.12] sufficient by itself for the strong convergence of [7.11] for arbitrary non-expansive mapping $T : C \to C$?

Bauschke (1996) proposed cyclic Halpern's algorithm for producing a sequence which strongly converges to the common fixed point of a finite set of non-expansive mappings. Moudafi (2000) and Yamada (2001) proposed the generalization of Halpern's scheme [7.11] for solving variational inequalities with a strongly monotone operator on the set of fixed points of a non-expansive operator.

It is well known that the Krasnoselsky–Mann method only has weak convergence in general. So, there are some strongly convergent modifications of this method. One of the most popular among such modifications is the so-called CQ-method (also known as the hybrid method). This method was proposed by Nakajo and Takahashi (2003). The algorithm has the following formulation:

$$\begin{cases} x_1 = x \in C, \\ y_n = \alpha_n x_n + (1-\alpha_n) T x_n, \\ C_n = \{ z \in C : \|z - y_n\| \le \|z - x_n\| \}, \\ Q_n = \{ z \in C : (x_n - z, x_0 - x_n) \ge 0 \}, \\ x_{n+1} = P_{C_n \cap Q_n} x, \end{cases}$$

where (α_n) is a sequence of numbers from $[0,\alpha]$ for some $\alpha \in [0,1)$. They proved that the generated sequence (x_n) strongly converges to a point $z = P_{F(T)}x$.

7.4. Modified extragradient algorithm for variational inequalities

The widely known and popular extragradient Korpelevich algorithm (1976) for solving variational inequality [7.1] with Lipschitz continuous operator A is formulated as

$$\begin{cases} x_0 \in H, \\ y_n = P_C\left(x_n - \lambda A x_n\right), \\ x_{n+1} = P_C\left(x_n - \lambda A y_n\right), \end{cases}$$

where $\lambda \in (0, 1/L)$, and L is a Lipschitz constant of A. Recently (Censor et al. 2011b; Lyashko et al. 2011), the modification with a single projection to C was proposed for this algorithm. It is called the subgradient extragradient method and is formulated as

$$\begin{cases} x_0 \in H, \\ y_n = P_C\left(x_n - \lambda A x_n\right), \\ T_n = \{ z \in H : \left(x_n - \lambda A x_n - y_n, z - y_n\right) \le 0 \}, \\ x_{n+1} = P_{T_n}\left(x_n - \lambda A y_n\right), \end{cases}$$

with the same conditions for λ as in the original method. In the papers by Censor et al. (2011b) and Lyashko et al. (2011), a weak convergence of the algorithm-generated sequences (x_n) and (y_n) to a solution of variational inequality is proved.

Assumption of the knowing Lipschitz constant of A (or at least having estimation for it) is an obvious flaw of the subgradient extragradient method. Also, there are problems where the Lipschitz condition does not hold at all. At the same time, most of the works related to algorithms for variational inequalities are considering Lipschitz-continuous operators.

Below, we discuss the modification of the subgradient extragradient algorithm with a dynamic step size regulation for variational inequalities with monotone non-Lipschitz operator and prove convergence of this algorithm.

So, let us consider the following algorithm for [7.1]:

Algorithm 7.1. Set parameters $\sigma > 0$, $\tau \in (0,1)$, $\theta \in (0,1)$ and element $x_0 \in H$.
Iteration. For $x_n \in H$ calculating

$$y_n = P_C(x_n - \lambda_n A x_n),$$

where λ_n is obtained from the condition

$$\begin{cases} j(n) = \min\Big\{ j \geq 0 : \left\| A P_C\left(x_n - \sigma\tau^j A x_n\right) - A x_n \right\| \\ \qquad\qquad \leq \dfrac{\theta}{\sigma\tau^j} \left\| P_C\left(x_n - \sigma\tau^j A x_n\right) - x_n \right\| \Big\}, \\ \lambda_n = \sigma\tau^{j(n)}. \end{cases} \quad [7.13]$$

If $y_n = x_n$, then stop; in any other case, calculate

$$x_{n+1} = P_{T_n}(x_n - \lambda_n A y_n),$$

where

$$T_n = \{ z \in H : (x_n - \lambda_n A x_n - y_n, z - y_n) \leq 0 \}.$$

It is obvious that condition $y_n = x_n$ implies that $x_n \in C$ and is a solution of the variational inequality. Indeed, the equality

$$x_n = P_C(x_n - \lambda_n A x_n)$$

is equivalent to the inequality

$$(x_n - x_n + \lambda_n A x_n, y - x_n) \geq 0 \quad \forall y \in C.$$

Let us show that the procedure [7.13] always finishes after a finite number of steps. The next lemma holds:

LEMMA 7.6.– *Step size λ_n selection rule from [7.13] is correct, so*

$$j(n) < +\infty.$$

PROOF.– Let $x_n \in VI(A,C)$. Then

$$x_n = P_C(x_n - \sigma A x_n) \text{ and } j(n) = 0.$$

Assume $x_n \notin VI(A,C)$ and let for all $j \geq 1$ the next inequality holds:

$$\sigma\tau^j \left\| AP_C\left(x_n - \sigma\tau^j Ax_n\right) - Ax_n \right\| > \theta \left\| P_C\left(x_n - \sigma\tau^j Ax_n\right) - x_n \right\|. \quad [7.14]$$

If $x_n \in C$, then

$$\lim_{j\to\infty} \left\| P_C\left(x_n - \sigma\tau^j Ax_n\right) - x_n \right\| = 0.$$

From the uniform continuity of A on bounded sets follows

$$\lim_{j\to\infty} \left\| AP_C\left(x_n - \sigma\tau^j Ax_n\right) - Ax_n \right\| = 0.$$

So

$$\lim_{j\to\infty} \frac{\left\| P_C\left(x_n - \sigma\tau^j Ax_n\right) - x_n \right\|}{\sigma\tau^j} = 0. \quad [7.15]$$

Let

$$y_n^j = P_C\left(x_n - \sigma\tau^j Ax_n\right).$$

Then

$$\left(\frac{y_n^j - x_n}{\sigma\tau^j}, x - y_n^j \right) + \left(Ax_n, x - y_n^j \right) \geq 0 \quad \forall x \in C. \quad [7.16]$$

Going to the limit in [7.16] and using [7.15], we get

$$(Ax_n, x - x_n) \geq 0 \quad \forall x \in C,$$

so, $x_n \in VI(A,C)$ and we have got a contradiction.

If $x_n \notin C$, we have

$$\lim_{j\to\infty} \left\| P_C\left(x_n - \sigma\tau^j Ax_n\right) - x_n \right\| = \left\| P_C x_n - x_n \right\| > 0.$$

and

$$\lim_{j\to\infty} \sigma\tau^j \left\| AP_C\left(x_n - \sigma\tau^j Ax_n\right) - Ax_n \right\| = 0.$$

so we again come to a contradiction (see [7.14]).

REMARK 7.5.– We have proved Lemma 7.6 completely without the assumption of monotonicity of operator A.

Let us investigate the weak convergence of Algorithm 7.1.

At first, we need to prove an important inequality for the distance from algorithm-generated points to the set $VI(A,C)$.

LEMMA 7.7.– *For the sequences (x_n) and (y_n), which are produced by Algorithm 7.1, the following inequality holds:*

$$\|x_{n+1} - z\|^2 \leq \|x_n - z\|^2 - \left(1 - \theta^2\right)\|x_n - y_n\|^2, \qquad [7.17]$$

where $z \in VI(A,C)$.

PROOF.– Let $z \in VI(A,C)$. Then

$$\|x_{n+1} - z\|^2 = \left\| P_{T_n}(x_n - \lambda_n Ay_n) - z \right\|^2$$

$$=\left\|P_{T_n}\left(x_n-\lambda_n Ay_n\right)-\left(x_n-\lambda_n Ay_n\right)+\left(x_n-\lambda_n Ay_n\right)-z\right\|^2$$

$$=\left\|\left(x_n-\lambda_n Ay_n\right)-z\right\|^2+\left\|P_{T_n}\left(x_n-\lambda_n Ay_n\right)-\left(x_n-\lambda_n Ay_n\right)\right\|^2$$

$$+2\left(P_{T_n}\left(x_n-\lambda_n Ay_n\right)-\left(x_n-\lambda_n Ay_n\right),\left(x_n-\lambda_n Ay_n\right)-z\right).$$

From the inequality,

$$2\left\|P_{T_n}\left(x_n-\lambda_n Ay_n\right)-\left(x_n-\lambda_n Ay_n\right)\right\|^2$$

$$+2\left(P_{T_n}\left(x_n-\lambda_n Ay_n\right)-\left(x_n-\lambda_n Ay_n\right),\left(x_n-\lambda_n Ay_n\right)-z\right)$$

$$=2\left(\left(x_n-\lambda_n Ay_n\right)-P_{T_n}\left(x_n-\lambda_n Ay_n\right),z-P_{T_n}\left(x_n-\lambda_n Ay_n\right)\right)\le 0,$$

follows that for all $n\ge 1$, we have

$$\left\|x_{n+1}-z\right\|^2\le\left\|\left(x_n-\lambda_n Ay_n\right)-z\right\|^2$$

$$-\left\|P_{T_n}\left(x_n-\lambda_n Ay_n\right)-\left(x_n-\lambda_n Ay_n\right)\right\|^2$$

$$=\left\|\left(x_n-\lambda_n Ay_n\right)-z\right\|^2-\left\|x_{n+1}-\left(x_n-\lambda_n Ay_n\right)\right\|^2$$

$$=\left\|x_n-z\right\|^2-\left\|x_n-x_{n+1}\right\|^2+2\lambda_n\left(z-x_{n+1},Ay_n\right).$$

Monotonicity of the mapping A, inclusion $z\in VI\left(A,C\right)$ and Lemma 7.1 imply

$$0\le\left(Ay_n-Az,y_n-z\right)=\left(Ay_n,y_n-z\right)-\underbrace{\left(Az,y_n-z\right)}_{\ge 0}$$

$$\le\left(Ay_n,y_n-z\right)=\left(Ay_n,y_n-x_{n+1}\right)+\left(Ay_n,x_{n+1}-z\right).$$

which means

$$\left(Ay_n,z-x_{n+1}\right)\le\left(Ay_n,y_n-x_{n+1}\right).$$

So, we have

$$\|x_{n+1}-z\|^2 \le \|x_n-z\|^2 - \|x_n-x_{n+1}\|^2 + 2\lambda_n\left(Ay_n, y_n-x_{n+1}\right).$$

Also

$$\|x_{n+1}-z\|^2 \le \|x_n-z\|^2 - \|(x_n-y_n)+(y_n-x_{n+1})\|^2$$

$$+2\lambda_n\left(Ay_n, y_n-x_{n+1}\right) = \|x_n-z\|^2 - \|x_n-y_n\|^2 - \|y_n-x_{n+1}\|^2$$

$$+2\left(x_n-\lambda_n Ay_n-y_n, x_{n+1}-y_n\right).$$

Because of $x_{n+1}\in T_n$, we have

$$\left(x_n-\lambda_n Ay_n-y_n, x_{n+1}-y_n\right) = \underbrace{\left(x_n-\lambda_n Ax_n-y_n, x_{n+1}-y_n\right)}_{\le 0}$$

$$+\lambda_n\left(Ax_n-Ay_n, x_{n+1}-y_n\right) \le \lambda_n\left(Ax_n-Ay_n, x_{n+1}-y_n\right).$$

And we came to the inequality

$$\|x_{n+1}-z\|^2 \le \|x_n-z\|^2 - \|x_n-y_n\|^2 - \|y_n-x_{n+1}\|^2$$

$$+2\lambda_n\left(Ax_n-Ay_n, x_{n+1}-y_n\right). \quad [7.18]$$

The last term in [7.18] can be estimated as

$$2\lambda_n\left(Ax_n-Ay_n, x_{n+1}-y_n\right) \le 2\lambda_n\|Ax_n-Ay_n\|\cdot\|x_{n+1}-y_n\|$$

$$\le 2\theta\|x_n-y_n\|\cdot\|x_{n+1}-y_n\| \le \theta^2\|x_n-y_n\|^2 + \|x_{n+1}-y_n\|^2. \quad [7.19]$$

Using estimation [7.19] for [7.18], we obtained the desired result inequality [7.17].

Using the alternative estimation of $2\lambda_n\left(Ax_n - Ay_n, x_{n+1} - y_n\right)$ from [7.18], we get another useful inequality

$$\|x_{n+1} - z\|^2 \le \|x_n - z\|^2 - (1-\theta)\|x_n - y_n\|^2 - (1-\theta)\|x_{n+1} - y_n\|^2, \qquad [7.20]$$

where $z \in VI(A,C)$.

Now, we can formulate one of the main results.

THEOREM 7.1.– *Sequences* (x_n) *and* (y_n), *which are generated by Algorithm 7.1, converge weakly to some point* $z \in VI(A,C)$.

PROOF.– It follows from [7.17] that the sequence (x_n) has a Fejer property relative to the set $VI(A,C)$, which means

$$\|x_{n+1} - z\|^2 \le \|x_n - z\|^2 \quad \forall n \ge 1 \ \ \forall z \in VI(A,C).$$

Especially, the sequence (x_n) is bounded.

Let us fix the number $N \ge 1$ and consider inequalities [7.17] for $n = 1, 2 \ldots N$. The sum of these inequalities gives us

$$\|x_{N+1} - z\|^2 \le \|x_1 - z\|^2 - \left(1-\theta^2\right)\sum_{n=1}^{N}\|x_n - y_n\|^2. \qquad [7.21]$$

Inequality [7.21] implies the convergence of the next series

$$\sum_n \|x_n - y_n\|^2.$$

So, we have

$$\lim_{n\to\infty}\|x_n - y_n\| = 0. \qquad [7.22]$$

Let us consider subsequence (x_{n_k}) which weakly converges to some point $z \in H$. Then, $y_{n_k} \xrightarrow{w} z$ and $z \in C$. Let us show that $z \in VI\left(A,C\right)$.

Two variants are possible:

1) sequence $\left(\lambda_{n_k}\right)$ does not converge to zero;

2) $\lim_{k\to\infty} \lambda_{n_k} = 0$.

Let us consider the first variant. We can assume that $\lambda_{n_k} \geq \lambda$ for all big enough k and some $\lambda > 0$. Then

$$\left(y_{n_k} - x_{n_k} + \lambda_{n_k} Ax_{n_k}, x - y_{n_k}\right) \geq 0 \quad \forall x \in C.$$

Using the inequality above and monotonicity of mapping A, we get

$$0 \leq \frac{\left(y_{n_k} - x_{n_k} + \lambda_{n_k} Ax_{n_k}, x - y_{n_k}\right)}{\lambda_{n_k}} = \frac{\left(y_{n_k} - x_{n_k}, x - y_{n_k}\right)}{\lambda_{n_k}}$$

$$+\left(Ax_{n_k}, x_{n_k} - y_{n_k}\right) + \left(Ax_{n_k}, x - x_{n_k}\right)$$

$$\leq \frac{\left(y_{n_k} - x_{n_k}, x - y_{n_k}\right)}{\lambda_{n_k}} + \left(Ax_{n_k}, x_{n_k} - y_{n_k}\right) + \left(Ax, x - x_{n_k}\right).$$

Passing to the limit and using [7.22], we get

$$\left(Ax, x - z\right) \geq 0 \quad \forall x \in C.$$

So, $z \in VI\left(A, C\right)$.

Let us deal with the second variant. Let $\lim_{k\to\infty} \lambda_{n_k} = 0$. Set

$$z_{n_k} = P_C\left(x_{n_k} - \mu_{n_k} Ax_{n_k}\right),$$

where $\mu_{n_k} = \lambda_{n_k}\tau^{-1} = \sigma\tau^{j(n_k)-1} > \lambda_{n_k} > 0$. Using Lemma 7.5, we have

$$\left\|x_{n_k} - z_{n_k}\right\| \leq \frac{1}{\tau}\left\|x_{n_k} - y_{n_k}\right\| \to 0.$$

So, the sequence $\left(z_{n_k}\right)$ is bounded and $z_{n_k} \xrightarrow{w} z$. From the uniform continuity of mapping A on bounded sets, it follows

$$\left\|Ax_{n_k} - Az_{n_k}\right\| \to 0.$$

At the same time, the inequality

$$\mu_{n_k}\left\|Az_{n_k} - Ax_{n_k}\right\| > \theta\left\|z_{n_k} - x_{n_k}\right\|$$

implies asymptotical behavior

$$\frac{\left\|x_{n_k} - z_{n_k}\right\|}{\mu_{n_k}} \to 0. \qquad [7.23]$$

So, we have

$$\left(z_{n_k} - x_{n_k} + \mu_{n_k} Ax_{n_k}, x - z_{n_k}\right) \geq 0 \quad \forall x \in C.$$

This leads to the estimation

$$0 \leq \frac{\left(z_{n_k} - x_{n_k}, x - z_{n_k}\right)}{\mu_{n_k}} + \left(Ax_{n_k}, x_{n_k} - z_{n_k}\right) + \left(Ax, x - x_{n_k}\right). \qquad [7.24]$$

Passing to the limit in [7.24] and using [7.23], we get

$$\left(Ax, x - z\right) \geq 0 \quad \forall x \in C,$$

so again, $z \in VI\left(A, C\right)$.

Let us show that $x_n \xrightarrow{w} z$. Which, in turn, will imply $y_n \xrightarrow{w} z$. Let us prove by contradiction. Suppose there is a sequence $\left(x_{m_k}\right)$ such that $x_{m_k} \xrightarrow{w} z'$ and $z \neq z'$. It is obvious that $z' \in VI\left(A, C\right)$. Using Lemma 7.2 twice, we get

$$\lim_{n\to\infty}\|x_n - z\| = \liminf_{k\to\infty}\left\|x_{n_k} - z\right\| < \liminf_{k\to\infty}\left\|x_{n_k} - z'\right\| = \lim_{n\to\infty}\|x_n - z'\|$$

$$= \liminf_{k\to\infty}\left\|x_{m_k} - z'\right\| < \liminf_{k\to\infty}\left\|x_{m_k} - z\right\| = \lim_{n\to\infty}\|x_n - z\|,$$

which is impossible. So, $z = z'$.

REMARK 7.6.– Weak limit $z \in VI(A,C)$ of the Fejer sequence (x_n) produced by Algorithm 7.1 has the next property (Bauschke and Combettes 2011): $P_{VI(A,C)}x_n \to z$. And if the set $VI(A,C)$ is an affine manifold, then $x_n \to P_{VI(A,C)}x_0$ (Bauschke and Combettes 2011). Asymptotics [7.22] can be improved – we can prove that $\liminf_{n\to\infty} \sqrt{n}\|x_n - y_n\| = 0$. Really, if we assume the contrary, we get $\|x_n - y_n\| \geq \mu n^{-1/2}$ for some $\mu > 0$ and any big enough n. So, the series $\sum_n \|x_n - y_n\|^2$ diverges, and we have a contradiction.

7.5. Modified extragradient algorithm for variational inequalities and operator equations with a priori information

Let us investigate the variant of the algorithm for finding a solution of variational inequality [7.1] which is also a fixed point of some provided mapping.

Let $S: H \to H$ be a quasi-non-expansive mapping with the set of fixed points $F(S) = \{x \in H : Sx = x\}$. Let us assume that the mapping $I - S$ is demiclosed at zero. Also, let

– (A3) $VI(A,C) \cap F(S) \neq \varnothing$.

Let $g: H \to (-\infty, +\infty)$ be a convex differentiable function. If the set $D = \{x \in H:\ g(x) \leq 0\}$ is not empty, then it can be interpreted as a set of fixed point quasi-non-expansive mapping

$$Sx = \begin{cases} x - \dfrac{g(x)}{\|\nabla g(x)\|^2}\nabla g(x), & \text{if } x \notin D, \\ x, & \text{if } x \in D, \end{cases}$$

where $\nabla g(x) \in H$ is a derivative of g at the point $x \in H$ (Vasin and Eremin 2005). For mapping $I-S$ to be demiclosed, it is sufficient for g to be bounded on any bounded set (Vasin and Eremin 2005).

Let us consider the next algorithm in order to find an element of the set $VI(A,C) \cap F(S)$.

Algorithm 7.2. Set parameters $\sigma > 0$, $\tau \in (0,1)$, $\theta \in (0,1)$, element $x_0 \in H$ and a sequence $(\delta_n) \subseteq [a,b] \subseteq (0,1)$.

Iteration. For $x_n \in H$, calculate

$$y_n = P_C(x_n - \lambda_n A x_n),$$

where λ_n is obtained from the condition

$$\begin{cases} j(n) = \min\Big\{ j \geq 0 : \left\| AP_C\left(x_n - \sigma\tau^j A x_n\right) - A x_n \right\| \\ \qquad \leq \dfrac{\theta}{\sigma\tau^j} \left\| P_C\left(x_n - \sigma\tau^j A x_n\right) - x_n \right\| \Big\}, \\ \lambda_n = \sigma\tau^{j(n)}. \end{cases}$$

And then calculate

$$x_{n+1} = \delta_n x_n + (1-\delta_n) S P_{T_n}(x_n - \lambda_n A y_n),$$

where

$$T_n = \{ z \in H : (x_n - \lambda_n A x_n - y_n, z - y_n) \leq 0 \}.$$

REMARK 7.7.– A method for finding elements of the set $VI(A,C) \cap F(S)$ for a non-expansive mapping S and a Lipschitz – continuous monotone mapping A with a fixed step $\lambda \in (0, 1/L)$, where $L > 0$ is a Lipschitz constant of A was proposed in the paper by Censor et al. (2011b). Algorithm 7.2 is a modification of that method with the dynamic step regulation procedure.

The next theorem holds.

THEOREM 7.2.– *Sequences* (x_n) *and* (y_n), *generated by Algorithm 7.2, converge weakly to a point* $z \in VI(A,C) \cap F(S)$.

PROOF.– Let

$$z_n = P_{T_n}(x_n - \lambda_n A y_n).$$

Because mapping S is non-expansive, for all points $z \in VI(A,C) \cap F(S)$, we have

$$\begin{aligned}
&\|x_{n+1} - z\|^2 \le \|\delta_n (x_n - z) + (1-\delta_n)(Sz_n - z)\|^2 \\
&= \delta_n \|x_n - z\|^2 + (1-\delta_n)\|Sz_n - z\|^2 - \delta_n(1-\delta_n)\|x_n - Sz_n\|^2 \\
&\le \delta_n \|x_n - z\|^2 + (1-\delta_n)\|z_n - z\|^2 - \delta_n(1-\delta_n)\|x_n - Sz_n\|^2.
\end{aligned} \quad [7.25]$$

Using [7.20] for [7.25] to estimate the term $(1-\delta_n)\|z_n - z\|^2$, we come to the inequality

$$\begin{aligned}
&\|x_{n+1} - z\|^2 \le \|x_n - z\|^2 - (1-\delta_n)(1-\theta)\|x_n - y_n\|^2 \\
&-(1-\delta_n)(1-\theta)\|z_n - y_n\|^2 - \delta_n(1-\delta_n)\|x_n - Sz_n\|^2.
\end{aligned} \quad [7.26]$$

From [7.26] it follows that the sequence (x_n) has a Fejer property relative to the set $VI(A,C) \cap F(S)$, so

$$\|x_{n+1} - z\|^2 \le \|x_n - z\|^2 \quad \forall n \ge 1 \quad \forall z \in VI(A,C) \cap F(S).$$

So, the sequence (x_n) is bounded. And the next inequalities hold

$$\|x_n - y_n\|^2 + \|z_n - y_n\|^2 \le \frac{\|x_n - z\|^2 - \|x_{n+1} - z\|^2}{(1-\delta_n)(1-\theta)},$$

$$\|x_n - Sz_n\|^2 \le \frac{\|x_n - z\|^2 - \|x_{n+1} - z\|^2}{\delta_n(1-\delta_n)}.$$

It follows that

$$\lim_{n\to\infty} \|x_n - y_n\| = \lim_{n\to\infty} \|z_n - y_n\| = \lim_{n\to\infty} \|x_n - Sz_n\| = 0. \quad [7.27]$$

Let us consider a subsequence (x_{n_k}), which is weakly convergent to a point $z \in H$. Then, (y_{n_k}) and (z_{n_k}) weakly converge to z and $z \in C$. Similar to the Proof of Theorem 7.1, we get $z \in VI(A, C)$.

It is left to show that $z \in F(S)$. Because of

$$\|z_n - Sz_n\| \le \|z_n - y_n\| + \|y_n - x_n\| + \|x_n - Sz_n\|,$$

equation [7.27] implies

$$\lim_{n\to\infty} \|z_n - Sz_n\| = 0.$$

Mapping $I - S$ is demiclosed at zero. So, from $z_{n_k} \overset{w}{\to} z$ and $z_{n_k} - Sz_{n_k} \to 0$, we get $z \in F(S)$.

Again, similarly to the Proof of Theorem 7.1, we can show that $x_n \overset{w}{\to} z$. Then, from [7.27], we get $y_n \overset{w}{\to} z$, which was required to prove.

Now, let us consider the operator equation with a priori information in the form of a set of fixed points of a quasi-non-expansive operator $T : H \to H$:

$$Ax = 0, \quad x \in F(T). \quad [7.28]$$

Similar problems were investigated in Vasin and Eremin (2005).

Algorithm 7.2 for the problem [7.28] takes the next form.

Algorithm 7.3. Set parameters $\sigma > 0$, $\tau \in (0,1)$, $\theta \in (0,1)$, element $x_0 \in H$ and a sequence $(\delta_n) \subseteq [a,b] \subseteq (0,1)$.

Iteration. For $x_n \in H$, calculate

$$y_n = x_n - \lambda_n A x_n,$$

where λ_n is calculated by the condition

$$\begin{cases} j(n) = \min\left\{ j \geq 0 : \left\| A\left(x_n - \sigma\tau^j A x_n\right) - A x_n \right\| \leq \theta \left\| A x_n \right\| \right\}, \\ \lambda_n = \sigma\tau^{j(n)}. \end{cases}$$

And then calculate

$$z_n = x_n - \lambda_n A y_n,$$
$$x_{n+1} = \delta_n x_n + (1-\delta_n) T z_n.$$

Partial case of Theorem 7.2 is the next result:

THEOREM 7.3.– *Let mapping $A : H \to H$ be monotone and uniformly continuous on bounded sets. Let mapping $T : H \to H$ be quasi-non-expansive with $I - T$ demiclosed at zero. And let us assume $A^{-1}0 \cap F(T) \neq \varnothing$. Then, sequences (x_n), (y_n) and (z_n), generated with Algorithm 7.3, converge weakly to a point $z \in A^{-1}0 \cap F(T)$.*

7.6. Strongly convergent modified extragradient algorithm

In this section, a strongly convergent modified extragradient method with dynamic step size calculation is proposed for solving variational inequalities with monotone operators acting in the Hilbert space. Regarding the operators, we, as before, do not require Lipschitz continuity. Variants of the method for variational inequalities and operator equations with a priori information about the solution given as a set of fixed points of the quasi-non-expansive operator are also studied. The most important theoretical result is theorems on the strong convergence of methods.

For the regularization of the modified extragradient algorithm, a simple Halpern scheme (Halpern 1967) is used, which essentially coincides with the iterative regularization scheme (Bakushinskii and Goncharskii 1989). Regularization with the help of projection CQ scheme (Nakajo and Takahashi 2003) and the Takahashi–Takeuchi–Kubota scheme (Takahashi et al. 2008) is also investigated.

7.6.1. *Algorithm variant for variational inequalities*

The next method is proposed for solving [7.1].

Algorithm 7.4. Set numerical parameters $\sigma > 0$, $\tau \in (0,1)$, $\theta \in (0,1)$, element $x_0 \in H$ and a sequence $(\alpha_n) \subseteq (0,1)$ such that $\lim_{n\to\infty} \alpha_n = 0$, $\sum_{n=1}^{\infty} \alpha_n = +\infty$.

Iteration. For $x_n \in H$, calculate

$$y_n = P_C(x_n - \lambda_n A x_n),$$

where λ_n is obtained by the condition

$$\begin{cases} j(n) = \min\Big\{ j \ge 0 : \left\| A P_C\left(x_n - \sigma\tau^j A x_n\right) - A x_n \right\| \le \\ \qquad \le \dfrac{\theta}{\sigma\tau^j} \left\| P_C\left(x_n - \sigma\tau^j A x_n\right) - x_n \right\| \Big\}, \\ \lambda_n = \sigma\tau^{j(n)}. \end{cases} \quad [7.29]$$

And then calculate

$$x_{n+1} = \alpha_n x_0 + (1-\alpha_n) P_{T_n}(x_n - \lambda_n A y_n),$$

where

$$T_n = \{ z \in H : (x_n - \lambda_n A x_n - y_n, z - y_n) \le 0 \}.$$

Let us prove a strong convergence of Algorithm 7.4. At first, we need to emphasize that step size λ_n selection procedure [7.29] always stops after a finite number of steps. So, $j(n) < +\infty$ (see Lemma 7.6).

Let us denote

$$z_n = P_{T_n}(x_n - \lambda_n A y_n). \quad [7.30]$$

The next Lemma holds.

LEMMA 7.8.– *For sequences* (x_n), (y_n) *and* (z_n), *generated by Algorithm 7.4, the next inequality holds*

$$\|z_n - z\|^2 \le \|x_n - z\|^2 - (1-\theta)\|x_n - y_n\|^2 - (1-\theta)\|z_n - y_n\|^2, \quad [7.31]$$

where $z \in VI(A, C)$.

PROOF.– Repeating the reasoning of Lemma 7.7, we come to inequality

$$\begin{aligned}\|z_n - z\|^2 &\le \|x_n - z\|^2 - \|x_n - y_n\|^2 - \|y_n - z_n\|^2 \\ &+2\lambda_n(Ax_n - Ay_n, z_n - y_n).\end{aligned} \quad [7.32]$$

The term $2\lambda_n(Ax_n - Ay_n, z_n - y_n)$ in [7.32] can be estimated as

$$\begin{aligned}&2\lambda_n(Ax_n - Ay_n, z_n - y_n) \le 2\lambda_n\|Ax_n - Ay_n\|\|z_n - y_n\| \\ &\le 2\theta\|x_n - y_n\|\|z_n - y_n\| \le \theta\|x_n - y_n\|^2 + \theta\|z_n - y_n\|^2.\end{aligned} \quad [7.33]$$

Using [7.33] in [7.32], we obtain the required inequality [7.31].

LEMMA 7.9.– *For sequences* (x_n), (y_n) *and* (z_n), *generated by Algorithm 7.4, the next inequality holds*

$$\begin{aligned}&\|x_{n+1} - z\|^2 - \|x_n - z\|^2 + (1-\theta)\|x_n - y_n\|^2 \\ &+(1-\theta)\|z_n - y_n\|^2 \le 2\alpha_n(x_0 - z, x_{n+1} - z).\end{aligned}$$

where $z \in VI(A, C)$.

PROOF.– Let $z \in VI(A,C)$. Using obvious inequality

$$\|a+b\|^2 \le \|a\|^2 + 2(b,a+b)$$

we obtain

$$\begin{aligned} \|x_{n+1}-z\|^2 &= \|\alpha_n(x_0-z)+(1-\alpha_n)(z_n-z)\|^2 \\ &\le (1-\alpha_n)^2\|z_n-z\|^2 + 2\alpha_n(x_0-z,x_{n+1}-z) \\ &\le \|z_n-z\|^2 + 2\alpha_n(x_0-z,x_{n+1}-z). \end{aligned} \quad [7.34]$$

Using [7.31] in [7.34], we obtain the required estimation.

Let us prove the boundedness of the algorithm-generated sequences.

LEMMA 7.10.– *Sequences* (x_n), (y_n) *and* (z_n), *generated by Algorithm 7.4 are bounded.*

PROOF.– Let $z \in VI(A,C)$. Then

$$\begin{aligned} \|x_{n+1}-z\| &= \|\alpha_n x_0 + (1-\alpha_n)z_n - z\| = \|\alpha_n(x_0-z)+(1-\alpha_n)(z_n-z)\| \\ &\le \alpha_n\|x_0-z\| + (1-\alpha_n)\|z_n-z\|. \end{aligned}$$

Using inequality from Lemma 7.8, we get

$$\|x_{n+1}-z\| \le \alpha_n\|x_0-z\| + (1-\alpha_n)\|x_n-z\| \le \max\{\|x_0-z\|, \|x_n-z\|\}.$$

So,

$$\|x_{n+1}-z\| \le \max\{\|x_0-z\|, \|x_1-z\|\} \quad \forall n \ge 1.$$

Which means that the sequence (x_n) is bounded.

Boundedness (y_n) and (z_n) follows from the boundedness of (x_n) and Lemma 7.8.

The next theorem holds.

THEOREM 7.4.– *Let the set $C \subseteq H$ be convex and closed, and the mapping $A: H \to H$ be monotone and uniformly continuous at bounded sets. Assume $VI(A,C) \neq \varnothing$. Then, the sequences (x_n), (y_n) and (z_n), which are generated by Algorithm 7.4, converge (strongly) to the point $\overline{z} = P_{VI(A,C)} x_0$.*

PROOF.– Let us consider element $\overline{z} = P_{VI(A,C)} x_0$. From Lemma 7.10, it follows that there is such number $M > 0$, that

$$\left|\left(x_0 - \overline{z}, x_{n+1} - \overline{z}\right)\right| \leq M \quad \forall n \geq 1.$$

Then, from inequality of Lemma 7.9, we get the estimation

$$\left\|x_{n+1} - \overline{z}\right\|^2 - \left\|x_n - \overline{z}\right\|^2 + (1-\theta)\left\|x_n - y_n\right\|^2 + (1-\theta)\left\|z_n - y_n\right\|^2 \leq 2\alpha_n M. \quad [7.35]$$

Consider the numerical sequence $\left(\left\|x_n - \overline{z}\right\|\right)$. There are two variants possible:

a) There is a number $\overline{n} \geq 1$ such that $\left\|x_{n+1} - \overline{z}\right\| \leq \left\|x_n - \overline{z}\right\|$ for all $n \geq \overline{n}$;

b) There is an increasing sequence of numbers (n_k) such that $\left\|x_{n_k+1} - \overline{z}\right\| > \left\|x_{n_k} - \overline{z}\right\|$ for all $k \geq 1$.

For variant a), there exists $\lim_{n\to\infty} \left\|x_n - \overline{z}\right\| \in [0, +\infty)$. We have $\left\|x_{n+1} - \overline{z}\right\|^2 - \left\|x_n - \overline{z}\right\|^2 \to 0$ and $\alpha_n \to 0$, so if $n \to \infty$, then

$$\left\|x_n - y_n\right\| \to 0, \quad [7.36]$$

$$\left\|z_n - y_n\right\| \to 0. \quad [7.37]$$

Boundedness of (x_n) implies the existence of subsequence $\left(x_{n_k}\right)$ which weakly converges to a point $w \in H$. From [7.36] it follows that $y_{n_k} \xrightarrow{w} w$. So, $w \in C$. Let us show that $w \in VI(A,C)$.

Again, only two variants are possible:

1) Numerical sequence $\left(\lambda_{n_k}\right)$ not converging to zero;

2) $\lim_{k\to\infty} \lambda_{n_k} = 0$.

Let us consider the first variant. We can assume $\lambda_{n_k} \geq \lambda$ for all big enough k and some $\lambda > 0$. We have

$$\left(y_{n_k} - x_{n_k} + \lambda_{n_k} Ax_{n_k}, x - y_{n_k}\right) \geq 0 \quad \forall x \in C.$$

This inequality together with monotonicity of the mapping A gives us an estimation

$$0 \leq \frac{\left(y_{n_k} - x_{n_k} + \lambda_{n_k} Ax_{n_k}, x - y_{n_k}\right)}{\lambda_{n_k}} = \frac{\left(y_{n_k} - x_{n_k}, x - y_{n_k}\right)}{\lambda_{n_k}}$$

$$+\left(Ax_{n_k}, x_{n_k} - y_{n_k}\right) + \left(Ax_{n_k}, x - x_{n_k}\right)$$

$$\leq \frac{\left(y_{n_k} - x_{n_k}, x - y_{n_k}\right)}{\lambda_{n_k}} + \left(Ax_{n_k}, x_{n_k} - y_{n_k}\right) + \left(Ax, x - x_{n_k}\right).$$

Passing to the limit and taking into account [7.36], we get

$$\left(Ax, x - w\right) \geq 0 \quad \forall x \in C.$$

So, $w \in VI\left(A, C\right)$.

Let us consider variant 2). Let $\lim_{k\to\infty} \lambda_{n_k} = 0$, and let

$$w_{n_k} = P_C\left(x_{n_k} - \mu_{n_k} Ax_{n_k}\right),$$

where $\mu_{n_k} = \lambda_{n_k} \tau^{-1} = \sigma\tau^{j(n_k)-1} > \lambda_{n_k} > 0$. Using Lemma 7.5, we get

$$\left\|x_{n_k} - w_{n_k}\right\| \leq \frac{1}{\tau}\left\|x_{n_k} - y_{n_k}\right\| \to 0.$$

In particular, the sequence $\left(w_{n_k}\right)$ is bounded and weakly converges to w. From uniform continuity of A on bounded sets, it follows that

$$\left\|Ax_{n_k} - Aw_{n_k}\right\| \to 0.$$

And from the inequality

$$\mu_{n_k}\left\|Aw_{n_k} - Ax_{n_k}\right\| > \theta\left\|w_{n_k} - x_{n_k}\right\|$$

we get the next asymptotic

$$\frac{\left\|x_{n_k} - w_{n_k}\right\|}{\mu_{n_k}} \to 0. \qquad [7.38]$$

Also, we have

$$\left(w_{n_k} - x_{n_k} + \mu_{n_k} Ax_{n_k}, x - w_{n_k}\right) \geq 0 \quad \forall x \in C.$$

This implies the estimation

$$0 \leq \frac{\left(w_{n_k} - x_{n_k}, x - w_{n_k}\right)}{\mu_{n_k}} + \left(Ax_{n_k}, x_{n_k} - w_{n_k}\right) + \left(Ax, x - x_{n_k}\right).$$

Passing to the limit and using [7.38], we get

$$\left(Ax, x - w\right) \geq 0 \quad \forall x \in C,$$

which gives us $w \in VI\left(A, C\right)$.

Now, let us prove

$$\limsup_{n\to\infty}\left(x_0 - \overline{z}, x_{n+1} - \overline{z}\right) \leq 0. \qquad [7.39]$$

Let us consider such subsequence $\left(x_{n_k}\right)$ that

$$\lim_{k\to\infty}\left(x_0 - \overline{z}, x_{n_k} - \overline{z}\right) = \limsup_{n\to\infty}\left(x_0 - \overline{z}, x_{n+1} - \overline{z}\right).$$

We can assume that $x_{n_k} \overset{w}{\rightarrow} w \in VI(A,C)$. Then

$$\lim_{k\to\infty}\left(x_0-\overline{z},x_{n_k}-\overline{z}\right)=\left(x_0-\overline{z},w-\overline{z}\right)$$

$$=\left(x_0-P_{VI(A,C)}x_0,w-P_{VI(A,C)}x_0\right)\le 0,$$

which finally implies [7.39].

Now from [7.39], the estimation

$$\begin{aligned}
&\|x_{n+1}-\overline{z}\|^2=\|\alpha_n(x_0-\overline{z})+(1-\alpha_n)(z_n-\overline{z})\|^2\\
&\le(1-\alpha_n)^2\|z_n-\overline{z}\|^2+2\alpha_n(x_0-\overline{z},x_{n+1}-\overline{z})\\
&\le(1-\alpha_n)\|x_n-\overline{z}\|^2+2\alpha_n(x_0-\overline{z},x_{n+1}-\overline{z}),
\end{aligned}$$

and Lemma 7.3 follows that $\|x_n-\overline{z}\|\to 0$. And from equations [7.36] and [7.37], we get $\|y_n-\overline{z}\|\to 0$ and $\|z_n-\overline{z}\|\to 0$.

Let us investigate variant b). We consider the sequence of numbers (m_k) with the next properties (see Lemma 7.4):

i) $m_k\to+\infty$;

ii) $\left\|x_{m_k+1}-\overline{z}\right\|\ge\left\|x_{m_k}-\overline{z}\right\|$ for all $k\ge n_1$;

iii) $\left\|x_{m_k+1}-\overline{z}\right\|\ge\|x_k-\overline{z}\|$ for all $k\ge n_1$.

From the inequality of Lemma 7.9 and ii), we have

$$\begin{aligned}
&(1-\theta)\left\|x_{m_k}-y_{m_k}\right\|^2+(1-\theta)\left\|z_{m_k}-y_{m_k}\right\|^2\\
&\le 2\alpha_{m_k}\left(x_0-\overline{z},x_{m_k+1}-\overline{z}\right)\le 2\alpha_{m_k}M.
\end{aligned}$$

This, in turn, implies

$$\lim_{k\to\infty}\left\|x_{m_k}-y_{m_k}\right\|=\lim_{k\to\infty}\left\|z_{m_k}-y_{m_k}\right\|=0.$$

Using similar reasoning, we show that partial weak limits of sequences $\left(x_{m_k}\right)$ and $\left(y_{m_k}\right)$ belong to the set $VI(A,C)$. Also, from the inequality

$$\begin{aligned}
&\left\|x_{m_k+1}-x_{m_k}\right\|=\left\|\alpha_{m_k}x_0+\left(1-\alpha_{m_k}\right)z_{m_k}-x_{m_k}\right\|\\
&\le\alpha_{m_k}\left\|x_0-x_{m_k}\right\|+\left(1-\alpha_{m_k}\right)\left\|z_{m_k}-x_{m_k}\right\|\\
&\le\alpha_{m_k}\left\|x_0-x_{m_k}\right\|+\left(1-\alpha_{m_k}\right)\left(\left\|z_{m_k}-y_{m_k}\right\|+\left\|y_{m_k}-x_{m_k}\right\|\right),
\end{aligned}$$

we have

$$\left\|x_{m_k+1}-x_{m_k}\right\|\to 0.$$

From the equality

$$\left(x_0-\overline{z},x_{m_k+1}-\overline{z}\right)=\left(x_0-\overline{z},x_{m_k}-\overline{z}\right)+\left(x_0-\overline{z},x_{m_k+1}-x_{m_k}\right)$$

we obtain

$$\overline{\lim_{k\to\infty}}\left(x_0-\overline{z},x_{m_k+1}-\overline{z}\right)=\overline{\lim_{k\to\infty}}\left(x_0-\overline{z},x_{m_k}-\overline{z}\right).$$

So, as in a previous part of the proof, we get

$$\limsup_{k\to\infty}\left(x_0-\overline{z},x_{m_k+1}-\overline{z}\right)\le 0.$$

And then

$$\begin{aligned}
&\left\|x_{m_k+1}-\overline{z}\right\|^2\le\left(1-\alpha_{m_k}\right)\left\|x_{m_k}-\overline{z}\right\|^2+2\alpha_{m_k}\left(x_0-\overline{z},x_{m_k+1}-\overline{z}\right)\\
&\le\left(1-\alpha_{m_k}\right)\left\|x_{m_k+1}-\overline{z}\right\|^2+2\alpha_{m_k}\left(x_0-\overline{z},x_{m_k+1}-\overline{z}\right).
\end{aligned}$$

Taking into account iii), we get

$$\|x_k - \bar{z}\|^2 \le \|x_{m_k+1} - \bar{z}\|^2 \le 2(x_0 - \bar{z}, x_{m_k+1} - \bar{z}).$$

So

$$\limsup_{k\to\infty} \|x_k - \bar{z}\|^2 \le 2 \limsup_{k\to\infty} (x_0 - \bar{z}, x_{m_k+1} - \bar{z}) \le 0.$$

This leads to

$$\lim_{n\to\infty} \|x_n - \bar{z}\| = 0,$$

and, at last, to the final result: $\lim_{n\to\infty} \|y_n - \bar{z}\| = \lim_{n\to\infty} \|z_n - \bar{z}\| = 0$.

Let us consider one more variant of the strongly convergent modified extragradient algorithm for solving variational inequality [7.1].

Algorithm 7.5. Set numerical parameters $\sigma > 0$, $\tau \in (0,1)$, $\theta \in (0,1)$ and element $x_0 \in H$.

Iteration. For $x_n \in H$, calculate

$$y_n = P_C(x_n - \lambda_n A x_n),$$

where λ_n is obtained from the condition

$$\begin{cases} j(n) = \min\Big\{ j \ge 0 : \left\| A P_C(x_n - \sigma\tau^j A x_n) - A x_n \right\| \\ \qquad\qquad \le \dfrac{\theta}{\sigma\tau^j} \left\| P_C(x_n - \sigma\tau^j A x_n) - x_n \right\| \Big\}, \\ \lambda_n = \sigma\tau^{j(n)}. \end{cases}$$

Calculate

$$z_n = P_{T_n}(x_n - \lambda_n A y_n),$$

where

$$T_n = \{z \in H : (x_n - \lambda_n A x_n - y_n, z - y_n) \le 0\}.$$

Calculate

$$x_{n+1} = P_{C_n \cap Q_n} x_0,$$

where

$$C_n = \{z \in H : \|z_n - z\| \le \|x_n - z\|\},$$
$$Q_n = \{z \in H : (x_n - z, x_0 - x_n) \ge 0\}.$$

Sets Q_n, C_n – closed subspaces. Calculating the projection of a point to the intersection $C_n \cap Q_n$ is a simple problem, which can be solved in a closed form (Bauschke and Combettes 2011).

REMARK 7.8.– Algorithm 7.5 is created by introducing the dynamic step size adjustment procedure to the scheme of the hybrid method (Nadezhkina and Takahashi 2006a), which in turn is a regularization of the Korpelevich method with projective CQ-scheme for approximating fixed points of a non-expansive operator, proposed by Japanese mathematicians Nakajo and Takahashi (2003).

The next theorem holds.

THEOREM 7.5.– *Let the set* $C \subseteq H$ *be convex and closed, and the mapping* $A : H \to H$ *be monotone and uniformly continuous at bounded sets. Assume* $VI(A,C) \ne \varnothing$. *Then, the sequences* (x_n), (y_n) *and* (z_n), *which are generated by Algorithm 7.5, strongly converge to the point* $\bar{z} = P_{VI(A,C)} x_0$.

PROOF.– Let us consider element $z = P_{VI(A,C)} x_0$. For the sequences (x_n), (y_n) and (z_n), which are generated by Algorithm 7.5, the next inequality holds

$$\|z_n - z\|^2 \le \|x_n - z\|^2 - (1-\theta)\|x_n - y_n\|^2 - (1-\theta)\|z_n - y_n\|^2. \quad [7.40]$$

Inequality [7.40] implies $z \in C_n$ for all $n \ge 1$. So, $VI(A,C) \subseteq C_n$ for all $n \ge 0$.

With mathematical induction, we can show that for every $n \geq 0$ holds $VI(A,C) \subseteq C_n \cap Q_n$. For $n = 0$, we have $Q_n = H$. So, $VI(A,C) \subseteq C_0 \cap Q_0$. Let for some $k \geq 1$ holds $VI(A,C) \subseteq C_k \cap Q_k$. Then, there exists a unique point $x_{k+1} \in C_k \cap Q_k$ such that $x_{k+1} = P_{C_k \cap Q_k} x_0$. Equation $x_{k+1} = P_{C_k \cap Q_k} x_0$ implies $(x_{k+1} - z, x_0 - x_{k+1}) \geq 0$ for all $z \in C_k \cap Q_k$. Inclusion $VI(A,C) \subseteq C_k \cap Q_k$ implies $VI(A,C) \subseteq Q_{k+1}$. As a result, $VI(A,C) \subseteq C_{k+1} \cap Q_{k+1}$.

Let us show that the sequence (x_n) is bounded. There exists a unique point $\bar{z} \in VI(A,C)$ such that $\bar{z} = P_{VI(A,C)} x_0$. $x_{n+1} = P_{C_n \cap Q_n} x_0$ implies

$$\|x_{n+1} - x_0\| \leq \|z - x_0\| \quad \forall z \in C_n \cap Q_n.$$

Because of $\bar{z} \in VI(A,C) \subseteq C_n \cap Q_n$, we have

$$\|x_{n+1} - x_0\| \leq \|\bar{z} - x_0\|. \qquad [7.41]$$

And [7.41] implies that (x_n) is bounded.

Let us prove that

$$\lim_{n \to \infty} \|x_{n+1} - x_n\| = 0. \qquad [7.42]$$

$x_{n+1} \in C_n \cap Q_n \subseteq Q_n$ and $x_n = P_{Q_n} x_0$ implies

$$\|x_{n+1} - x_0\| \geq \|x_n - x_0\| \quad \forall n \geq 1.$$

Numerical sequence $(\|x_n - x_0\|)$ bounded and non-decreasing. So, there exists a finite limit $\lim_{n \to \infty} \|x_n - x_0\|$. At the same time, as $x_{n+1} \in Q_n$, we have

$$(x_n - x_{n+1}, x_0 - x_n) \geq 0$$

and

$$
\begin{aligned}
\|x_n - x_{n+1}\|^2 &= \|(x_n - x_0) - (x_{n+1} - x_0)\|^2 \\
&= \|x_n - x_0\|^2 - 2(x_n - x_0, x_{n+1} - x_0) + \|x_{n+1} - x_0\|^2 \\
&= \|x_{n+1} - x_0\|^2 - \|x_n - x_0\|^2 - 2(x_n - x_{n+1}, x_0 - x_n) \\
&\leq \|x_{n+1} - x_0\|^2 - \|x_n - x_0\|^2,
\end{aligned}
$$

which implies [7.42]

As $x_{n+1} \in C_n$, so $\|z_n - x_{n+1}\| \leq \|x_n - x_{n+1}\|$. And this implies

$$\|z_n - x_n\| \leq \|z_n - x_{n+1}\| + \|x_{n+1} - x_n\| \leq 2\|x_n - x_{n+1}\| \to 0. \qquad [7.43]$$

Using [7.40], we get

$$
\begin{aligned}
\|x_n - y_n\|^2 &\leq \frac{\|x_n - z\|^2 - \|z_n - z\|^2}{(1-\theta)} \\
&= \frac{(\|x_n - z\| - \|z_n - z\|)(\|x_n - z\| + \|z_n - z\|)}{(1-\theta)} \\
&\leq \frac{(\|x_n - z\| + \|z_n - z\|)}{(1-\theta)}\|x_n - z_n\| = O(\|x_n - z_n\|),
\end{aligned}
$$

$$
\begin{aligned}
\|z_n - y_n\|^2 &\leq \frac{\|x_n - z\|^2 - \|z_n - z\|^2}{(1-\theta)} \\
&= \frac{(\|x_n - z\| - \|z_n - z\|)(\|x_n - z\| + \|z_n - z\|)}{(1-\theta)} \\
&\leq \frac{(\|x_n - z\| + \|z_n - z\|)}{(1-\theta)}\|x_n - z_n\| = O(\|x_n - z_n\|),
\end{aligned}
$$

where $z \in VI(A, C)$. Equation [7.43] implies

$$\|x_n - y_n\| \to 0, \qquad [7.44]$$

$$\|z_n - y_n\| \to 0. \qquad [7.45]$$

Sequence (x_n) is bounded, so there exists a subsequence (x_{n_k}) which weakly converges to a point $w \in H$. Equation [7.44] implies $y_{n_k} \xrightarrow{w} w$. So, $w \in C$. With the reasoning, similar to Theorem 7.4, we can show that $w \in VI(A,C)$.

For $\bar{z} = P_{VI(A,C)}x_0$, inequality [7.41] implies

$$\|x_0 - \bar{z}\| \le \|x_0 - w\| \le \liminf_{k\to\infty} \|x_0 - x_{n_k}\| \le \limsup_{k\to\infty} \|x_0 - x_{n_k}\| \le \|x_0 - \bar{z}\|.$$

So, we have got $\lim_{k\to\infty} \|x_0 - x_{n_k}\| = \|x_0 - w\| = \|x_0 - \bar{z}\|$. This implies $x_{n_k} \to w = \bar{z}$. So, $x_n \to \bar{z}$. Equations [7.44] and [7.45] imply $y_n \to \bar{z}$ and $z_n \to \bar{z}$, which was required to prove.

The next strongly convergent algorithm for solving variational inequality [7.1] can be constructed using the Takahashi–Takeuchi–Kubota method (Takahashi et al. 2008).

Algorithm 7.6. Set numerical parameters $\sigma > 0$, $\tau \in (0,1)$, $\theta \in (0,1)$ and element $x_0 \in H$.

Iteration. For $x_n \in H$, calculate

$$y_n = P_C(x_n - \lambda_n A x_n),$$

where λ_n is calculated using the condition

$$\begin{cases} j(n) = \min\{ j \ge 0 : \left\| AP_C\left(x_n - \sigma\tau^j Ax_n\right) - Ax_n \right\| \\ \qquad\qquad \le \dfrac{\theta}{\sigma\tau^j} \left\| P_C\left(x_n - \sigma\tau^j Ax_n\right) - x_n \right\| \Big\}, \\ \lambda_n = \sigma\tau^{j(n)}. \end{cases}$$

Calculate

$$z_n = P_{T_n}(x_n - \lambda_n A y_n),$$

where

$$T_n = \{z \in H : (x_n - \lambda_n A x_n - y_n, z - y_n) \leq 0\}.$$

Calculate

$$x_{n+1} = P_{C_{n+1}} x_0,$$

where

$$C_{n+1} = \{z \in C_n : \|z_n - z\| \leq \|x_n - z\|\}, \ C_0 = H.$$

REMARK 7.9.– The main disadvantage of the scheme above is the increasing complexity of convex sets C_n, on which x_0 is projected.

The next theorem holds.

THEOREM 7.6.– *Let conditions (A1) and (A2) be satisfied. Then, sequences* (x_n), (y_n) *and* (z_n), *produced by Algorithm 7.6, strongly converge to a point* $\bar{z} = P_{VI(A,C)} x_0$.

PROOF.– For the sequences (x_n), (y_n) and (z_n), which are produced by Algorithm 7.6, inequality [7.40] holds. Let us show that this algorithm generates a chain of nested sets

$$H = C_0 \supseteq C_1 \supseteq \ldots \supseteq C_n \supseteq \ldots \supseteq VI(A,C).$$

It is obvious that $VI(A,C) \subseteq C_0 = H$. Let us assume $VI(A,C) \subseteq C_n$. And let $z \in VI(A,C)$. Equation [7.40] implies

$$\|z_n - z\| \leq \|x_n - z\|.$$

So, $z \in C_{n+1}$, and this implies $VI(A,C) \subseteq C_{n+1} \subseteq C_n$ for all $n \geq 1$.

Let us show that there exists $\lim_{n\to\infty} \|x_n - x_0\| \in [0,+\infty)$. As

$$x_n = P_{C_n} x_0, \quad VI(A,C) \subseteq C_n,$$

so for all $z \in VI(A,C)$, the next inequality holds

$$\|x_n - z\|^2 \le \|x_0 - z\|^2 - \|x_n - x_0\|^2.$$

This implies

$$\|x_n - x_0\|^2 \le \|x_0 - z\|^2 - \|x_n - z\|^2 \le \|x_0 - z\|^2.$$

In turn, this implies that the sequence $(\|x_n - x_0\|)$ is bounded from above. Since $C_{n+1} \subseteq C_n$, so $\|x_{n+1} - x_0\| \ge \|x_n - x_0\|$. Then, the sequence $(\|x_n - x_0\|)$ is bounded from above and non-decreasing, and there exists finite limit $\lim_{n\to\infty} \|x_n - x_0\|$.

Let us show that the sequence (x_n) is fundamental. For every $m \ge 1$, we have (with account to $C_{n+m} \subseteq C_n$)

$$\|x_{n+m} - x_n\|^2 = \left\|x_{n+m} - P_{C_n} x_0\right\|^2 \le \|x_0 - x_{n+m}\|^2 - \|x_n - x_0\|^2.$$

This implies that (x_n) is the Cauchy sequence. So,

$$x_n \to \bar{z} \in H, \ n \to \infty.$$

Let us show that $\bar{z} \in VI(A,C)$. Since $x_{n+1} \in C_{n+1}$, so

$$\|z_n - x_{n+1}\| \le \|x_n - x_{n+1}\|.$$

This implies

$$\|z_n - x_n\| \le \|z_n - x_{n+1}\| + \|x_n - x_{n+1}\| \le 2\|x_n - x_{n+1}\| \to 0.$$

Using inequality [7.40], for all $z \in VI(A,C)$, we get

$$\|x_n - y_n\|^2 = O(\|x_n - z_n\|), \quad \|z_n - y_n\|^2 = O(\|x_n - z_n\|),$$

which, in turn, leads to

$$\lim_{n\to\infty} \|x_n - y_n\| = \lim_{n\to\infty} \|y_n - z_n\| = 0.$$

The above asymptotics imply that $y_n \to \overline{z}$ and $z_n \to \overline{z}$ if $n \to \infty$. So, $\overline{z} \in C$. With a reasoning, similar to Theorem 7.4, it can be shown that $\overline{z} \in VI(A,C)$.

Since $x_n = P_{C_n} x_0$ and $VI(A,C) \subseteq C_n$, so

$$(x_n - x_0, z - x_n) \geq 0 \quad \forall z \in VI(A,C).$$

Passing to the limit in the inequality above, we get

$$(\overline{z} - x_0, z - \overline{z}) \geq 0 \quad \forall z \in VI(A,C),$$

so $\overline{z} = P_{VI(A,C)} x_0$.

As can be seen from the reasoning above, sequences (y_n) and (z_n) are strongly converging to $\overline{z} = P_{VI(A,C)} x_0$.

7.6.2. *Variant for problems with a priori information*

Let us consider quasi-non-expansive mapping $S : H \to H$ with fixed set points set $F(S)$. Let us also consider the problem of finding

$$x \in VI(A,C) \cap F(S). \tag{7.46}$$

Let us assume that mapping $I - S$ is demiclosed at zero, and condition (A3) holds. The next algorithm can be used to find solutions of [7.46].

Algorithm 7.7. Set numerical parameters $\sigma > 0$, $\tau \in (0,1)$, $\theta \in (0,1)$, element $x_0 \in H$, sequence $(\delta_n) \subseteq [a,b] \subseteq (0,1)$ and sequence $(\alpha_n) \subseteq (0,1)$ such that $\lim_{n\to\infty} \alpha_n = 0$, $\sum_{n=1}^{\infty} \alpha_n = +\infty$.

Iteration. For $x_n \in H$, calculate

$$y_n = P_C(x_n - \lambda_n A x_n),$$

where λ_n is calculated using the condition

$$\begin{cases} j(n) = \min\left\{ j \geq 0 : \left\| AP_C\left(x_n - \sigma\tau^j Ax_n\right) - Ax_n \right\| \right. \\ \qquad\qquad \left. \leq \dfrac{\theta}{\sigma\tau^j} \left\| P_C\left(x_n - \sigma\tau^j Ax_n\right) - x_n \right\| \right\}, \\ \lambda_n = \sigma\tau^{j(n)}. \end{cases}$$

Calculate

$$x_{n+1} = \alpha_n x_0 + (1-\alpha_n)\left(\delta_n x_n + (1-\delta_n) SP_{T_n}(x_n - \lambda_n A y_n)\right),$$

where

$$T_n = \left\{ z \in H : (x_n - \lambda_n A x_n - y_n, z - y_n) \leq 0 \right\}.$$

For analysis of Algorithm [7.7], we continue using notations from [7.30].

LEMMA 7.11.– *For the sequences (x_n), (y_n) and (z_n), generated by Algorithm 7.7, the next inequality holds*

$$\begin{aligned} &\|x_{n+1} - z\|^2 - \|x_n - z\|^2 + (1-\delta_n)(1-\theta)\|x_n - y_n\|^2 \\ &+ (1-\delta_n)(1-\theta)\|z_n - y_n\|^2 + \delta_n(1-\delta_n)\|x_n - Sz_n\|^2 \\ &\leq 2\alpha_n(x_0 - z, x_{n+1} - z). \end{aligned}$$

where $z \in VI(A,C) \cap F(S)$.

PROOF.– Let $z \in VI(A,C) \cap F(S)$. Then

$$\begin{aligned}\|x_{n+1}-z\|^2 &= \|\alpha_n(x_0-z)+(1-\alpha_n)(\delta_n x_n+(1-\delta_n)Sz_n-z)\|^2 \\ &\le (1-\alpha_n)^2\|\delta_n x_n+(1-\delta_n)Sz_n-z\|^2+2\alpha_n(x_0-z,x_{n+1}-z) \\ &\le \|\delta_n(x_n-z)+(1-\delta_n)(Sz_n-z)\|^2+2\alpha_n(x_0-z,x_{n+1}-z) \\ &= \delta_n\|x_n-z\|^2+(1-\delta_n)\|Sz_n-z\|^2-\delta_n(1-\delta_n)\|x_n-Sz_n\|^2 \\ &+2\alpha_n(x_0-z,x_{n+1}-z) \\ &\le \delta_n\|x_n-z\|^2+(1-\delta_n)\|z_n-z\|^2-\delta_n(1-\delta_n)\|x_n-Sz_n\|^2 \\ &+2\alpha_n(x_0-z,x_{n+1}-z).\end{aligned} \quad [7.47]$$

Estimating the term $(1-\delta_n)\|z_n-z\|^2$ in [7.47] with Lemma 7.8, we obtain the inequality

$$\begin{aligned}\|x_{n+1}-z\|^2 &\le \|x_n-z\|^2-(1-\delta_n)(1-\theta)\|x_n-y_n\|^2 \\ &-(1-\delta_n)(1-\theta)\|z_n-y_n\|^2-\delta_n(1-\delta_n)\|x_n-Sz_n\|^2 \\ &+2\alpha_n(x_0-z,x_{n+1}-z),\end{aligned}$$

which was required to prove.

LEMMA 7.12.– *Sequences (x_n), (y_n) and (z_n), generated by Algorithm 7.7, are bounded.*

PROOF.– Let $z \in VI(A,C) \cap F(S)$. We have

$$\begin{aligned}\|x_{n+1}-z\| &= \|\alpha_n(x_0-z)+(1-\alpha_n)(\delta_n(x_n-z)+(1-\delta_n)(Sz_n-z))\| \\ &\le \alpha_n\|x_0-z\|+(1-\alpha_n)\delta_n\|x_n-z\|+(1-\alpha_n)(1-\delta_n)\|Sz_n-z\| \\ &\le \alpha_n\|x_0-z\|+(1-\alpha_n)\delta_n\|x_n-z\|+(1-\alpha_n)(1-\delta_n)\|z_n-z\|.\end{aligned}$$

Using inequality [7.31], we get

$$\|x_{n+1}-z\| \le \alpha_n\|x_0-z\|+(1-\alpha_n)\|x_n-z\| \le \max\{\|x_0-z\|,\|x_n-z\|\}.$$

So, $\|x_{n+1}-z\| \leq \max\{\|x_0-z\|, \|x_1-z\|\}$ for all $n \geq 1$, and the sequence (x_n) is bounded. With inequality [7.31], this also implies that sequences (y_n) and (z_n) are also bounded.

The next theorem holds.

THEOREM 7.7.– *Let $C \subseteq H$ be convex and closed, and mapping $A: H \to H$ is monotone and uniformly continuous at bounded sets. Let mapping $S: H \to H$ be quasi-non-expansive and mapping $I-S$ is demiclosed at zero. Assume $VI(A,C) \cap F(S) \neq \varnothing$. Then, sequences (x_n), (y_n) and (z_n), generated by Algorithm 7.7, strongly converge to a point $\bar{z} = P_{VI(A,C) \cap F(S)} x_0$.*

PROOF.– Let us consider element $\bar{z} = P_{VI(A,C) \cap F(S)} x_0$. Lemma 7.12 implies that there is such $M > 0$ that $|(x_0 - \bar{z}, x_{n+1} - \bar{z})| \leq M \quad \forall n \geq 1$. Using inequality from Lemma 7.11, we get the estimation

$$\begin{aligned} &\|x_{n+1}-\bar{z}\|^2 - \|x_n - \bar{z}\|^2 + (1-\delta_n)(1-\theta)\|x_n - y_n\|^2 \\ &+(1-\delta_n)(1-\theta)\|z_n - y_n\|^2 + \delta_n(1-\delta_n)\|x_n - Sz_n\|^2 \leq 2\alpha_n M. \end{aligned} \quad [7.48]$$

For the numerical sequence $(\|x_n - \bar{z}\|)$, we have two possible variants:

a) There exists such number $\bar{n} \geq 1$ that $\|x_{n+1} - \bar{z}\| \leq \|x_n - \bar{z}\|$ for all $n \geq \bar{n}$;

b) There exists an increasing sequence of numbers (n_k) such that $\|x_{n_k+1} - \bar{z}\| > \|x_{n_k} - \bar{z}\|$ for all $k \geq 1$.

Let us consider variant a). In this case, there is a $\lim_{n \to \infty} \|x_n - \bar{z}\| \in [0, +\infty)$. As $\|x_{n+1} - \bar{z}\|^2 - \|x_n - \bar{z}\|^2 \to 0$ and $\alpha_n \to 0$, so for $n \to \infty$, we have

$$\|x_n - y_n\| \to 0, \quad [7.49]$$

$$\|z_n - y_n\| \to 0, \quad [7.50]$$

$$\|x_n - Sz_n\| \to 0. \quad [7.51]$$

Sequence (x_n) is bounded, so there is a subsequence $\left(x_{n_k}\right)$ which weakly converges to a point $w \in H$. Equation [7.49] implies that (y_n) converges weakly to $w \in H$. So, we have $w \in C$. With a reasoning, similar to the Proof of Theorem 7.4, we get $w \in VI(A,C)$. So, it is left to prove that $w \in F(S)$. Because of

$$\|z_n - Sz_n\| \leq \|z_n - y_n\| + \|y_n - x_n\| + \|x_n - Sz_n\|,$$

and with [7.49]–[7.51], we have $\lim_{n\to\infty} \|z_n - Sz_n\| = 0$. Mapping $I - S$ is demiclosed at 0, so $z_{n_k} \overset{w}{\rightarrow} w$ together with $\lim_{k\to\infty} \left\|z_{n_k} - Sz_{n_k}\right\| = 0$ implies $w \in F(S)$.

Similar to the Proof of Theorem 7.4, we get

$$\limsup_{n\to\infty} \left(x_0 - \bar{z}, x_{n+1} - \bar{z}\right) \leq 0. \qquad [7.52]$$

From [7.52], Lemma 7.3 and the next inequality

$$\begin{aligned}&\|x_{n+1} - \bar{z}\|^2 \leq (1-\alpha_n)^2 \|\delta_n x_n + (1-\delta_n) Sz_n - \bar{z}\|^2 \\ &+2\alpha_n \left(x_0 - \bar{z}, x_{n+1} - \bar{z}\right) \\ &\leq (1-\alpha_n)\|x_n - \bar{z}\|^2 + 2\alpha_n \left(x_0 - \bar{z}, x_{n+1} - \bar{z}\right)\end{aligned}$$

we get $\|x_n - \bar{z}\| \to 0$. And with [7.49] and [7.50], we get $\|y_n - \bar{z}\| \to 0$ and $\|z_n - \bar{z}\| \to 0$.

Let us look at the variant b). We can consider the sequence of numbers (m_k) with the next properties:

i) $m_k \to +\infty$;

ii) $\left\|x_{m_k+1} - \bar{z}\right\| \geq \left\|x_{m_k} - \bar{z}\right\|$ for all $k \geq n_1$;

iii) $\left\|x_{m_k+1} - \bar{z}\right\| \geq \|x_k - \bar{z}\|$ for all $k \geq n_1$;

From ii), it follows that

$$\left(1-\delta_{m_k}\right)(1-\theta)\left\|x_{m_k}-y_{m_k}\right\|^2+\left(1-\delta_{m_k}\right)(1-\theta)\left\|z_{m_k}-y_{m_k}\right\|^2$$
$$+\delta_{m_k}\left(1-\delta_{m_k}\right)\left\|x_{m_k}-Sz_{m_k}\right\|^2\leq 2\alpha_{m_k}\left(x_0-\bar{z},x_{m_k+1}-\bar{z}\right)\leq 2\alpha_{m_k}M.$$

Which, in turn, implies

$$\lim_{k\to\infty}\left\|x_{m_k}-y_{m_k}\right\|=\lim_{k\to\infty}\left\|z_{m_k}-y_{m_k}\right\|=\lim_{k\to\infty}\left\|x_{m_k}-Sz_{m_k}\right\|=0.$$

A similar reasoning gives us that weak partial limits of sequences $\left(x_{m_k}\right)$, $\left(y_{m_k}\right)$ and $\left(z_{m_k}\right)$ are in the set $VI(A,C)\cap F(S)$.

Inequality

$$\left\|x_{m_k+1}-x_{m_k}\right\|=\left\|\alpha_{m_k}\left(x_0-x_{m_k}\right)+\left(1-\alpha_{m_k}\right)\left(1-\delta_{m_k}\right)\left(Sz_{m_k}-x_{m_k}\right)\right\|$$
$$\leq\alpha_{m_k}\left\|x_0-x_{m_k}\right\|+\left(1-\alpha_{m_k}\right)\left(1-\delta_{m_k}\right)\left\|Sz_{m_k}-x_{m_k}\right\|$$

implies $\left\|x_{m_k+1}-x_{m_k}\right\|\to 0$. And the proof is obtained with the reasoning similar to the Proof of Theorem 7.4.

And now let us consider the operator equation with a priori information in the form of a fixed point set of quasi-non-expansive mapping $T:H\to H$:

$$Ax=0,\quad x\in F(T). \qquad [7.53]$$

Algorithm 7.7 for the problem [7.53] becomes:

Algorithm 7.8. Set numerical parameters $\sigma>0$, $\tau\in(0,1)$, $\theta\in(0,1)$, element $x_0\in H$, sequence $(\delta_n)\subseteq[a,b]\subseteq(0,1)$ and sequence $(\alpha_n)\subseteq(0,1)$ such that

$$\lim_{n\to\infty}\alpha_n=0,\ \sum_{n=1}^{\infty}\alpha_n=+\infty.$$

Iteration. For $x_n\in H$, calculate

$$y_n=x_n-\lambda_n Ax_n,$$

where λ_n is obtained from the condition

$$\begin{cases} j(n) = \min\left\{ j \geq 0 : \left\| A\left(x_n - \sigma\tau^j Ax_n\right) - Ax_n \right\| \leq \theta \left\| Ax_n \right\| \right\}, \\ \lambda_n = \sigma\tau^{j(n)}. \end{cases}$$

Calculate

$$z_n = x_n - \lambda_n Ay_n,$$
$$x_{n+1} = \alpha_n x_0 + (1-\alpha_n)\left(\delta_n x_n + (1-\delta_n) Tz_n\right).$$

The next theorem is a special case of Theorem 7.7.

THEOREM 7.8.– *Let mapping* $A : H \to H$ *be monotone and uniformly continuous on bounded sets. Let mapping* $T : H \to H$ *be quasi-non-expansive and* $I - T$ *is demiclosed at zero. Assume* $A^{-1}0 \cap F(T) \neq \varnothing$. *Then, sequences* (x_n), (y_n) *and* (z_n), *produced by Algorithm 7.8, strongly converge to point* $\bar{z} = P_{A^{-1}0 \cap F(T)} x_0$.

7.7. References

Antipin, A.S. (1976). On a convex programming method using a symmetrical modification of the Lagrange function. *Matecon.*, 12(6), 1164–1173.

Aubin, J.P. and Ekeland, I. (1984). *Applied Nonlinear Analysis*. John Wiley & Sons, New York.

Baiocchi, C. and Capello, A. (1984). *Variational and Quasivariational Inequalities: Application to Free Boundary Problems*. John Wiley & Sons, New York.

Bakushinskii, A.B. and Goncharskii, A.V. (1989). *Iterative Methods for Solving Ill-posed Problems*. Nauka, Moscow.

Bauschke, H.H. (1996). The approximation of fixed points of compositions of nonexpansive mappings in Hilbert space. *J. Math. Anal. Appl.*, 202, 150–159.

Bauschke, H.H. and Combettes, P.L. (2011). *Convex Analysis and Monotone Operator Theory in Hilbert Spaces*. Springer, New York.

Bensoussan, A. (1974). Points de Nash dans le cas de fontionnelles quadratiques et jeux différentiels linéaires à N personnes. *SIAM J. Control*, 12, 460–499.

Brezis, H. (1972). Problèmes unilateraux. *J. Math. Pures Appl.*, 51, 1–168.

Browder, F.E. (1965a). Fixed-point theorems for noncompact mappings in Hilbert space. *Proc. Natl. Acad. Sci. USA*, 53, 1272–1276.

Browder, F.E. (1965b). Nonexpansive nonlinear operators in a Banach space. *Proc. Natl. Acad. Sci.*, 54, 1041–1044.

Burachik, R.S., Lopes, J.O., Svaiter, B.F. (2005). An outer approximation method for the variational inequality problem. *SIAM J. Control Optim.*, 43(6), 2071–2088.

Censor, Y., Gibali, A., Reich, S. (2011a). Strong convergence of subgradient extragradient methods for the variational inequality problem in Hilbert space. *Optim. Methods Softw.*, 26, 827–845.

Censor, Y., Gibali, A., Reich, S. (2011b). The subgradient extragradient method for solving variational inequalities in Hilbert space. *JOTA*, 148, 318–335.

Combettes, P.L. (1999). Strong convergence of block-iterative outer approximation methods for convex optimization. *SIAM J. Control Optim.*, 38(2), 538–565.

Dafermos, S. (1982). Traffic equilibria and variational inequalities. *Transport. Sci.*, 16, 231–240.

Denisov, S.V. and Semenov, V.V. (2011). Proximal algorithm for bilevel variational inequalities: Strong convergence. *J. Num. Appl. Math.*, 3, 27–32.

Denisov, S.V., Semenov, V.V., Chabak, L.M. (2015). Convergence of the modified extragradient method for variational inequalities with non-Lipschitz operators. *Cybern. Syst. Anal.*, 51(5), 757–765.

Duveau, G. and Lions, J.-L. (1980). *Inequalities in Mechanics and Physics*. Nauka, Moscow.

Ekeland, I. and Temam, R. (1976). *Convex Analysis and Variational Problems*. North-Holland, Amsterdam.

Eremin, I.I. and Mazurov, V.D. (1979). *Non-Stationary Processes of Mathematical Programming*. Nauka, Moscow.

Fichera, G. (1964). Problemi Elastostatici con Vincoli Unilaterali, il Problema di Signorini con Ambigue Condizioni al Contorno. *Mem. Accad. Naz. Lincei*, VIII(7), 91–140.

Glowinski, R., Lions, J.-L., Trémolières, R. (1981). *Numerical Analysis of Variational Inequalities*. North-Holland, Amsterdam.

Göhde, D. (1965). Zum Prinzip der kontraktiven Abbildung. *Math. Nachr.*, 30, 251–258.

Goldstein, A.A. (1964). Convex programming in Hilbert space. *Bull. AMS*, 70, 709–710.

Halpern, B. (1967). Fixed points of nonexpanding maps. *Bull. AMS*, 73, 957–961.

Harker, P.T. and Pang, J.-S. (1990). Finite-dimensional variational inequality and nonlinear complementarity problems: A survey of theory, algorithms and applications. *Math. Program.*, 48, 161–220.

Hartman, P. and Stampacchia, G. (1966). On some nonlinear elliptic differential functional equations. *Acta Math.*, 115, 153–188.

Haugazeau, Y. (1968). Sur les inéquations variationnelles et la minimisation de fonctionnelles convexes. PhD Thesis, Université de Paris.

Iusem, A.N. and Nasri, M. (2011). Korpelevich's method for variational inequality problems in Banach spaces. *JOGO*, 50, 59–76.

Iusem, A.N. and Svaiter, B.F. (1997). A variant of Korpelevich's method for variational inequalities with a new search strategy. *Optimization*, 42, 309–321.

Ivanenko, V.I. and Melnik, V.S. (1988). *Variational Methods in Control Problems for Systems with Distributed Parameters*. Nauk. Dumka, Kyiv.

Khobotov, E.N. (1987). Modification of the extra-gradient method for solving variational inequalities and certain optimization problems. *USSR Comput. Math. Math. Phys.*, 27(5), 120–127.

Kinderlehrer, D. and Stampacchia, G. (2000). *An Introduction to Variational Inequalities and their Applications*. SIAM, Philadelphia.

Kirk, W.A. (1965). A fixed point theorem for mappings which do not increase distances. *Am. Math. Monthly*, 72, 1004–1006.

Konnov, I.V. (1993). Combined relaxation methods for finding equilibrium points and solving related problems. *Izv. VUZov. Math.*, 2, 46–53.

Konnov, I.V. (2001). *Combined Relaxation Methods for Variational Inequalities*. Springer-Verlag, Berlin.

Korpelevich, G.M. (1976). An extragradient method for finding saddle points and for other problems. *Matecon.*, 12(4), 747–756.

Krasnoselsky, M.A. (1955). Two remarks about the method of successive approximations. *UMN*, 10(1), 123–127.

Levitin, E.S. and Polyak, B.T. (1966). Constrained minimization methods. *USSR Comput. Math. Math. Phys.*, 6(5), 1–50.

Lions, J.-L. (1968). *Contrôle optimal de systèmes gouvernés par des équations aux dérivées partielles*. Dunod, Paris.

Lions, J.-L. (1969). *Some Methods of Solving Non-Linear Boundary Value Problems*. Dunod-Gauthier-Villars, Paris.

Lions, P.-L. (1977). Approximation de points fixes de contractions. *C. R. Acad. Sci. Paris, Ser. A-B.*, 284, A1357–A1359.

Lions, J.-L. and Stampacchia, G. (1967). Variational inequalities. *Commun. Pure Appl. Math.*, XX, 493–519.

Lyashko, S.I., Semenov, V.V., Voitova, T.A. (2011). Low-cost modification of Korpelevich's methods for monotone equilibrium problems. *Cybern. Syst. Anal.*, 47(4), 631–639.

Mainge, P.-E. (2008). Strong convergence of projected subgradient methods for nonsmooth and nonstrictly convex minimization. *Set-Valued Anal.*, 16, 899–912.

Malitsky, Yu.V. (2015). Projected reflected gradient methods for monotone variational inequalities. *SIAM J. Optim.*, 25, 502–520.

Malitsky, Yu.V. and Semenov, V.V. (2014). An extragradient algorithm for monotone variational inequalities. *Cybern. Syst. Anal.*, 50(2), 271–277.

Malitsky, Yu.V. and Semenov, V.V. (2015). A hybrid method without extrapolation step for solving variational inequality problems. *JOGO*, 61(1), 193–202.

Mann, W.R. (1953). Mean value methods in iteration. *Proc. AMS*, 506–510.

Marcotte, P. (1991). Application of Khobotov's algorithm to varational inequalities and network equilibrium problems. *Indorm. Systems Oper. Res.*, 29, 258–270.

Moreau, J.J. (1967). Fonctionnelles convexes. *Séminaire : équations aux dérivées partielles*, Collège de France, Paris.

Moudafi, A. (2000). Viscosity approximation methods for fixed-points problems. *J. Math. Anal. Appl.*, 241(1), 46–55.

Nadezhkina, N. and Takahashi, W. (2006a). Strong convergence theorem by a hybrid method for nonexpansive mappings and Lipschitz-continuous monotone mappings. *SIAM J. Optim.*, 16, 1230–1241.

Nadezhkina, N. and Takahashi, W. (2006b). Weak convergence theorem by an extragradient method for nonexpansive mappings and monotone mappings. *JOTA*, 128, 191–201.

Nagurney, A. (1999). *Network Economics: A Variational Inequality Approach*. Kluwer, Dordrecht.

Nakajo, K. and Takahashi, W. (2003). Strong convergence theorems for nonexpansive mappings and nonexpansive semigroups. *J. Math. Anal. Appl.*, 279, 372–379.

Nurminski, E.A. (2008). The use of additional diminishing disturbances in Fejer models of iterative algorithms. *Comput. Math. and Math. Phys.*, 48(12), 2154–2161.

Opial, Z. (1967). Weak convergence of the sequence of successive approximations for nonexpansive mappings. *Bull. AMS*, 73, 591–597.

Patriksson, M. (1999). *Nonlinear Programming and Variational Inequality Problems: A Unified Approach*. Kluwer, Dordrecht.

Popov, L.D. (1980). A modification of the Arrow-Hurwicz method for search of saddle points. *Math. Notes Acad. Sci. USSR*, 28(5), 845–848.

Semenov, V.V. (2013). Two methods of approximation of the fixed point of the Fejerian operator. *J. Num. Appl. Math.*, 1, 46–56.

Semenov, V.V. (2014a). A strongly convergent splitting method for systems of operator inclusions with monotone operators. *J. Autom. Inf. Sci.*, 46(5), 45–56.

Semenov, V.V. (2014b). Strongly convergent algorithms for variational inequality problem over the set of solutions the equilibrium problems. In *Continuous and Distributed Systems, Solid Mechanics and Its Applications, vol. 211*, Zgurovsky, M.Z. and Sadovnichiy, V.A. (eds). Springer International Publishing, Cham.

Solodov, M.V. (2001). A class of globally convergent algorithms for pseudomonotone variational inequalities. In *Complementarity: Applications, Algorithms and Extensions*, Ferris, M.C., Mangasarian, O.L., Pang, J.S. (eds). Kluwer, Dordrecht.

Solodov, M.V. and Svaiter, B.F. (1999a). A hybrid projection-proximal point algorithm. *J. Conv. Anal.*, 6(1), 59–70.

Solodov, M.V. and Svaiter, B.F. (1999b). A new projection method for variational inequality problems. *SIAM J. Control Optim.*, 37, 765–776.

Solodov, M.V. and Svaiter, B.F. (2000). Forcing stong convergence of proximal point iterations in a Hilbert space. *Math. Program.*, 87, 189–202.

Stampacchia, G. (1965). Le problème de Dirichlet pour les équations elliptiques du second ordre à coefficients discontinus. *Ann. Inst. Fourier*, 15(1), 189–257.

Suzuki, T. (2007). A sufficient and necessary condition for Halpern-type strong convergence to fixed points of nonexpansive mappings. *Proc. AMS*, 135, 99–106.

Takahashi, W., Takeuchi, Y., Kubota, R. (2008). Strong convergence theorems by hybrid methods for families of nonexpansive mappings in Hilbert spaces. *J. Math. Anal. Appl.*, 341, 276–286.

Tseng, P. (2000). A modified forward-backward splitting method for maximal monotone mappings. *SIAM J. Optim.*, 38, 431–446.

Vainberg, M.M. (1974). *Variational Method and Method of Monotone Operators in the Theory of Nonlinear Equations*. John Wiley & Sons, New York.

Vasin, V.V. and Eremin, I.I. (2005). *Operators and Iterative Processes of the Fejer Type: Theory and Applications*. Regular and Chaotic Dynamics, Moscow-Izhevsk.

Verlan, D.A., Semenov, V.V., Chabak, L.M. (2015). A strongly convergent modified extragradient method for variational inequalities with non-Lipschitz operators. *J. Autom. Inf. Sci.*, 47(7), 31–46.

Wittmann, R. (1992). Approximation of fixed points of nonexpansive mappings. *Arch. Math.*, 58, 486–491.

Xu, H.K. (2002a). Another control condition in an iterative method for nonexpansive mappings. *Bull. Austral. Math. Soc.*, 65, 109–113.

Xu, H.K. (2002b). Iterative algorithms for nonlinear operators. *J. LMS*, 2, 240–256.

Yamada, I. (2001). The hybrid steepest descent method for the variational inequality problem over the intersection of fixed point sets of nonexpansive mappings. *Stud. Comput. Math.*, 8, 473–504.

Zykina, A.V. and Melenchuk, N.V. (2010). Two-step extragradient method for variational inequalities. *Izv. VUZov. Math.*, 9, 82–85.

8

On Multivariate Algorithms of Digital Signatures on Secure El Gamal-Type Mode

Vasyl USTIMENKO[1,2]

[1] *Royal Holloway, University of London, United Kingdom*

[2] *Maria Curie-Sklodowska University, Lublin, Poland*

The intersection of noncommutative and multivariate cryptography contains studies of cryptographic applications of subsemigroups and subgroups of affine Cremona semigroups defined over finite commutative ring K with the unit. We will consider special subsemigroups (platforms) in a semigroup of all endomorphisms of $K[x_1, x_2, ..., x_n]$.

Efficiently computed homomorphisms between such platforms can be used in post-quantum key exchange protocols when correspondents elaborate common transformation of $(K^*)^n$. The security of these schemes is based on a complexity of decomposition problem for an element of a semigroup into a product of given generators.

We suggest three such protocols (with a group and with two semigroups as platforms) for their usage with multivariate digital signature systems. The usage of protocols allows us to convert public maps of these systems into private modes, i.e. one correspondent uses the collision map to safely transfer selected multivariate rule to their partner.

Computational Methods and Mathematical Modeling in Cyberphysics and Engineering Applications 1,
coordinated by Dmitri KOROLIOUK, Sergiy LYASHKO and Nikolaos LIMNIOS.

The "privatization" of former publicly given map allows the usage of digital signature system for which some cryptanalytic instruments were found (estimation of different attacks on rainbow oil and vinegar system, cryptanalytic studies LUOV) with the essentially smaller size of hashed messages. Transition of the basic multivariate map to safe El Gamal-type mode does not allow the usage of cryptanalytic algorithms for already broken Imai–Matsumoto cryptosystem or original oil and vinegar signature schemes proposed by J. Patarin.

Thus, even broken digital signature schemes can be used in combination with protocol execution during some restricted "trust interval" of polynomial size.

A minimal trust interval can be chosen as a dimension n of the space of hashed messages, i.e. the safely transported multivariate map has to be used at most n times. Before the end of this interval, correspondents have to start the session of multivariate protocol with a modified multivariate map.

The security of such algorithms rests not on properties of quadratic multivariate maps but on the security of the protocol for the map delivery and corresponding NP hard problem.

This research was supported by the Fellowship of British Academy for Researchers at Risk 2022.

8.1. On post-quantum, multivariate and non-commutative cryptography

Post-quantum cryptography[1] (PQC) is an answer to threats coming from a full-scale quantum computer able to execute Shor's algorithm. With this algorithm implemented on a quantum computer, currently used public key schemes, such as RSA and elliptic curve cryptosystems, are no longer secure. The U.S. NIST made a step towards mitigating the risk of quantum attacks by announcing the PQC standardization process post-quantum cryptography for a new public key algorithm. In June 2020, NIST published a list of candidates qualified to the third round of the standardization process. Some public key candidates are implemented, like candidate called Round 5 (see Andrzejczak (2019)) or the classic Mc Eliece algorithm (see McEliece (1978)).

1 Post-quantum cryptography: https://www.nist.gov/news-events/news/2022/07/pqc-standardization-process-announcing-four-candidates-be-standardized-plus.

Current candidates are developed within following directions of PQC: lattice-based systems, code-based cryptosystems, multivariate cryptography (see Ding and Petzoldt (2017) and further references), hash-based cryptography, studies of isogenies for superelliptic curves. There is the following alternative approach to public key cryptography. Instead of public encryption rule, correspondents can use protocol for elaboration of some common information which allows us to define encryption rule for one user and decryption instrument for their partner. We refer to such algorithms as cryptosystem of El Gamal type. Recall that El Gamal proposed such cryptosystem based on the classical Diffie–Hellman algorithm over multiplicative group F^*_p.

In March 2021, it was announced that the prestigious Abel prize will be shared by A. Wigderson and L. Lovasz. They contribute valuable applications of theory of expanding graphs to theoretical computer science (see Grzesik et al. (2018), Hoory et al. (2006) and further references). We have been working on applications of these graphs to cryptography. This paper is dedicated to the usage of geometrical expanders in the sense of Alon (1986) as tools for digital signatures.

In this publication, we continue to develop new cryptosystems with an alternative approach (Ustimenko 2017, 2018, 2019a, 2020; Ustimenko and Klisowski 2019) to multivariate public key cryptography based on the idea of the modified Diffie–Hellman-type protocols in terms of noncommutative cryptography for subsemigroups of endomorphisms of $K[x_1, x_2,\dots,x_n]$. Security of these algorithms rests on the complexity of word problem to decompose the given multivariate map into generators of affine Cremona semigroup $End(K[x_1, x_2,\dots,x_n])$ (see Wagner and Magyarik (1984)) for the first application of word problem in the case of group). Thus, we are working in the area of intersection of multivariate and noncommutative cryptographies.

Recall that multivariate cryptography (see Ding and Petzoldt 2017) uses polynomial maps of affine space K^n defined over a finite commutative ring K into itself as encryption tools. It exploits the complexity of finding a solution of a system of nonlinear equations from many variables. Multivariate cryptography uses, as encryption tools, nonlinear polynomial transformations of kind $x_1\rightarrow f_1(x_1, x_2,\dots,x_n)$, $x_2\rightarrow f_2(x_1, x_2,\dots,x_n)$, …, $x_n\rightarrow f_n(x_1, x_2,\dots,x_n)$ transforming affine space K^n, where $f_i \in K[x_1, x_2,\dots,x_n]$, $i=1,2,\dots,n$ are multivariate polynomials, usually given in a standard form, i.e. via a list of monomials in a chosen order (nonlinear endomorphisms of $K[x_1, x_2,\dots,x_n]$).

Noncommutative cryptography is an active area of cryptology where the cryptographic primitives and systems are based on algebraic structures like groups, semigroups and noncommutative rings (see Kahrobaei and Khan 2006; Maze et al. 2007; Kropholler et al. 2010; Fine et al. 2011; Ding and Petzoldt 2017; Kumar and Saini 2017; Lopez Ramos et al. 2019; Moldovyan and Moldovyan 2010; Myasnikov et al. 2011). It is important that this direction is well supported by cryptanalytic research (Ben-Zvi et al. 2018; Roman'kov 2016, Roman'kov 2019). Semigroup-based cryptography consists of general cryptographical schemes defined in terms of wide classes of semigroups and their implementations for chosen semigroup families (so-called platform semigroups).

Papers by Ustimenko (2017, 2018, 2019a, 2020) and Ustimenko and Klisowski (2019) contain some modifications of the Diffie–Hellman protocol when G is given as a subgroup of affine Cremona semigroup $S(K^n)$ over a finite commutative ring K of all polynomial transformations. These papers use the assumption that each element is given in its standard form of multivariate cryptography. To use semigroup operation, we have to compute the composition of transformations. This was an attempt to combine methods of noncommutative cryptography and multivariate cryptography.

Papers by Ustimenko (2017, 2018) suggest some usage of homomorphisms of subsemigroups of affine Cremona groups for protocols and cryptosystems which are not generalizations of the Diffie–Hellman algorithm and its El Gamal-type modifications. Some examples are given there. The implementations of these schemes with the evaluation of densities of involved polynomial transformations are described in Ustimenko (2019a, 2020) and Ustimenko and Klisowski (2019). Elements of graph-based stable subgroups used in Ustimenko and Klisowski (2019) can serve as encryption tools of stream ciphers (see Ustimenko et al. 2019b and further references).

8.2. On stable subgroups of formal Cremona group and privatization of multivariate public keys based on maps of bounded degree

Let $K[x_1, x_2, \ldots, x_n]$ be the commutive ring of all polynomials in variables $x_1, x_2, \ldots, x_n$ defined over a commutive ring K. Each endomorphism $F \in E_n(K)$ is uniquely determined by its values on formal generators x_i, $i=1,2,\ldots, n$. Symbol $End(K[x_1, x_2, \ldots, x_n])=E_n(K)$ stands for the semigroup of all endomorphisms of $K[x_1, x_2, \ldots, x_n]$. So, we can identify F with the formal rule of $x_1 \rightarrow f_1(x_1, x_2, \ldots, x_n)$, $x_2 \rightarrow f_2(x_1, x_2, \ldots, x_n)$, $\ldots$, $x_n \rightarrow f_n(x_1, x_2, \ldots, x_n)$, where $f_i \in K[x_1, x_2, \ldots, x_n]$. Element F naturally induces the transformation $\Delta(F)$ of affine space K^n given by the following rule $\Delta(F)$:$(\alpha_1, \alpha_2, \ldots, \alpha_n) \rightarrow (f_1 (\alpha_1, \alpha_2, \ldots, \alpha_n), f_2(\alpha_1, \alpha_2, \ldots, \alpha_n), \ldots, f_n(\alpha_1, \alpha_2, \ldots, \alpha_n))$ for each $(\alpha_1, \alpha_2, \ldots, \alpha_n) \in K^n$. Luigi Cremona (see Noether 1904) introduced $\Delta(E_n(K)) = CS(K^n)$ which is currently called *affine Cremona semigroup*. A group of all invertible transformations of $CS(K^n)$ with an inverse from $CS(K^n)$ is known as *affine Cremona group* $CG(K^n)$ (shortly known as *Cremona group*: see for instance Šafarevič (1982) and Bodnarchuk (2001)).

We refer to infinite $E_n(K)$ as *formal affine Cremona semigroup*. Density of the map F is the total number of monomial terms in all f_i.

It is a known fact that typically, the degree of iteration of the "random" multivariate polynomial transformation grows exponentially. This is due to the fact that the composition of two "random" elements of $CS(K^n)$ of degree d in majority cases will have a degree d^2.

In particular, this can be seen in the one-dimensional case (i.e. for $n = 1$); the exponential growth of the degree of iterations of a nonlinear polynomial is unavoidable. However, it turns out that for $n\psi >\psi 1$ the degree of some special transformations may grow significantly slower. There are discovered examples of transformations with a degree that may grow polynomially (see Ostafe 2010; Ostafe and Shparlinski 2010) or be even no greater than some established constant d. The last fact means the existence of stable subsemigroups/subgroups of formal Cremona groups of degree d, i.e. subsemigroups of endomorphisms of degree of at least d. The first example of families of stable groups was introduced in Ustimenko (1998, 2001) in terms of the well-known extremal graphs $D(n, q)$ and their analogs $D(n,K)$ defined over arbitrary commutative ring K. However, the importance of their stability was later realized (Ustimenko 2007a, 2007b). The formal proof of the fact that these groups are stable of degree 3 is presented in Wroblewska (2008). Obviously, the semigroup of transformations of degree 1 and general affine group of bijective transformations of degree 1 are examples of stable subsemigroups and subgroups of $E_n(K)$ of degree 1. Currently, nonlinear families of stable semigroups and groups are known for each parameter $d \geq 2$; they have various applications to cryptography (see Ustimenko 2019a, 2019b, 2020, and further references). All these results are obtained via studies of "symbolic walks" on algebraic graphs, i.e. graphs defined by a system of algebraic equations. The following statement follows from the results of Ustimenko (2020).

THEOREM 8.1.– *For each natural number $n>1$ and each commutative ring K and $d \geq 2$, there is a family of noncommutative stable subgroups.*

$G_n(K) < E_n(K)$ of degree d and stable semigroups $S_n(K)<E_n(K)$ of degree d such that $G_n(K) <S_n(K)$.

Let ${}^iZ= \{{}^ig_1, {}^ig_2, ..., {}^ig_t\}$ be a sequence of sets of elements from $E_{n(i)}(K)$, where $n(i)>1$ is an increasing sequence of positive integers. We say that ${}^iZ$ is a *noncommutative system of stable Cremona generators of degree d* and rank *t* if

(1) $\Delta({}^ig_k{}^ig_j) \neq \Delta({}^ig_j{}^ig_k)$ for arbitrary $k \neq j$.

(2) ${}^iSZ= <{}^ig_1, {}^ig_2, ..., {}^ig_t>$ are stable semigroups of degree *d*.

PROPOSITION 8.1.– *For each commutative ring K, sequence $n(i)=i, i\geq 2$ and each value of parameters d and t, there is a noncommutative system of stable Cremona generators of degree d and rank t.*

We say that ${}^iZ$ is a *regular noncommutative system* of stable Cremona generators if $n(i)=i$ for each value of *i*.

Let $n(i)$, $m(i)$, $m(i) \leq n(i)$ be two increasing sequences of natural numbers and ${}^iZ$, ${}^iZ^l$ are corresponding stable systems of growing periods of degrees *d* and *d'* $(d' \leq d)$ and rank *t*, $t>1$.

We say that ${}^iZ' =\{{}^ig'_1, {}^ig'_2, ..., {}^ig'_t\}$ *is a quotient of stable Cremona system* ${}^iZ$ if the rule $\varphi({}^ig_j)= {}^ig'_j$, $j=1, 2,..., t$ defines computationally tame homomorphism of semigroup ${}^iSZ$ onto ${}^iSZ^l$, i.e. homomorphism computable in time $O(n_i^{\alpha})$ for some positive constant α. We refer to ${}^iZ$ as *a stable cover of noncommutative system of stable Cremona generators.*

In Ustimenko (2019b), the following statement was proven.

THEOREM 8.2.– *For each finite commutative ring K and natural numbers d, $d>0$ and t, $t \geq 2$, there is an increasing sequence n(i) of natural numbers and noncommutative system of stable Cremona generators ${}^iZ =\{{}^ig_1, {}^ig_2, ..., {}^ig_t\}$ of degree d and rank t which has a regular quotient ${}^iZ'$.*

We say that the stable Cremona system of degree *d* of elements has *enveloping family* of stable subsemigroup $EZ^i(K)$ of degree *d* if $E_{(i)}(K)>EZ^i(K)>SZ^i(K)$.

8.3. Multivariate Tahoma protocol for stable Cremona generators and its usage for multivariate encryption algorithms

The word *tahoma* stands here for "tame homomorphism".

It is worth noting that Tahoma is the name of a mountain in North America and is a popular font in text processing.

Let us assume that Alice selects a noncommutative system $Z(K)$ of *stable Cremona generators* of degree d and rank t with quotient $Z'(K)$ such that there is an enveloping family $EZ(K)$ of $Z(K)$ and enveloping family $EZ^1(K)$ of $Z'(K)$.

Alice chooses parameter i and bijective affine transformation T, $deg(T)=1$ and T', $deg(T')=1$ acting on $(K)^{n(i)}$ and $(K)^{m(i)}$. She selects elements E and ${}^1E$ from $EZ_{n(i)}(K)$ and $EZ'_{m(i)}(K)$. Alice takes generators $g_1, g_2, \ldots, g_t$ of $SZ_i(K)$ and the corresponding images $g'_1, g'_2, \ldots, g'_t$ in the $SZ'_i(K)$.

So, she forms $a_j = TEg_jE^{-1}T^{-1}$, $j=1,2,\ldots,t$ and $b_j = T'E'g'_j(E')^{-1}(T')^{-1}$, $j=1,2,\ldots,t$ written in the standard form of $E_{n(i)}(K)$ and $E_{m(i)}(K)$.

Alice sends (a_j, b_j) and $j=1,2,\ldots,t$ to Bob. He takes an alphabet $\{z_1, z_2, \ldots, z_t\}$ and selects a word $w(z_1, z_2, \ldots, z_t)$, $=z_{i(1)}^{\alpha(1)} z_{i(2)}^{\alpha(2)} \ldots z_{2i(l)}^{\alpha(l)}$, where $\alpha(j)>0$, $j=1,2,\ldots,l$, $l>1$, $i(s)\in\{1,2,\ldots,t\}$, $i(j)\neq i(j+1)$ for $j=1,2,\ldots,t-1$.

Bob computes $b=w(b_1, b_2,\ldots,b_t)$ and keeps it safely for himself. He forms $a=w(a_1, a_2, \ldots a_t)$ and sends this element of $E_{n(i)}(K)$ to Alice.

She uses the following restoration process to get $w(b_1, b_2,\ldots,b_t)$. Alice computes $E^{-1}T^{-1}aTE=c$. She uses tame homomorphism φ corresponding to noncommutative system Z and its quotient Z^1 and computes $\varphi(c)=c'$. Secondly, she computes $b=w(b_1, b_2,\ldots,b_t)$ as $T'E'c^1(E')^{-1}(T')^{-1}$.

REMARK 8.1.– *Adversary has to decompose the available multivariate map $a=w(a_1, a_2)$ from $E_{n(i)}$ into a word in the given generators $a_1, a_2, \ldots, a_t$ written in their standard form. So, security rests on the word problem in semigroup $E_{n(i)}(K)$ (or stable semigroup $<a_1, a_2, \ldots, a_t>$).*

It is worth noting that due to this algorithm, correspondents Alice and Bob can safely elaborate collision quadratic transformation of $(K)^{m(i)}$ with the chosen dimension $m(i)$. In the case of a regular quotient $m(i)=i$.

So, correspondents have an algorithm to safely elaborate the stable collision map of the selected degree d acting of free module K^d of the arbitrarily chosen dimension. This option raises the following question.

Do we need multivariate public key?

One of the directions of security research is CLASSICAL MULTIVARIATE CRYPTOGRAPHY (see books Koblitz 1998; Ding et al. 2006) which present examples of old public keys in the form of a family of quadratic elements ^nF of $E_n(F_q)$ which induces a bijective map $\Delta(^nF)$. The map ^nF is given publicly.

So, each public user (and adversary as well) can create as many pairs $(x, y=\Delta(^nF)(x))$, where x is chosen as a plaintext from the plainspace $F_q{}^n$, and obtain the corresponding ciphertext y as they want.

Thus, due to the above protocol, there is no need to give F to the public. Alice can select some family which is a current or former (already broken) candidate to a safe pubic key. So, she has a private algorithm to solve the equation $F(x)=b$.

She can deformate ^nF by bijective elements ^1T_n and ^2T_n from $AGL_n(F_q)$ to use elements $G_n={}^1T_n\ {}^nF\ {}^2T_n$. Secondly, Alice and Bob execute MULTIVARIATE TAHOMA PROTOCOL to elaborate the common element $B_n(x)$ from $E_n(F_q)$.

We assume that G_n and B_n are given in the forms of tuples $(g_1, g_2, ..., g_n)$ and $(b_1, b_2, ..., b_n)$, where g_i and b_i are elements of $F_q[x_1, x_2, ..., x_n]$. Finally, Alice sends $(g_1+b_1, g_2+b_2, ..., g_n+b_n)$ to Bob. He uses his knowledge on $B(x)$ and publicly known addition rule to restore $G(x)$. So, Bob uses $G(x)$ as an encryption function. Alice uses her private algorithm of finding the reimage of $F(x)$ for the decryption.

CONCLUSION 1.– *No need to give multivariate rule $F(x)$ of the bounded degree for public audience. There is a safe way to transport the rule to your partner.*

The next question is: how long are correspondents able to keep G in private mode?

REMARK 8.2.– *Assume that correspondents use multivariate stable protocol and Alice safely transports quadratic multivariate encryption map G on F_q^n (or transformation of constant degree d).*

Bob sends Alice several elements of the kind $G(p_i)$, $i=1,2, ..., l$ via open channel during their communication.

If parameter l is "large enough" ($l=O(n^2)$ for $d=2$), then ADVERSARY can use their cyberterrorist tools and intercept some *pairs* (p_i, c_i) where p_i is a plaintext and c_i is a corresponding ciphertext. *If they intercept* $O(n^2)$ such pairs then they can approximate G with costly polynomial algorithm (cryptanalytic linearization method), which requires $O(n^5)$ elementary operation for $d=2$.

Such activity allows ADVERSARY to become in the same position with user Bob. They get G from the private storage of correspondents. Similarly to the case of having public rule adversary can compute $G(p)$ for any chosen p.

Of course, adversary has to break "public rule"' G as well, i.e. to find out the way of computation of its reimage.

Alice and Bob could be smart enough to use reasonable trust interval restricting parameter l (*number of messages*) by some expression $C(n)$ such that $C(n) \leq den(G)/(2n)$ (half of the *average density of* G_i, $i=1,2,\ldots,n$).

At the end of trust interval, Alice may select other conjugates of generators of Singer system, correspondents conduct a new session of the protocol. Alice safely delivers a new quadratic encryption rule for Bob.

CONCLUSION 2.– *After the transportation of the encryption multivariate rule F(x) to your partner, there is an option of its periodical modification.*

This scheme allows privatization of multivariate public rules, i.e. transition of the rule to noncommutative El Gamal Gamal-type cryptosystem which uses protocols of noncommutative cryptography which hide the encryption algorithm from adversary.

PRACTICAL ASPECTS.– Instead of quasipublic rules, Alice can create invertible quadratic multivariate map G of density $O(n)$ (density is the number of monomial terms in all G_i, $i=1,2,\ldots,n$) which allows us to compute its reimage for $O(n)$ elementary steps. Note that $O(n)$ is a speed of reading of string from M^n, where M is selected finite alphabet.

Some of the presented cryptographic privatization schemes are already implemented on the level of prototype models (see Ustimenko and Klisowski 2019; Ustimenko 2020).

8.4. On multivariate digital signature algorithms and their privatization scheme

It is commonly admitted that multivariate cryptography turned out to be more successful as an approach to build signature schemes primarily because multivariate schemes provide the shortest signature among post-quantum algorithms. Such signatures use the system of nonlinear polynomial equations

$$^1p(x_1,x_2,\ldots,x_n) = {}^1p_{i,j}\cdot x_ix_j + {}^1p_i\cdot x_i + {}^1p_0$$

$$^2p(x_1,x_2,\ldots,x_n) = {}^2p_{i,j}\cdot x_ix_j + {}^2p_i\cdot x_i + {}^2p_0$$

$$\ldots$$

$$^mp(x_1,x_2,\ldots,x_n) = {}^mp_{i,j}\cdot x_ix_j + {}^mp_i\cdot x_i + {}^mp_0$$

where ${}^kp_{i,j}$, ${}^kp_i$ are elements of selected commutative ring K.

The quadratic multivariate cryptography map consists of two bijective affine transformations, S and T of dimensions n and m, and a quadratic element P' of kind xi →ip of formal Cremona group, where ip are written above elements of K[x1, x2,...,xn]. We denote via Δ(P') -1(y) some reimage of y=Δ(P(x)). The triple Δ({\displaystyle (S^{-1},{P'}^{-1},T^{-1})}S) -1, Δ(P') -1, Δ({\displaystyle (S^{-1},{P'}^{-1},T^{-1})}T) -1 is the private key also known as the trapdoor. {\displaystyle P'\colon F^{m}\rightarrow F^{n}}.

The public key is the composition *S, P'* and *T* which is by assumption hard to invert without the knowledge of the trapdoor. Signatures are generated using the private key and are verified using the public key as follows. The message is hashed to a vector y via a known hash function. The signature is $\Delta(T)^{-1}(\Delta(P')^{-1})(\Delta(S)^{-1})$. The receiver of the signed document must have the public key P in possession. He computes the hash y and checks that the signature x fulfils $\Delta(P)(y)=x$.

EXAMPLE.– Assume we have two groups of variables $z_1, z_2, \ldots, z_r$ and $z'_1, z'_2, \ldots, z_{n-r}$ such that the substitution $x_1=z_1$, $x_2=z_2,\ldots$, $x_r=z_r$, $x_{r+1}=z'_1$, $x_{r+2}=z'_2,\ldots$, $x_n=z'_{n-r}$ converts every single element ^ip to expression in the form $\Sigma\gamma_{ijk}z_jz'_k+\Sigma\lambda_{ijk}z'_jz'_k+\Sigma\varsigma_{ij}z_j+\Sigma\varsigma'_{ij}z'_j+\sigma_i$. In this situation, we have to sign a given message y and the result is a valid signature x. The coefficients, γ_{ijk}, λ_{ijk}, ς_{ij}, ς'_{ij} and σ_i must be chosen secretly. The vinegar variables z'_i are chosen randomly (or pseudorandomly). The resulting linear equation system is solved for the oil variables z_i.

The unbalanced oil and vinegar (UOV) scheme described above is a modified version of the oil and vinegar scheme designed by J. Patarin. Both are digital signature protocols. They are algorithms of multivariate cryptography. The security of this signature scheme is based on an *NP*-hard mathematical problem. To create and validate signatures, a minimal quadratic equation system must be solved. Solving m equations with n variables is *NP*-hard. While the problem is easy if m is either essentially larger or essentially smaller than n,[1] importantly for cryptographic purposes, the problem is thought to be difficult in the average case when m and n are nearly equal, even when using a quantum computer. Multiple signature schemes have been devised based on multivariate equations with the goal of achieving quantum resistance. We assume that parameter n can be selected in a free way and parameters n and m are connected via linear equation $\alpha n+\beta m+b$ where $\alpha\neq 0,\beta\neq 0$. So $m=)(n)$. We take integer k, which is $\geq max(n, m)$, $k=O(n)$ and commutative ring $K[x_1,x_2,...,x_n, x_{n+1}, x_{n+2},..., x_k]$ where x_i, $i=1,2,...,n$ are variables of public equations ${}^jp(x_1,x_2, \ldots, x_n)$, $j=1,2, ...,m$ and $x_{n+1}, x_{n+2},\ldots,x_k$ are formal variables. Furthermore, one of the suggested schemes below can be used.

SCHEME 8.1.– Let us assume that Alice selects a noncommutative system $Z(K)$ of *stable Cremona generators* of degree d =2 and rank t, $t>1$ with regular quotient $Z'(K)$ such that there is an enveloping family $EZ(K)$ of $Z(K)$ and enveloping family $EZ^1(K)$ of $Z'(K)$. Let us assume that $Z(K)$ corresponds to the sequence of increasing integers $n(i)$ and i coincides with the number of total variables k introduced above. So, Alice takes generators ${}^kg_1, {}^kg_2,..., {}^kg_t$ and uses multivariate *tahoma protocol.* She forms $a_j \in E_{n(k)}(K)$ and $b_j \in E_k(K)$. *She* sends them to Bob. Correspondents execute the protocol and elaborate common collision element $h=(h_1, h_2,...,h_k) \in E_k(K)$. Alice (or Bob) selects polynomials ${}^ip$ for the multivariate digital signature system (m.d.s.). They take h' with coordinates $h'_1={}^ip+h_i$, $i=1,2,...m$ and $h'_i= h_i +f_i$, $i=m+1, m+2,...,k$ where f_i are quadratic "pseudorandom" elements of $K[x_1, x_2,..., x_k]$.

It is noteworthy that users have to know formats n and m of hashed vector and a signature. So Bob (or Alice) restores (${}^1p$, ${}^2p$,..., ${}^mp$) and correspondents can use the multivariate digital signature system on private El Gamal mode based on the chosen multivariate tahoma protocol.

SCHEME 8.2.– Alice (or Bob) can use the above-described scheme with $d=3$. For $f \in K[x_1, x_2, \ldots x_k]$; we define its quadratic restriction $r(f)$ as the sum of monomial terms of degree at most 2. Let $D=d/d_{x1}+ d/d_{x2}, \ldots + d/d_{xk}$ be the differentiation operator for elements of $K[x_1, x_2, \ldots x_k]$. We define deformation $def(f)$ of f as $D(f)+r(f)$ (see Ustimenko (2018) for the idea of deformated collision maps). So, correspondents use scheme 8.1 to elaborate common cubical collision map h. They continue the remaining steps of scheme 8.1 with quadratic $def(h)$ instead of h.

8.5. Examples of stable cubical groups

8.5.1. *Simplest graph-based example*

The following family of stable groups is already used in some algorithms of symmetric cryptography and protocols of commutative and noncommutative cryptography (see Ustimenko 2013, 2019c; and further references). Let K be a commutative ring. We define $A(n, K)$ as a bipartite graph with the point set $P=K^n$ and line set $L=K^n$ (two copies of a Cartesian power of K are used). We will use brackets and parentheses to distinguish tuples from P and L. So $(p)=(p_1, p_2, \ldots, p_n) \in P_n$ and $[l]=[l_1, l_2, \ldots, l_n] \in L_n$. The incidence relation $I=A(n,K)$ (or corresponding bipartite graph I) is given by condition $pI\ l$ if and only if the equations of the following kind hold.

$p_2 - l_2=l_1p_1$, $p_3 - l_3= p_1 l_2$, $p_4 - l_4 = l_1p_3$, $p_5 - l_3 = p_1 l_4$, …, $p_n - l_n= p_1 l_{n-1}$ for odd n and $p_n - l_n = l_1 p_{n-1}$ for even n.

Let us consider the case of finite commutative ring K, $|K|=m$. As it instantly follows from the definition, the order of our bipartite graph $A(n, K)$ is $2m^n$. The graph is m-regular. In fact, the neighbor of a given point p is given by the above equations, where parameters $p_1, p_2,\ldots, p_n$ are fixed elements of the ring and symbols $l_1, l_2,\ldots, l_n$ are variables. It is easy to see that the value for l_1 could be freely chosen. This choice uniformly establishes values for $l_2, l_3, \ldots, l_n$. So, each point has precisely m neighbors. In a similar way, we observe the neighborhood of the line, which also contains m neighbors. We introduce the color $\rho(p)$ of the point p and the color $\rho(l)$ of line l as parameters p_1 and l_1, respectively.

Graphs $A(n, K)$ with coloring ρ belong to the class of Γ *linguistic graphs* considered in Ustimenko (2004). Linguistic graph $\Gamma= \Gamma(K)$ is defined over commutative ring K as a bipartite graph with partition sets L=K^n and P=K^k and color sets K^s and K^r, respectively. Projection ρ of point $x=(x_1, x_2, \ldots, x_n)$, or line $y=[y_1, y_2, \ldots, y_t]$, on the tuple of their first s and r coordinates, respectively, defines colors of vertices. Each vertex has a unique neighbor of selected color. So, $n+r=t+s$. The incidence of linguistic graphs is given by a system of polynomial equation over the ring K.

In the case of linguistic graph $\Gamma(K)$ with $s=r=1$ the path consisting of its vertices $v_0, v_1, v_2, \ldots, v_k$ is uniquely defined by initial vertex v_0, and colors $\rho(v_i)$, $i=1, 2,\ldots, k$ of other vertices from the path. We can consider graph $\Gamma=\Gamma'(K[x_1, x_2, \ldots, x_n])$ defined by the same with Γ equations but over the commutative ring $K[x_1, x_2, \ldots, x_n])$.

So the following symbolic computation can be defined. Take the *symbolic point* $x=(x_1, x_2, ..., x_n)$, where x_i are generic variables of $K[x_1, x_2, ..., x_n]$ and *symbolic string* C which is a tuple of polynomials $f_1, f_2, ..., f_k$ from $K[x_1]$ with even parameter k. Form the path of vertices $v_0=x$, v_1 such that v_1Iv_o and $\rho(v_1)=f_1(x_1)$, v_2 such that v_2Iv_1 and $\rho(v_2)=f_2(x_1)$, ..., v_k such that v_kIv_{k-1} and $\rho(v_k)=f_k(x_1)$. We choose parameter k as even number. So v_k is the point from the partition set $K[x_1, x_2, ..., x_n]^n$ of the graph Γ'.

We note that the computation of each coordinate of v_i depending on variables $x_1, x_2, ..., x_n$ and polynomials $f_1, f_2, ..., f_k$ needs only arithmetical operations of addition and multiplication. As it follows from the definition of linguistic graph final vertex v_k (point) has coordinates $(h_1(x_1), h_2(x_1, x_2), h_3(x_1, x_2, x_3), ..., h_n(x_1, x_2, ..., x_n))$, where $h_1(x_1)=f_k(x_1)$. Let us consider the map $H= \Gamma\hat{}\eta(C)$: $x_i\rightarrow h_i(x_1, x_2, ..., x_n)$, $i=1, 2, ..., n$ which corresponds to symbolic string C. Assume that the equation $b=f_k(x_1)$ has exactly one solution. Then, the map $H:x_i\rightarrow h_i(x_1, x_2, ..., x_n)$, $i=1, 2, ..., n$ is a bijective transformation. In the case of finite parameter k and finite densities of $f_i(x_1)$, $i=1, 2, ..., n$, the map H also has finite density. If all parameters $\deg(f_i(x_1))$ are finite, then the map H has a linear degree in variable n. The idea of symbolic computation (see Ustimenko 2013 and further references) is the following one.

Let us consider the totality $St=St(K)$ of all symbolic strings with the product $(f_1, f_2, ..., f_r) \cdot (g_1, g_2, ..., g_s) = (f_1, f_2, ..., f_s, g_1(f_r), g_2(f_r), ...g_s(f_r))$. It is easy to see that $St(K)$ is a semigroup for which empty string serves as a unity.

We can check that the map ${}^{\Gamma}\eta=\eta$ *is a* homomorphism of semigroup $St(K)$ into Cremona semigroup $S(K^n)$ for each linguistic graph Γ with $r=s=1$ and point set K^n. We consider a subsemigroup $\sum=\sum(K)$ *of symbolic strings* C of kind $(x_1+a_i, x_2+a_2, ..., x_t+a_t)$ where parameter t is even. In the case of a linguistic graphs with $r=s=1$, we identify a symbolic string C with the corresponding tuple $(a_1, a_2, ..., a_t)$. Natural product of two strings given by tuples $C_1=(a_1, a_2, ..., a_t)$ and $C_2=(b_1, b_2, ..., b_m)$ is a string $C=C_1\circ C_2=(a_1, a_2, ..., a_t, b_1+a_t, b_2+a_t, ..., b_m+a_t)$. This product transforms $\sum$ to a semigroup. The map η' sending C to $\eta(C)$ is a homomorphism of $\sum$ into affine Cremona group $C(K^n)$. It is a restriction of ${}^{\Gamma}\eta$ onto $\sum(K)$. Let $C=(x_1\ a_1, x_1+a_2, ..., x_1+a_s)$ be a symbolic string from semigroup $\sum(K)$. We refer to $Rev(C)=(x_1-a_s+a_{s-1}, x_1-a_s+a_{s-2}, ..., x_1-a_s+a_1, x_1-a_s)$ as a reversing string for C. It is easy to see that *$\eta'(CRev(C))$ is a unity* of Cremona semigroup.

In the case of linguistic graphs $\Gamma=A(n, K)$, the totality $G(n, K) = \eta'(\sum(K))$ is a stable subgroup of degree 3 (see Ustimenko 2013 and further references). We use notation ${}^n\eta'$ for the restriction of ${}^{\Gamma}\eta$, $\Gamma=A(n,K)$ onto $\sum$. We assume that $a_0=0$ and say that transformation $\eta'(C)$ is irreducible if $a_i\neq a_{i+2}$, $i=1, 2, ..., t-2$. If $a_1\neq a_{t-1}$, and $a_2\neq a_t$, we say that irreducible symbolic string C and corresponding transformation $\eta'(C)$ are standard elements. We have a natural homomorphism $G(n+1, K)$ onto $G(n, K)$ induced by the homomorphism Δ from $A(n+1, K)$ onto $A(n, K)$ sending point $(x_1, x_2, \ldots, x_n, x_{n+1})$ to $(x_1, x_2, \ldots, x_n)$ and line $[x_1, x_2, \ldots, x_n, x_{n+1}]$ to $[x_1, x_2, \ldots, x_n]$. It means that there is a well-defined projective limit $A(K)$ of graphs $A(n, K)$ and groups $G(K)$ of groups $G(n, K)$ when n is growing to infinity. In fact, in the case of $K=F_q$, $q>2$, infinite graph $A(F_q)$ is a tree.

It means that group $G(F_q)$ is a group of walks of even length on q-regular tree starting in zero point with natural addition of them. A standard symbolic string C defines transformation ${}^n\eta'(C)$ in each group $G(n, K)$, $n \geq 2$ and $G(K)$. An irreducible transformation $\eta'(C)$ from $G(K)$ has an infinite order.

We are going to use the family of maps introduced below.

Let $\Delta=\Delta n, k, n>k$ be a canonical homomorphism of $A(n,K)$ onto $A(k,K)$ corresponding to the procedure of deleting of coordinates with indices $k+1, k+2, \ldots, n$. This map defines the canonical homomorphism $\mu=\mu(n, k)$ of group $G(n, K)$ onto $G(k, K)$. Let us consider the diagram

Σ(K)

/↓

G(k, K)← G(n, K)

where vertical arrow corresponds to homomorphism ${}^n\eta'$ from $\Sigma(K)$, skew line corresponds to ${}^k\eta'$ and horizontal arrow stands for $\mu(n,k)$, $n>k$. It is easy to see that this diagram is a commutative one.

As was noted in Ustimenko (2013), subgroups $G(n, K)$ of $E_n(K)$ form a family of stable cubical maps. So, correspondents can take pair $G(n, K)$ and $G(k, K)$ with $n(k)=k+\gamma$ where parameter γ is a positive constant or a positive linear function in variable k.

Alice can use the above-defined computationally tame homomorphism $\mu=\mu(n,k)$, $n>k$ of groups $G(n,K)$ and $G(k,K)$.

She considers family of subgroups $G_k=G(n,K)$, $n=n(k)$, k-2,3 and family $G'_k=G(k,K)$,k=2,3,…,n.

She selects different strings $w_1, w_2, \ldots w_t$ of even length of semigroup $\Sigma(K)$ such that $w_i\, w_j \neq w_j w_i$ for different i and from $\{1,2,\ldots,t\}$. This condition implies that

${}^{s}\eta(w_i w_j) \neq {}^{s}\eta(w_j w_i)$ for $s \geq 2$. After the check of noncommutativity of generators, Alice takes generators ${}^{i}g_j={}^{n(i)}\eta\ (w_j),\ j=1,\ 2,\ \ldots,\ t,\ i \geq 2$ which form a Cremona system of stable noncommutative generators corresponding to sequence $n(2),n(3),\ldots$. Let us denote this system as Z. Note that ${}^{i}SZ = <{}^{i}g_1,\ {}^{i}g_2,\ldots,\ {}^{i}g_t>$ is a subgroup of G_i. So, the $G_i(K)$ forms an enveloping family of Z.

Similarly, Alice considers generators ${}^{i}g'_j = {}^{i}\eta\ (w_j)$, $j=2, 3, \ldots,t$, $i\geq 2$ of system Z'. Note that ${}^{i}SZ' = <{}^{i}g'_1, {}^{i}g'_2,\ldots, {}^{i}g'_t>$ is a subgroup of $G'_i(K)$.

So, the $G'_i(K),\ i=2,3,\ldots$ is an enveloping family of Z'.

It is easy to see that the system Z' is a quotient of Z defined by tame homomorphisms which move ${}^{i}g_j$ to ${}^{i}g'_j$. So, Alice takes $i=k$ and starts Stable Tahoma Protocol with Z and Z' and described enveloping families. So, she conducts steps of the algorithm, generates pairs $a_i,\ b_i,\ i=1,2,\ldots,t$ and sends them to Bob. Thus, correspondents generate the collision element h in $E\_k(K)$.

Alice (or Bob) selects a multivariate map for digital signatures *(MDS)* to Bob (or Alice). They use *def(h)* and safely deliver the map to a partner.

8.5.2. *Other stable subgroups defined via linguistic graphs*

Let us consider more general graph-based constructions of semigroups of formal Cremona semigroup $E_n(K)$.

Element $x_1 \rightarrow f_i(x_1,\ x_2,\ \ldots,\ f_n),\ i=1,2,\ldots,n$ of this semigroup will be identified with the tuple of elements $(f_1, f_2,\ldots, f_n)$, $f_i \in K[x_1,\ x_2,\ldots,x_n]$ when it is convenient.

Let us consider a totality ${}^{s}BS(K)$ of sequences of kind $u=(H_0,\ G_1,\ G_2,\ H_3,H_4,G_5,\ G_6,\ldots,\ H_{t-1},\ H_t),\ t=4i$, where $H_k \in E_s(K),\ G_j \in E_s((K)$. We refer to ${}^{s}BS(K)$ as a totality of free symbolic strings of rank s. We define a product of u with $u'=(H'_0,\ G'_1,\ G'_2,\ H'_3,\ H'_4,\ G'_5,\ G'_6,\ldots,\ H'_{l-1},\ H_l)$ as $w=(H_0,\ G_1,\ G_2,\ H_3,\ H_4,\ G_5,\ G_6,\ldots,\ H_{t-1},\ H'_0(H_t),G'_1(H_t),\ G'_2(H_t),\ H'_3(H_t),\ H'_4(H_t),\ G'_5(H_t),\ G'_6(H_t),\ \ldots,\ H'_{l-1}(H_t),\ H'_l(H_t))$. Note that the compositions of maps is computed in $E_s(K)$.

It is easy to see that this operation transforms ${}^{s}BS(K)$ into the semigroup with the unity element (H_0), where E_0 is an identity transformation from $S(K^s)$. Elements of kind $(H_0,\ G_1,\ G_2,\ H_3,\ H_4)$ are generators of the semigroup. This subsemigroup has some similarity with subsemigroup of special chains in the free product $E_s(K)\cdot E_s(K)$. We refer to ${}^{s}BS(K)$ as a *semigroup of free regular strings of dimension s*.

Let us assume that H_t of the above-written $u \in {}^sBS(K)$ is a bijective map and its inverse is a polynomial map (in the case of infinite ring K). Then, we can consider a reverse linguistic string $Rev(u)= (H_{t-1}(H_t^{-1}), G_{t-2}(H_t^{-1}), G_{t-3},(H_t^{-1}), H_{t-4}(H_t^{-1}),H_{t-5}{}^1(H_t), \ldots,G_2(H_t^{-1}), G_1(H_t^{-1}), H_0(H_t^{-1}), H_t^{-1})$ and refer to u as a reversible string. Let ${}^sBR(K)$ stand for the semigroup of reversible strings.

Let K be a finite commutative ring. We refer to an incidence structure with a point set $P=P_{s,m}=K^{s+m}$ and a line set $L=L_{r,m}=K^{r+m}$ as linguistic incidence structure I_m if point $x=(x_1, x_2,\ldots, x_s, x_{s+1}, x_{s+2}, \ldots, x_{s+m})$ is incident to line $y=[y_1, y_2, \ldots, y_r, y_{r+1}, y_{r+2}, \ldots, y_{r+m}]$ if and only if the following relations hold:

$$a_1x_{s+1}+b_1y_{r+1}=f_1(x_1, x_2,\ldots, x_s, y_1, y_2, \ldots, y_r)$$

$$a_2x_{s+2}+b_2y_{r+2}=f_2(x_1, x_2,\ldots, x_s, x_{s+1}, y_1, y_2, \ldots, y_r, y_{r+1})$$

$$\ldots$$

$$a_mx_{s+m}+b_my_{r+m}=f_m(x_1, x_2,\ldots, x_s, x_{s+1},\ldots, x_{s+m}, y_1, y_2, \ldots, y_r, y_{r+1},\ldots, y_{r+m})$$

where a_j, and b_j, $j=1,2,\ldots,m$ are not zero divisors, and f_j are multivariate polynomials with coefficients from K. Brackets and parentheses allow us to distinguish points from lines (see Ustimenko 2020).

The color $\rho(x)=\rho((x))$ $(\rho(y)=\rho([y]))$ of point $\boldsymbol{x}$ (line $[y]$) is defined as a projection of an element (x) (respectively, $[y]$) from a free module on its initial s (relatively r) coordinates. As it follows from the definition of linguistic incidence structure for each vertex of incidence graph, there exists the unique neighbor of a chosen color.

We refer to $\rho((x))=(x_1, x_2,\ldots, x_s)$ for $(x)=(x_1, x_2,\ldots, x_{s+m})$ and $\rho([y])=(y_1, y_2, \ldots, y_r)$ for $[y]=[y_1, y_2, \ldots, y_{r+m}]$ as the color of the point and the color of the line, respectively. For each $b\in K^r$ and $p=(p_1, p_2,\ldots, p_{s+m})$, there is the unique neighbor of the point $[l]=N_b(p)=N((p),b)$ with the color b. Similarly for each $c\in K^s$ and line $l=[l_1, l_2,\ldots, l_{r+m}]$, there is the unique neighbor of the line $(p)= N_c([l])=N([l],b)$ with the color c. We refer to the operator of taking the neighbor of vertex accordingly chosen color as a sliding operator. On the sets P and L of points and lines of linguistic graph, we define jump operators ${}^1J={}^1J_b(p)=J((p),b)=(b_1, b_2,\ldots,b_s, p_1, p_2,\ldots, p_{s+m})$, where $(b_1, b_2,\ldots,b_s)\in K^s$ and ${}^2J={}^2J_b([l])=J([l],b)=[b_1, b_2,\ldots,b_r, l_1, l_2,\ldots, l_{r+m}]$, where $(b_1, b_2,\ldots,b_r)\in K^r$. We refer to tuple (s, r, m) as a type of the linguistic graph $I=I(K)$.

Note that we can consider the same set of above equations with coefficients from K for variables x_i and y_i from the extension K' of K and define graph ${}^{K'}I={}^{K'}I(K)$. Let

$s=r$ and $K'=K[x_1, x_2,..., x_n]$, $n=m+s$. We consider induced subgraph in I' of all vertices of ${}^{K'}I$ with colors from $K[x_1, x_2,..., x_s]$ (tuples of $K[x_1, x_2,..., x_s]^s$).

We form the sequence of vertices (walk with jumps) of graph I' with the usage of string u from free linguistic semigroup ${}^sBS(K)$.

We take the initial point $(x)=(x_1, x_2,..., x_s, x_{s+1}, x_{s+2},..., x_{s+m})$ formed by the generic variables of K' and consider a skating chain

$$(x),J((x),H_0)=({}^1x),N(({}^1x),G_1)=[{}^2x],J([{}^2x],G_2)=[{}^3x],N([{}^3x],H_3)=({}^4x),J(({}^4x),H_4)= ({}^5x),\ldots, J([{}^{t-2}x],G_{t-2})=[{}^{t-1}x],N([{}^{t-1}x],H_{t-1})=({}^tx),J(({}^tx),H_t)=({}^tx).$$

Let $({}^tx)$ be the tuple $(H_t, F_2, F_3,\ldots,F_n)$, where $F_i \in K[x_1, x_2,\ldots, x_n]$. We define ${}^I\Psi(u)$, $I=I(K)$ as the map $(x_1, x_2,\ldots, x_n)\rightarrow(H_t, F_2, F_3,\ldots,F_n)$ and refer to it as a chain transition of point variety.

The statement written below follows from the definition of the map.

LEMMA 8.1.– *The map* $\psi={}^I\psi$: ${}^sBS(K)\rightarrow E_n(K)$ *is a homomorphism of semigroups,* $\psi({}^sBR(K))$*is a group* (Ustimenko 2019b).

We refer to ${}^I\psi({}^sBS(K))={}^ICT(K)$ as a *chain transition semigroup* of linguistic graph $I(K)$ and to map ψ as *a linguistic compression map.* Note that in the case of the finite commutative ring, homomorphism composition $\Delta\psi$ *of homomorphism* Δ *and* ψ maps infinite semigroup into a finite set of elements of $\Delta({}^ICT(K))$.

We define subsemigroup ${}^sGS(K)$ of symbolic ground strings as a totality of bipartite strings $u=(H_0, G_1, G_2, H_3, H_4, G_5, G_6,\ldots, H_{t-1}, H_t)$ in ${}^sBS_r(K)$ with $H_0=E_0$, $G_1=G_2$, $H_3=H_4$,$G_5=G_6$,…, $H_{t-1}=H_t$ where E_0 is a unit of $E_n(K)$ and refer to ${}^I\psi({}^sGS(K))={}^IGCT(K)$ as a semigroup of ground chain transitions on a linguistic graph I.

In the case of linguistic graph $A(n,K)$ of type $(1,1,n\text{-}1)$, we can consider a subgroup $St(K)$ of elements of ${}^IGS(K)$ with coordinates of type x_1+t, $t\in K$ and identify ${}^{A(n,K)}\psi\ (St(K))$ with the above-introduced group of cubical endomorphisms $GA(n, K)$.

We can consider a subgroup $LSt(K)$ of elements ${}^IBS(K)$ *with* coordinates of type x_1+t, $t\in K$ and construct extension $Ext(GA(n,K))$ of $G(n,K)$ as ${}^{A(n,K)}\psi\ (LSt(K))$. As shown in the paper Ustimenko (2018), the elements of the stable group $Ext(AG(n,K))$ are also cubical endomorphisms.

In fact, the first family $GD(n,K)$ of stable groups of degree 3 was introduced as a group of transformative of bijective stream ciphers of multivariate nature defined via linguistic graphs $D(n, K)$ of type $(1,1,n-1)$ in Ustimenko (1998) The implementation of this cipher in the case $K=F_q$ is described in Ustimenko (2001). Graphs $D(n, F_q)$ and their connected components were introduced in Lazebnik and Ustimenko (1993a, 1993b) and Lazebnik and Ustimenko (1995) as a family of graphs of large girth of extremal graph theory (see also Lazebnik et al. 1995, 1996). Connections between graphs $D(n,K)$ and $A(n,K)$ are discussed in Ustimenko (2013).

8.5.3. *Special homomorphisms of linguistic graphs and corresponding semigroups*

Let $I(K)$ be the linguistic graph over commutative ring K defined in section 8.5.2 and $M = \{m1, m2,..., md\}$ be a subset of $\{1, 2, ..., m\}$ (set of indices for equations). Assume that equations indexed by elements from M of the following kind:

$$a_{m1}x_{m1} - b_{m1}y_{m1} = f_{m1}(x_1, x_2, \ldots, x_s, y_1, y_2, \ldots, y_r)$$

$$a_{m2}x_{m2} - b_{m2}y_{m2} = f_{m2}(x_1, x_2, \ldots, x_s, x_{m1}, y_1, y_2, \ldots, y_r, y_{m1})$$

...

$$a_{md}x_{md} - b_{md}y_{md} = f_{md}(x_1, x_2, \ldots, x_s, x_{m1}, x_{m2}, \ldots, x_{m\,d-1}, y_1, y_2, \ldots, y_r, y_{m1}, y_{m2}, \ldots, y_{m\,d-1})$$

define other linguistic incidence structure I_M. Then, the natural projections δ_1: $(x) \rightarrow (x_1, x_2, \ldots, x_s, x_{m1}, x_{m2}, \ldots, x_{md})$ and δ_2: $[y] \rightarrow [y_1, y_2, \ldots, y_r, y_{m1}, y_{m2}, \ldots, y_{md}]$ of free modules define the natural homomorphism δ of incidence structure I onto I_M. We will use the same symbol ρ for the coloring of linguistic graph I_M.

It is clear that δ is the color preserving homomorphism of incidence structures (bipartite graphs). We refer to δ as a symplectic homomorphism and graph I_M as a symplectic quotient of linguistic graph I. In the case of linguistic graphs defined by an infinite number of equations, we may consider symplectic quotients defined by an infinite subset M (see Ustimenko et al. (2019b), where symplectic homomorphism was used for the cryptosystem construction).

LEMMA 8.2.– *A symplectic homomorphism $\acute{\eta}$ of linguistic graph I of type (r, s, m) onto I' defined over commutative ring K induces the semigroup homomorphism $\acute{\eta}^*$ of ${}^{I}CT(K)$ into ${}^{I'}CT(K)$ and the following diagram is commutative:*

$^{s}BS_r(K) \rightarrow {}^{I}CT(K)$

$\downarrow /$

$^{I'}CT(K)$

where horizontal and vertical arrows correspond to the linguistic compression homomorphisms $^{I}\psi$ and $^{I'}\psi$ and symbol $/$ corresponds to η^*.

If S is a stable subsemigroup of $^{I}CT(K)$ of degree d, then $\dot{\eta}^*(S)$ is also a stable subsemigroup (or subgroup). The degree of $\dot{\eta}^*(S)$ is bounded above by d. We will search for subsemigroup X *of* $^{s}BS_r(K)$ and linguistic graphs $I(K)$ such that $\Psi(X)$ is a stable subsemigroups of $^{I}CT(K)$.

8.5.4. *Example of stable subsemigroups of arbitrary degree*

We define double Schubert graph $DS(k,K)$ over commutative ring K as an incidence structure defined as a disjoint union of partition sets $PS=K^{k(k+1)}$ consisting of points which are tuples of kind $x=(x_1, x_2, \ldots, x_k, x_{11}, x_{12}, \ldots, x_{kk})$ and $LS=K^{k(k+1)}$ consisting of lines which are tuples of kind $y=[y_1, y_2, \ldots, y_k, y_{11}, y_{12}, \ldots, y_{kk}]$, where x is incident to y, if and only if $x_{ij} - y_{ij} = x_i\, y_j$ for $i=1, 2, \ldots, k$ and $j=1, 2, \ldots, k$. It is convenient to assume that the indices of kind i,j are placed for tuples of $K^{k(k+1)}$ in the lexicographical order.

The term double Schubert graph is chosen because points and lines of $DS(k, F_q)$ can be treated as subspaces of $F_q^{(2k+1)}$ of dimensions $k+1$ and k, which form two largest Schubert cells. Recall that the largest Schubert cell is the largest orbit of group of unitriangular matrices acting on a variety of subsets of given dimensions. We will consider these connection in details in the next section.

We define the color of point $x=(x_1, x_2, \ldots, x_k, x_{11}, x_{12}, \ldots, x_{kk})$ from PS as tuple $(x_1\, x_2, \ldots, x_k)$ and the color of a line $y=[y_1, y_2, \ldots, y_k, y_{11}, y_{12}, \ldots, y_{kk}]$ as the tuple $(y_1, y_2, \ldots, y_k)$. For each vertex v of $DS(k, K)$, there is the unique neighbor $y=N_a(v)$ of a given color $a=(a_1, a_2, \ldots, a_k)$. It means the graphs $DS(k, K)$ form a family of linguistic graphs.

Let us consider the subsemigroup $^{k}Y(d, K)$ of $^{k}BS(K)$ consisting of strings $u=(H_0, G_1, G_2, H_3, H_4, G_5, G_6, \ldots, H_{t-1}, H_t)$ such that the maximum of parameters $\deg(H_0)+\deg(G_1)$, $\deg(G_2)+\deg(H_3)$, $\deg(H_4)+\deg(G_5)$, $\deg(G_6)+\deg(H_7)$, $\deg(G_{t-2})+\deg(H_{t-1})$, $\deg(H_t)=1$ equals d, $d>1$.

THEOREM 8.3.– *Let I(K) be an incidence relation of double Schubert graph DS(k, K). Then,* $^{I}\psi(^{k}Y(d, K))={}^{k}U(d,K)$ *form a family of stable semigroups of degree d* (see Ustimenko 2019b).

The proof is based on the fact that the chain transition u from $^{k}U(d, K)$ moves $x_{i,j}$ into expression $x_{i,j}+T(u)$, where $T(u)$ is a linear combination of products $f\epsilon K[x_1, x_2,..., x_k]$, $g\epsilon K[y_1, y_2,..., y_k]$ where $deg(f)+deg(g)\leq d$.

New semigroup $^{k}U(d, K)$ consists of transformations of the free module K^t, $t=(k+1)k$. If $d=2$, then $^{k}U(d, K)$ contains semigroups of quadratic transformation defined in Ustimenko (2020), which consists of ground chain transitions.

Let J be a subset of the Cartesian square of $M=\{1,2,...,k\}$. We can identify its element (i,j) with the index ij of double Schubert graph $DS(k,K)$.

PROPOSITION 8.2.– *Each subset J of* M^2 *defines symplectic homomorphism* δ_J *of DS(k, K) onto linguistic graph* $DS_J(k,K)$.

It is easy to see that in the case of empty set J, the image of the map is a complete bipartite graph with the vertex set K^kUK^k.

COROLLARY 8.1.– *Let I(J, K)) be an incidence relation of linguistic graph* $DS_J(k, K)$. *Then,* $^{I(J,K)}\psi(^{k}Y(d, K))={}^{k}U_J(d,K)$ *form a family of stable semigroups of degree d.*

Second protocol of safe delivery of MDS.

Let f: $N\rightarrow R$ be the real function in natural variable and $[,]'$ stands for ceiling function, i.e. $[f(n)]'$ is closest to $f(n)$ parameter n' such that $n'\geq n$.

Alice considers a family of $^{r(n)}Y(d, K))$, $d=2$ or $d=3$, where $r(i)=[i^{1/2}]'$, i=2,3,... So the point set $^{r(i)}DS(d, K))$ is a free module of dimension $[i^{1/2}]'+([i^{1/2}]')^2$ which is at least $[i^{1/2}+i]'$.

For each I, she can select the strings u(1)=u(1,i), u(2,i), …, u(t,i), t ≥ 2 of kind u(k,i)=($^{k,i}H_0$,, $^{k,i}G_1$, $^{k,i}G_2$, $^{k,i}H_3$, $^{k,i}H_4$, $^{k,i}G_5$, $^{k,i}G_6$,…, $^{k,i}H_{t-1}$, $^{k,i}H_t$), k=1,2,…,t such that $^{k,i}H_t\,{}^{j,i}H_t\neq{}^{ji}H_t\,{}^{k,i}H_t$ for distinct k and j.

The last condition ensures that for $^{I}\psi$(u(k,i))=a(k,i), k=1,2,…,t conditions a(k,i)a(j,i)≠a(j, i)a(k,i) holds if j ≠k.

So cubical endomorphisms $a(l,i)\epsilon E_{r(i)}$, $l=1,2,...,t$, $i=2,3,...$ form stable cubical noncommutative Cremona system Z corresponding to the sequence $r(i)$. Semigroups

${}^{r(i)}U(d, K)$ form enveloping family of Z. Alice can take ${}^{r(i)}DS(d, K))$ and subset $J(i)$ which defines an incidence system $I(J, K))$ such that $|J(i)|=i-r(i)$. So the point set of $I(J(i), K)$ is K^i.

Symplectic homomorphism of ${}^{r(i)}DS(d, K))$ onto $I(J(i),K)$ induces homomorphism $\phi(i,J(i))$ *of semigroup* ${}^{r(i)}U(d, K)$ onto ${}^{i}U_{J(i)}(d,K)$. It is easy to see that $\phi(i,J(i))(a(k,i))=a'(k,i)$ form the quotient Z' of the system Z with enveloping family ${}^{i}U_{J(i)}(d,K)$.

So, Alice selects the multivariate digital signature system with parameters m and n. She takes $i=k$ which exceed $max(m, n)$ for parameters m and n *of MDC*. She generates $a(k,j)$, $j=1,2,...,t$ from ${}^{r(k)}U(d, K)$ and considers subset $J(i)$. Alice uses invertible elements of enveloping semigroups ${}^{r(k)}U(d, K)$ and ${}^{k}U_{J(k)}(d,K)$ together with bijective affine transformations of free modules $K^{r(k)}$ and K^k. So, she uses multivariate tahoma protocol, generates pairs (a_i, b_i) and sends them to Bob.

Correspondents execute the protocol and generate the collision map $h=(h_1, h_2,..., h_k)$ from $E_k(K)$.

In the case of $d=2$, Alice (or Bob) uses scheme 1 of section 8.3. So they send $h_i+{}^{i}p_i$, $i=1,2,...,m$ to the partner who restores ${}^{i}p$. If $d=3$, Alice (or Bob) uses scheme 2 of section 8.3 with the computation of $def(h)$.

8.6. Conclusion

We consider an option to keep multivariate public map of digital signature system on "secure mode of El Gamal type". So the combination of multivariate Tahoma protocol based on the selected platform with the multivariate digital signature defined over the chosen group field $K=F_q$ is used. The base is in fact triple which consists of two stable subsemigroups $S_1(q)$ and $S_2(q)$ of $E_{p(n)}$ and $E_n(F_q)$ and homomorphism between them. Open size n gives the format of used hashed vectors. This number is known to adversary.

According to the cryptographic conventions, the whole used combination has to be known up to some hidden formal parameters which form the key of the algorithm.

Thus, adversary knows ground field F_q, dimensions n and $p(n)$, pairs of generators (a_i, b_i), $i=1,2$, where $a_i \in E_{p(n)}(F_q)$ and $b_i \in E_n(q)$. So he knows the degree of stable subsemigroups $<a_1, a_2,...,a_t>$ an $<b_1, b_2,...b_t>$, can use homomorphism which moves a_i to b_i but does not know an alternative efficient way of homomorphism computation.

Adversary is informed that correspondents use protocol in a periodic way; the collision map is used exactly one time and multivariate map is used for signatures $\leq n$ times. From session to session, correspondents can change sets of generators without the change of the platform.

They know the type of multivariate map $P=SP'T$, S and T of dimensions n and m, and a quadratic element P' of kind $x_i \rightarrow {}^i p$, $i=1,2,...m$, but do not know multivariate map P itself. It means that the adversary knows that correspondents are followers of one selected method. So correspondents are followers of the Imai–Matsumoto method, original oil and vinegar method by J. Patarin, Rainbow UOV, or lifted UOV algorithms.

Note that cryptanalysis for the first two methods is known only in the case when P is given public. Adversary knows that correspondents are able to change maps S and T and internal parameters of P' in each of these four cases. For the simplicity, we assume that the number of equations m is known to adversary.

Thus, adversary can try to break the tahoma multivariate protocol based on known NP hard problem (work decomposition of element of formal Cremona group into a product of given generators presented in the standard form of multivariate cryptography). Of course, if they break the protocol, the adversary gets the multivariate digital signature in its public form. They have to periodically compute such a decomposition of obtained element into generators to get the corresponding collision map. So in the case of the already broken classical Imai–Matsumoto and oil and vinegar schemes, the adversary can forge the signature. In the case of rainbow and lifted UOV, the adversary has to find a solution for the related crypt analytic problem. For the current state of the cryptanalysis of these schemes, the reader should see Ding et al. (2019, 2020) and Shuhei et al. (2020).

Breaking the WORD PROBLEM is currently unsolvable post-quantum problem, so we discuss the remaining options for adversary. Assume that the adversary intercepts all pairs of kind (hash vector/corresponding signature) during the session. These n pairs are not sufficient to approximate the quadratic multivariate map of the signature. So the security of the entire algorithm rests entirely on the security of the protocol.

We present several combined algorithm (three families of platforms and four types of general digital signatures). Time execution of the combination is a sum of the already investigated protocol and signature scheme.

8.7. References

Alon, N. (1986). Eigenvalues, geometric expanders, sorting in rounds and Ramsey Theory. *Combinatorica*, 6(3), 207–219.

Andrzejczak, M. (2019). The low-area FPGA design for the post-quantum cryptography proposal Round5. In *Proceedings of the Federated Conference on Computer Science and Information Systems (FedCSIS)*. Cryptography and Security Systems, Leipzig.

Anshel, I., Anshel, M., Goldfeld, D. (1999). An algebraic method for public-key cryptography. *Math. Res. Lett.*, 6(3–4), 287–291.

Ben-Zvi, A., Kalka, A., Tsaban, B. (2018). Cryptanalysis via algebraic span. In *Advances in Cryptology – CRYPTO 2018 – 38th Annual International Cryptology Conference*, Shacham, H. and Boldyreva, A. (eds). Santa Barbara, August 19–23, 10991, 255–274. Springer, Cham.

Blackburn, S.R. and Galbraith, S.D. (1999). Cryptanalysis of two cryptosystems based on group actions. *Advances in Cryptology – ASIACRYPT '99*, 1716, 52–61. Springer, Berlin.

Bodnarchuk, Y. (2001). Every regular automorphism of the affine Cremona group is inner. *J. Pure Appl. Algebra*, 157(1), 115–119.

Cao, Z. (2012). *New Directions of Modern Cryptography*. CRC Press, Taylor & Francis Group, Boca Raton.

Ko, C.K.H., Lee, S.J., Cheon, J.H., Han, J.W., Kang, J.S., Park, C. (2000). New public-key cryptosystem using braid groups. In *Advances in Cryptology – CRYPTO 2000*. Santa Barbara, 1880, 166–183. Springer, Berlin.

Ding, J. and Petzoldt, A. (2017). Current state of multivariate cryptography. *IEEE Security & Privacy*, 15(4), 28–36.

Ding, J., Gower, J.E., Schmidt, D.S. (2006). *Multivariate Public Key Cryptosystems*, 2nd edition. Springer, New York.

Ding, J., Deaton, J., Schmidt, K., Vishakha, Zhang, Z. (2019). Cryptanalysis of the lifted unbalanced oil vinegar signature scheme. Paper, Cryptology ePrint Archive, 2019/1490.

Ding J., Deaton J., Vishakha, Yang, B.-Y. (2020). The nested subset differential attack: A practical direct attack against LUOV which forges a signature within 210 minutes. Paper, Cryptology ePrint Archive, 2020/967.

Fine B., Maggie H., Kahrobaej, D., Rosenberg G. (2011). Aspects of nonabelian group based cryptography: A survey and open problems. *arXiv:1103.4093*.

Grzesik, A., Král', D., Lovász, L.M. (2018). Elusive extremal graphs. *arXiv:1807.01141*.

Hoory N., Linial A., Wigderson A. (2006). Expander graphs and their applications. *Bull. Amer. Math Soc.*, 43, 439–561.

Kahrobaei, D. and Khan, B. (2006). A non-commutative generalization of ElGamal key exchange using polycyclic groups. In *IEEE GLOBECOM 2006 – 2006 Global Telecommunications Conference [4150920]*. doi: 10.1109/GLOCOM.2006.

Koblitz, N. (1998). *Algebraic Aspects of Cryptography*. Springer Berlin, Heidelberg.

Kropholler, P.H., Pride, S.J., Othman, W.A.M., Wong, K.B., Wong, P.C. (2010). Properties of certain semigroups and their potential as platforms for cryptosystems. *Semigroup Forum*, 81, 172–186.

Kumar, G. and Saini, H. (2017). Novel noncommutative cryptography scheme using extra special group. *Security and Communication Networks*, 9036382, 21. doi: 10.1155/2017/9036382.

Lazebnik, F. and Ustimenko, V. (1993a). New examples of graphs without small cycles and of large size. *Eur. J. Combinat.*, 14, 445–460.

Lazebnik, F. and Ustimenko, V. (1993b). Some algebraic constractions of dense graphs of large girth and of large size. *DIMACS Series in Discrete Mathematics and Theoretical Computer Science*, 10, 75–93.

Lazebnik, F. and Ustimenko, V. (1995). Explicit construction of graphs with arbitrary large girth and of large size. *Discrete Applied Mathematics*, 60, 275–228.

Lazebnik, F., Ustimenko, V., Woldar, A.J. (1995). A new series of dense graphs of high girth. *Bull. AMS*, 32(1), 73–79.

Lazebnik, F., Ustimenko, V., Woldar, A.J. (1996). A characterisation of the components of the graphs D(k, q). *Discr. Math.*, 157, 271–283.

Lopez Ramos, J.A., Rosenthal, J., Schipani, D., Schnyder, R. (2019). Group key management based on semigroup actions. *J. Algebra Appli.*, 16.

Maze, G., Monico, C., Rosenthal, J. (2007). Public key cryptography based on semigroup actions. *Adv. Math. Commun.*, 1(4), 489–507.

McEliece, R.J. (1978). A public-key cryptosystem based on algebraic coding theory. DSN Progress Report, 44, 114–116.

Moldovyan, D.N. and Moldovyan, N.A. (2010). A new hard problem over non-commutative finite groups for cryptographic protocols. In *International Conference on Mathematical Methods, Models, and Architectures for Computer Network Security, MMM-ACNS 2010: Computer Network Security*, 183–194.

Myasnikov, A.G., Shpilrain, V., Ushakov, A. (2011). *Non-commutative Cryptography and Complexity of Group-Theoretic Problems*. American Mathematical Society, Rhode Island.

NIST PQC Team (2022). PQC standardization process: Announcing four candidates to be standardized, plus fourth round candidates. *NIST* [Online]. Available at: https://www.nist.gov/news-events/news/2022/07/pqc-standardization-process-announcing-four-candidates-be-standardized-plus.

Noeter, M. (1904). Luigi Cremona. *Math. Ann.*, 59, 1–19.

Ostafe, A. (2010). Multivariate permutation polynomial systems and nonlinear pseudorandom number generators. *Finite Fields Appli.*, 16(3), 144–154.

Ostafe, A. and Shparlinski, I.E. (2010). On the degree growth in some polynomial dynamical systems and nonlinear pseudorandom number generators. *Math. Comput.*, 79(269), 501–511.

Roman'kov, V.A. (2016). A nonlinear decomposition attack. *Groups Complex. Cryptol*, 8(2) 197–207.

Roman'kov, V.A. (2019). An improved version of the AAG cryptographic protocol. *Groups, Complex., Cryptol*, 11, 35–42.

Sakalauskas, L., Tvarijonas, P., Raulynaitis, A. (2007). Key agreement protocol (KAP) using conjugacy and discrete logarithm problema in group representation level. *Informatica*, 8(1), 115–124.

Šafarevič, I.R. (1982). On some in_nite dimension groups II. *Mathematics of the USSR-Izvestiya*, 18(1), 185–194.

Shpilrain, V. and Ushakov, A. (2006). The conjugacy search problem in public key cryptography: Unnecessary and insufficient. *Applicable Algebra in Engineering, Communication and Computing*, 17(3–4), 285–289.

Shuhei, N., Yasuhiko, I., Wang, Y., Ding, J., Takagi, T. (2020). New complexity estimation on the rainbow-band-separation attack. Paper, Cryptology ePrint Archive, 2020/703.

Tsaban, B. (2015). Polynomial-time solutions of computational problems in noncommutative-algebraic cryptography. *J. Cryptol.*, 28(3), 601–622.

Ustimenko, V. (1998). Coordinatisation of trees and their quotients, in the Voronoj's impact on modern science. *Kyiv, Institute of Mathematics*, 2, 125–152.

Ustimenko, V. (2001). CRYPTIM: Graphs as tools for symmetric encryption. In *Proceedings of AAECC-14 Symposium on Applied Algebra, Algebraic Algorithms and Error Correction Codes*, November, 278–286.

Ustimenko, V. (2004). Maximality of affine group, hidden graph cryptosystem and graph's stream ciphers. *J. Algebra Discr. Math.*, 1(10), 51–65.

Ustimenko V. (2007a). On the extremal graph theory for directed graphs and its crypto-graphical applications. In *Advances in Coding Theory and Cryptography, Series on Coding and Cryptology*, Shaska, T., Huffman, W.C., Joener, D., Ustimenko, V. (eds). World Scientific Publishing, Singapore.

Ustimenko, V. (2007b). On the graph based cryptography and symbolic computations. *Serdica Journal of Computing.* doi: 10.55630/sjc.2007.1.131-156.

Ustimenko, V. (2013). On the extremal graph theory and symbolic computations. *Dopov. Nac. Akad. Nauk Ukr.*, 2, 42–49.

Ustimenko, V. (2017). On the families of stable transformations of large order and their cryptographical applications. *Tatra Mt. Math. Publ.*, 70, 107–117.

Ustimenko, V. (2018). On new symbolic key exchange protocols and cryptosystems based on hidden tame homomorphism. *Dopovidi. NAS of Ukraine*, 10, 26–36.

Ustimenko, V. (2019a). On desynchronised multivariate algorithms of El Gamal type for stable semigroups of affine Cremona group. *Theoretical and Applied Cybersecurity*, 1(1), 22–30.

Ustimenko, V. (2019b). On inverse protocols of post quantum cryptography based on pairs of noncommutative multivariate platforms used in tandem. Paper, Cryptology ePrint Archive, 897.

Ustimenko, V. (2019c). On semigroups of multiplicative cremona transformations and new solutions of post quantum cryptography. Paper, Cryptology ePrint Archive, 133.

Ustimenko, V. (2020). On the usage of postquantum protocols defined in terms of transformation semigroups and their homomophisms. *Theoretical and Applied Cybersecurity*, 1(2), 32–44.

Ustimenko, V. and Klisowski, M. (2019). On noncommutative cryptography with cubical multivariate maps of predictable density. In "*Intelligent Computing*", *Proceedings of the 8. 2019 Computing Conference*. London, 2, 998. Springer, Cham.

Ustimenko, V., Romanczuk-Polubiec, U., Wroblewska, A. (2019a). Expanding graphs of the Extremal Graph Theory and expanded platforms of Post Quantum Cryptography. In *Position Papers of the 2019 Federated Conference on Computer Science and Information Systems*, FedCSIS 2019. Leipzig, September 1–4, 41–46.

Ustimenko, V., Romanczuk-Polubiec, U., Wroblewska, A., Polak, M., Zhupa, E. (2019b). On the constructions of new symmetric ciphers based on non-bijective multivariate maps of prescribed degree. *Security Commun. Net.*, 2137561, 15. doi: 10.1155/2019/2137561.

Wagner, R. and Magyarik, M.R. (1984). A public-key cryptosystem based on the word n problem. In *Advances in cryptology. Proceedings of CRYPTO '84*. Santa Barbara, 19–22 August.

Wroblewska, A. (2008). On some properties of graph based public keys. *Alb. J. Math.*, 2(3), 229–234.

9

Metasurface Model of Geographic Baric Field Formation

Dmitri KOROLIOUK[1,2], Maksym ZOZIUK[1], Pavlo KRYSENKO[1] and Yuriy YAKYMENKO[1]

[1] *National Technical University of Ukraine Igor Sikorsky Kyiv Polytechnic Institute, Ukraine*

[2] *UNESCO Chair in Biotechnology and Bioethics, University of Rome Tor Vergata, Italy*

9.1. Introduction

In the task of studying and predicting weather phenomena, the main factor is that the air is very mobile, and therefore the entire atmosphere is changeable. In a matter of hours, meteorological parameters that determine weather conditions can change.

In this case, the state of the weather in a certain location, for a short period of time, largely depends on parameters such as the distribution of pressures and temperatures of the lower atmosphere (weather conditions).

Therefore, the charts with isobars and isotherms (lines connecting places with the same pressures and temperatures, respectively) form the basis of synoptic activity.

Figure 9.1 shows typical local-geographic baric fields.

Computational Methods and Mathematical Modeling in Cyberphysics and Engineering Applications 1, coordinated by Dmitri KOROLIOUK, Sergiy LYASHKO and Nikolaos LIMNIOS.

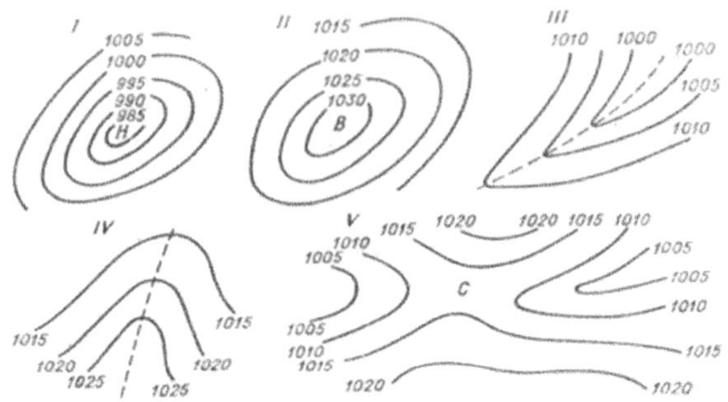

Figure 9.1. *Some baric field configurations*

In this chapter, we propose a metasurface model for the formation of locally inhomogeneous baric fields in the lower atmospheric layers.

The atmospheric baric field defines the spatial distribution of atmospheric pressure at certain altitude levels. The baric field is a scalar one, represented by surfaces of equal values of a given scalar (the value of atmospheric pressure, usually expressed in hectopascals), and on a plane – by the lines of equal values. For a scalar baric field, these are isobaric surfaces and isobars.

To trace the dynamics of 2D baric field values distribution according to the data of real observations, maps of baric topography are used (see Figure 9.2).

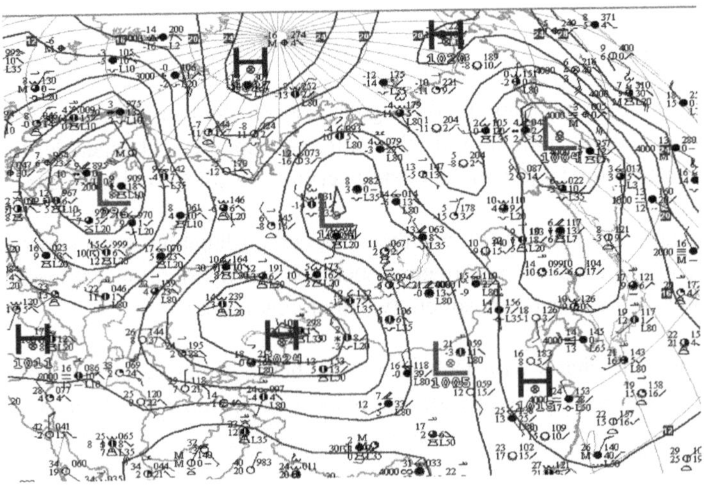

Figure 9.2. *Example of a topographic baric map. For a color version of this figure, see www.iste.co.uk/koroliouk/computational.zip*

9.2. The parametric scalar field model principle

If we consider the surface layer of the atmosphere as a metasurface on a large scale, then it is natural to try to transfer the corresponding research methods, taking into account the change in scale, as well as the initial and boundary conditions.

Since we are interested in the formation of locally inhomogeneous baric fields, we will consider surfaces of the order of 100–10,000 sq. Km, which is $2 \cdot 10^{-7} - 2 \cdot 10^{-5}$ of the total surface of the Earth globe.

That is, based on the topology of local regions as geodesic quasi-spherical sectors, it is natural to neglect the curvature of the Earth and consider a plane problem. Our constructions are based on the model for creating Chladni figures.

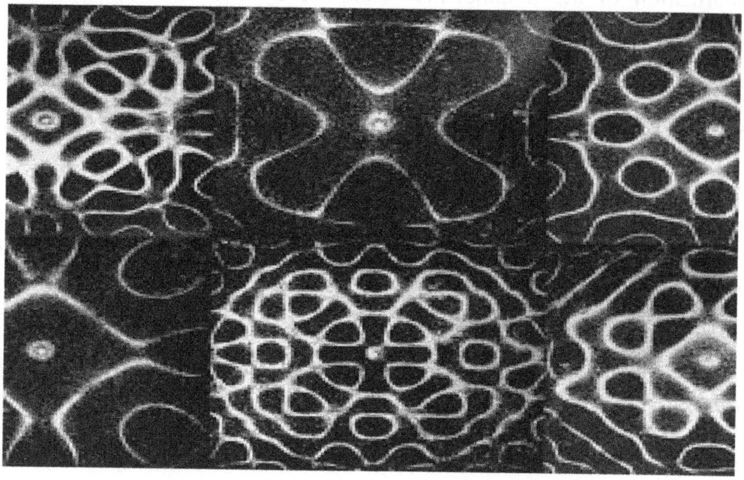

Figure 9.3. *Six configurations of Chladni figures*

In the 18th century, Ernst Chladni showed the original processes of creating an image on a plate, in the center of which there is a source of a periodic impulse (exciting action), as a result of which an acoustic standing wave is created (shown in Figure 9.3). A bulk material (e.g. sand) is placed on the surface of the plate, which, in accordance with the configuration of the standing wave, occupies the most energetically optimal location, and a surface redistribution of the bulk material is observed in the form of a specific surface figure.

9.3. Local isobaric scalar field model

The model is constructed using heuristic assumptions about how an all-around wave is converted to a standing wave. Consider the displacement of surface air masses under an external frequency acoustic action:

$$h(x, y, t) = \sin(2\pi f t) s(x, y), \qquad [9.1]$$

where f is the frequency of the acoustic wave and $s(x, y)$ is the displacement coefficient.

In this case, the standing wave provides a time-stationary displacement coefficient $s(x, y)$, which "normalizes" the periodic function $sin(f, t)$. That is, there is a factorization of the displacement function by two factors, where the first depends only on time, and the second only on coordinates.

It is known (Comer et al. 2004) that the displacement function h is represented by the solution of the wave equation:

$$\partial^2 h/\partial t^2 = c^2(\partial^2 h/\partial x^2 + \partial^2 h/\partial y^2), \qquad [9.2]$$

where $c^2 = 1/v^2$, and v is the velocity of the wavefront of the disturbance frequency f.

Since

$$\partial h/\partial t = s2\pi f \cos(2\pi f t), \qquad [9.3]$$

$$\partial^2 h/\partial t^2 = -s4\pi^2 f^2 \sin(2\pi f t), \qquad [9.4]$$

and also

$$\begin{cases} \partial^2 h/\partial x^2 = \sin(2\pi f t)\, \partial^2 s/\partial x^2 \\ \partial^2 h/\partial y^2 = \sin(2\pi f t)\, \partial^2 s/\partial y^2, \end{cases} \qquad [9.5]$$

where $s(x, y)$ is the displacement coefficient.

From equations [9.2] and [9.4], we obtain the following relation

$$-s(2\pi f t)^2 \sin(2\pi f t) = c^2 sin(2\pi f t)(\partial^2 s/\partial x^2 + \partial^2 s/\partial y^2),$$

whence we have

$$-(2\pi f/c)^2 s = (\partial^2 s/\partial x^2 + \partial^2 s/\partial y^2). \qquad [9.6]$$

Thus, the stationary wave model is reduced to an eigenvalue problem for the Laplace operator

$$\Delta s - \lambda s = 0, \lambda = -(2\pi f/c)^2, \quad [9.7]$$

in a rectangle with Dirichlet boundary conditions:

$$s\,(\pm 1, y) = 0, -1 \leq y \leq +1$$

$$s\,(x, \pm 1) = 0, -1 \leq x \leq +1. \quad [9.8]$$

The solution to this problem is known (Muller 2013):

$$s_{m,n}(x, y) = a \sin(\pi n x) \sin(\pi m y) + b \sin(\pi m x) \sin(\pi n y),$$

$$n, m = 1,2,3, \ldots \quad [9.9]$$

where a and b are the constants

$$n^2 + m^2 = K; K := (2f/c)^2. \quad [9.10]$$

The following relationship also takes place:

$$c\sqrt{n^2 + m^2}/2 = f. \quad [9.11]$$

Below, we present the simulated Chladni figures for an elementary area with the size 1×1.

In particular, for $n = m$:

$$s = 2(a + b)\sin(2\pi f x/c)\sin(2\pi f y/c). \quad [9.12]$$

That is, when the numbers at the frequency are equal, there is a simple dependence on the frequency, the parameters of the medium and the material.

9.4. Modeling Chladni figures based on the proposed model

In this section, the solutions of equation [9.12] are presented for some values of the parameters a, b, m, n.

The values of the parameters were chosen for practical interest, based on the effect of increasing the complexity of the pattern and the number of transitional lines of standing waves (inflection lines of the density gradient of redistributed air masses).

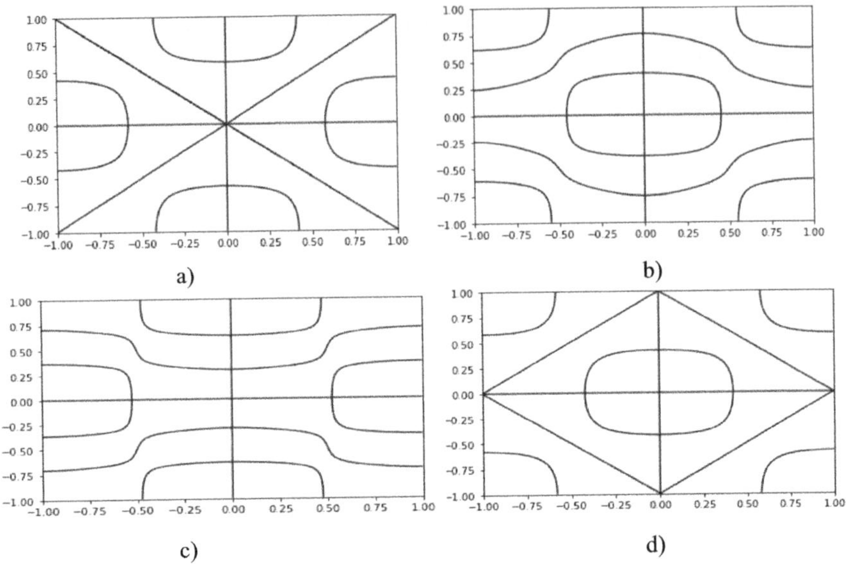

Figure 9.4. *Chladni figure with values: (a): a = 1; b = -1; m = 2; n = 3; (b): a = 2; b = 2; m = 2; n = 3; (c): a = 2; b = -4; m = 2; n = 3; (d): a = 2; b = 1; m = 2; n = 3*

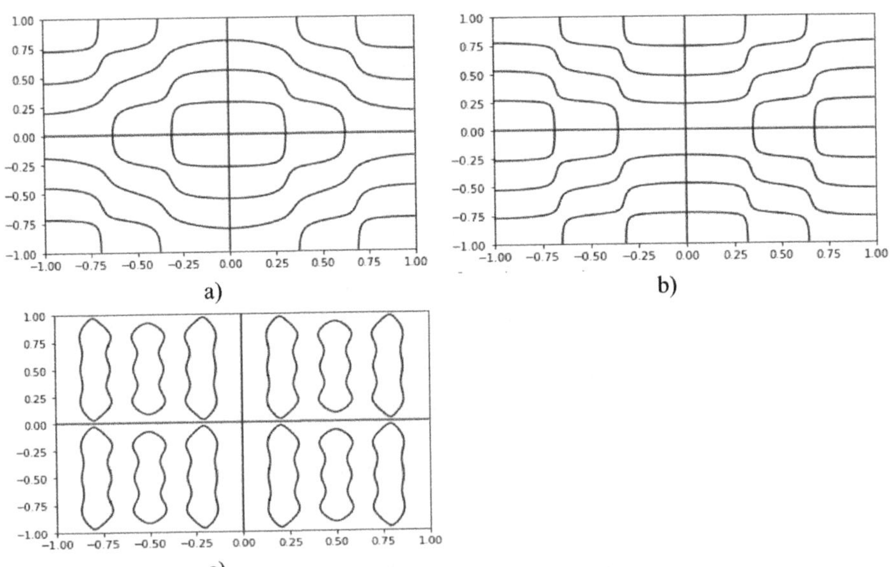

Figure 9.5. *Chladni figure with values: (a): a = -2; b = -1; m = 4; n = 3; (b): a = 4; b = -1; m = 4; n = 3; (c): a = -4; b = -1; m = 1; n = 7*

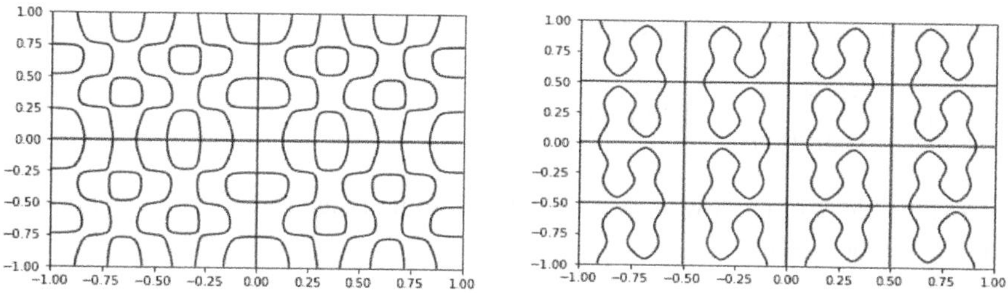

a) b)

Figure 9.6. *Chladni figure with values: (a): a = 4; b = -1; m = 4; n = 7; (b): a = 1; b = -3; m = 8; n = 2*

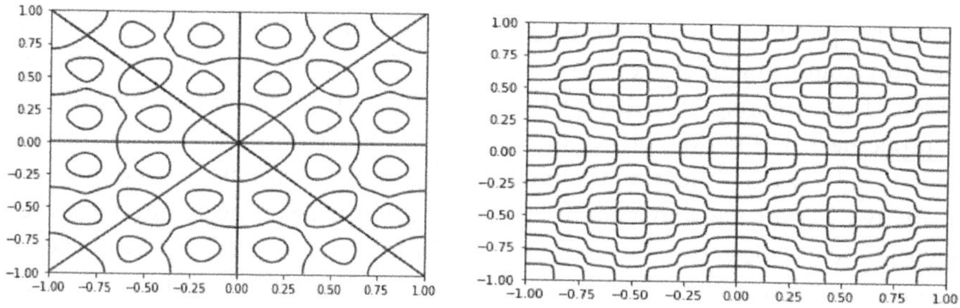

a) b)

Figure 9.7. *Chladni figure with values: (a): a = 1; b = -1; m = 8; n = 3; (b): a = -4; b = -1; m = 9; n = 7*

As shown in Figures 9.4–9.7, by varying the values of the parameters, we simulate different configurations of baric fields.

9.5. The frequency of forcing influences and the problem of its detection

Since the scale of formation of locally inhomogeneous baric fields was considered on elementary square surfaces of the order of 100–10,000 sq.km, we will assume that the side of the base square takes a value of about 10 km, and it corresponds to the wavelength of the frequency forcing action.

It is known that the frequency ν of an acoustic wave is expressed as follows:

$$\nu = c/\lambda, \qquad [9.13]$$

where c is the speed of propagation of the sound wave, and λ is the wavelength.

Then, in our case, the frequency:

$$\nu = \frac{340\ m/sec}{10{,}000\ m} = 0.034\ Hz$$

i.e. a value close to the lower limit of the infrasonic range (which is conventionally defined as 0.001 Hz).

It seems appropriate to measure the frequency and amplitude of infrasonic frequency driving forces using the method of a resonant vibrator (strings, horns, pipes) with a resonant frequency close to kilometer wavelengths or by microbarometric methods.

Another problem is related to the fact that the nature of the wave is acoustic. In order to capture low-frequency acoustic waves, first, a high sensitivity is needed. Second, strong filtering of such detectors is required. The resonant vibrator can have the properties of, for example, gravitational antennas, which are used to capture gravitational waves. Filtration is provided by the dimensions of the rod and sensitivity by cryogenic temperatures. Also, we can use a Michelson interferometer, which is used to detect gravitational waves. In the case of gravitational waves, a vacuum is used, where mirrors are directly located, which are deflected in the case of detection. In the case of infrasound, we can use, for example, not vacuum, but air pressure – infra-waves. To weed out the input stream, we can use sophisticated methods or huge mirrors (cut, which is weakly influenced by high-frequency acoustic waves, but sensitive to low-frequency ones).

We will not exclude more exotic assumptions that it is gravitational waves that are the source of infra-waves.

Let us present a model for detecting ultra-low-frequency infra-waves. This means we need very large detectors for high sensitivity and low temperatures with additional seismic noise isolation systems. Let us take these resonators and place them in places located at equal distances from each other. Depending on the number of such resonators, we will have different shapes described by them – a line, a triangle, a square, and so on. If one of the detectors detects an infra-wave, then after re-radiation it will propagate in all directions, which will be captured by other resonators. They, in turn, will also re-emit and so on. In the end, analyzing the data from all the detectors, we will get something like a Chladni figure, and we will be able to conclude whether infra-waves form baric fields, and whether they have enough power.

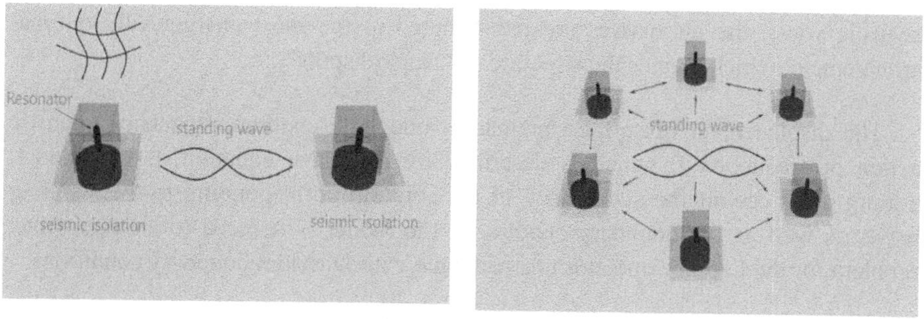

Figure 9.8. *Schematic variants of different topology of infrasound sensors network. For a color version of this figure, see www.iste.co.uk/koroliouk/computational.zip*

Obviously, with an increase in the number of resonators, the probability of fixing an infra-wave increases. This is due to the fact that after a series of reflections and re-radiation, the detectors on the resonators will record the signal of the standing wave amplitude amplified several times.

9.6. Conclusion

Meteorological phenomena are known to manifest themselves as local "leopard spots".

Infrasound acoustic waves are also known to influence the formation of downpours, tornadoes and other weather phenomena.

Our work aims to simulate the formation of such "leopard spots", as well as the role of infrasonic acoustic influences in this mechanism.

The formation of local inhomogeneities of baric scalar fields is of paramount importance in the problem of weather phenomena studies. We propose a new mathematical model and simulation of the formation of local baric fields as a numerical experiment.

The model describes the formation of locally inhomogeneous baric fields on a surface scale of the order of 100–10,000 sq. Km, which, due to the size of the globe, is a flat problem.

Our constructions are based on the principle of creating Chladni figures (Rossing 2005; Muller 2013), which consists of the fact that under the influence of a standing acoustic wave, the air masses are redistributed to the most energetically optimal arrangement, which creates the structure of "leopard spots".

The displacement of surface air masses under the external frequency acoustic action of frequency f, which is described by the wave equation, is considered. Taking into account the stationarity of the solution corresponding to the standing wave, as well as the boundary conditions, the model is reduced to an eigenvalue problem for the Laplace operator in a rectangle with Dirichlet boundary conditions.

Numerical modeling based on the proposed model, at different values of its parameters, illustrates the creation of figures of redistribution of air masses, which are very reminiscent of local configurations of baric fields, actually built on the basis of meteorological observations.

An empirical estimate of the frequency of an acting acoustic wave gives a value of the order of 0.01 Hz, that is, a value close to the lower limit of the infrasonic range.

The problem arises of determining the characteristics and monitoring the infrasonic acoustic waves and their correlations with the formation of local weather phenomena. Several models are proposed to detect ultra-low-frequency waves to create infrasonic observatories.

The physical phenomenon of Chladni's figures is described in the works of Chladni (1787), Faraday (1831), Rossing (1982), Comer et al. (2004). For acoustic waves, see Waller (1961), Rossing (2005), Muller (2013), and for electromagnetic waves, see Zozyuk et al. (2019). The fundamentals of mathematical and numerical simulation of Chladni figures are given in Kac (1966), Schwarz (1991), Driscoll (1997), Tikhonov et al. (1999). The statistical methods of experimental data processing are presented in Koroliouk (2015); Koroliouk et al. (2018). The various aspects and models of baric fields and their interactions are described in (Wilcox et al. (1974), Speranza (1975), Gabis and Troshichev (2000), Karakhanyan et al. (2011), Lam and Tinsley (2015), Veretenenko et al. (2018), Vyklyuk et al. (2018) and Tsurkis (2021). The presentation of infrasound measurements are reported in Garcés et al. (1998), Hedlin and Walker (2013), Liashchuk (2015) and Lyashchuk et al. (2015). Finally, some methods of gravitational wave detections are shown in Bassan (1994), Baggio et al. (2008), Punturo et al. (2010), Nappo (2012) and Canuel et al. (2018).

9.7. References

Baggio, L., Bignotto, M., Bonaldi, M., Camarda, M., Cerdonio, M., Conti, L., De Rosa, M., Falferi, P., Fattori, S., Fortini, P. et al. (2008). A joint search for gravitational wave bursts with AURIGA and LIGO. *AURIGA Collaboration, LIGO Scientific Collaboration*. arXiv:0710.0497 [gr-qc].

Bassan, M. (1994). Resonant gravitational wave detectors: A progress report. *Class. Quantum Grav.*, 11 A39 [Online]. Available at: http//https://iopscience.iop.org/article/10.1088/0264-9381/11/6A/004/pdf.

Canuel, B., Bertoldi, A., Amand, L., Pozzo di Borgo, E., Chantrait, T., Danquigny, C., Dovale Álvarez, M., Fang, B., Freise, A., Geiger, R. et al. (2018). Exploring gravity with the MIGA large scale atom interferometer. *Sci Rep.*, 8, 14064. doi: 10.1038/s41598-018-32165-z.

Chladni, E.F.F. (1787). *Entdeckungen über die Theorie des Klanges*. Breitkopf und Härtel, Leipzig.

Comer, J.R., Shepard, M.J., Henriksen, P.N., Ramsier, R.D. (2004). Chladni plates revisited. *Am. J. Phys.*, 72(10), 1345–1346.

Driscoll, T. (1997). Eigenmodes of isospectral drums. *SIAM Rev.*, 39(1), 1–17. doi: 10.1137/S0036144595285069.

Faraday, M. (1831). On a peculiar class of acoustical figures; and on certain forms assumed by groups of particles upon vibrating elastic surfaces. *Phil. Trans. Roy. Soc. Lond.*, 121, 299–340.

Gabis, I.P. and Troshichev, O.A. (2000). Influence of short-term changes in solar activity on baric field perturbations in the stratosphere and troposphere. *J. Atmos. Solar-Terrestrial Phy.*, 62(9), 725–735. doi: 10.1016/S1364-6826(00)00063-8.

Garcés, M.A., Hansen, R.A., Lindquist, K.G. (1998). Traveltimes for infrasonic waves propagating in a stratified atmosphere. *Geophysical Journal International*, 135(1), 255–263. doi: 10.1046/j.1365-246X.1998.00618.x.

Hedlin, M. and Walker, K. (2013). A study of infrasonic anisotropy and multipathing in the atmosphere using seismic networks. *Royal Soc.* doi: 10.1098/rsta.2011.0542.

Kac, M. (1966). Can one hear the shape of a drum? *Am. Math. Monthly*, 73(4), 1–23. doi: 10.2307/2313748.

Karakhanyan, A.A., Zherebtsov, G.A., Kovalenko, V.A., Molodykh, S.I. (2011). Role of atmospheric circulation in the baric field response to the Earth's crossing of the interplanetary magnetic field sector boundaries. *Geomagn. Aeron.*, 51, 1155–1159. doi: 10.1134/S0016793211080214.

Koroliouk, D. (2015). Binary statistical experiments with persistent nonlinear regression. *Theor. Probability Math. Statist., AMS*, 91, 71–80. doi: 10.1090/tpms/967.

Koroliouk, D. and Koroliuk, V.S. (2018). Filtration of stationary Gaussian statistical experiments. *J. Math. Sci.*, 229(1), 30–35. doi: 10.1007/s10958-018-3660-0.

Lam, M.M. and Tinsley, B.A. (2016). Solar wind-atmospheric electricity-cloud microphysics connections to weather and climate. *J. Atmos. Solar-Terres. Phys.* doi: 10.1016/j.jastp.2015.10.019.

Liashchuk, O.I. (2015). Using data of infrasound measurements in Ukraine for explosions and earthquakes identification. *JGD*, 18(1), 36–44. doi: 10.23939/jgd2015.01.036.

Lyashchuk, A., Andrushchenko, Y., Gordienko, Y., Karyagin, E., Kornienko, I. (2015). Possibility of the use of data of infrasonic monitoring for identification of the nature of seismic events. *Geofizicheskiy Zhurnal*, 37(6), 105–114. doi: 10.24028/gzh.0203-3100.v37i6.2015.111177.

Muller, T. (2013). Numerical Chladni figures. arXiv:1308.5523v1 [physics.ed-ph].

Nappo, C.J. (ed.) (2012). *An Introduction to Atmospheric Gravity Waves*. Elsevier, Amsterdam.

Punturo, M., Abernathy, M., Acernese, F., Allen, B., Andersson, N., Arun, K., Barone, F., Barr, B., Barsuglia, M., Beker, M. et al. (2010). The third generation of gravitational wave observations and their science reach. *Classical and Quantum Gravity*, 27(8).

Rossing, T.D. (2005). Comment on "Chladni plates revisited," by J.R. Comer, M.J. Shepard, P.N. Henriksen, and R.D. Ramsier [Am. J. Phys., 72(10), 1345–1346 (2004)]. *Am. J. Phys.*, 73(3), 283–283.

Schwarz, H.R. (1991). *Methode der finiten Elemente*. B.G. Teubner, Stuttgart.

Speranza, A. (1975). The formation of baric depressions near the Alps. *Ann. Geophys.*, 28(2–3), 177–221. doi: 10.4401/ag-4898.

Thomas, D.R. (1982). Chladni's law for vibrating plates. *Am. J. Phys.*, 50(3), 271–274.

Tikhonov, A.N. and Samarskiy, A.A. (1999). *Equations of Mathematical Physics*, 6th revised edition. Publishing House of Moscow State University, Moscow.

Tsurkis, I.Y. (2021). Two components of the baric tilt of elements of the earth surface: The case of a low-angle topography. *Seism. Instr.*, 57, 17–26. doi: 10.3103/S0747923921010102.

Veretenenko, S., Ogurtsov, M., Lindholm, M., Jalkanen, R. (2018). Galactic cosmic rays and low clouds: Possible reasons for correlation reversal. In *Cosmic Rays*, Szadkowski, Z. (ed.). InTech, London.

Vyklyuk, Y., Radovanović, M., Milovanovic, B., Leko, T., Milenkovic, M., Milošević, Z., Pešić, A.M., Jakovljević, D. (2018). Hurricane genesis modelling based on the relationship between solar activity and hurricanes II. *J. Atmos. Solar-Terres. Phys.*, 180, 159–164. doi: 10.1016/j.jastp.2017.09.008.

Waller, M.D. (1961). *Chladni Figures*. G. Bell, London.

Wilcox, J.M., Scherrer, P.H., Svalgaard, L., Roberts, W.O., Olson, R.H., Jenne, R.L. (1974). Influence of solar magnetic sector structure on terrestrial atmospheric vorticity. *J. Atmos. Sci.*, 31(2), 581–588 [Online]. Available at: https://journals.ametsoc.org/view/journals/atsc/31/2/1520-0469_1974_031_0581_iosmss_2_0_co_2.xml.

Zozyuk, M., Koroliouk, D., Moskaliuk, V., Yurikov, A., Yakymenko, Y. (2019). Creation of quasiperiodic surfaces under the action of vibrating dielectric matrices. *ELNANO-2020*, 224–229. doi: 10.1109/ELNANO50318.2020.9088821.

10

Simulation of the Electron–Hole Plasma State by Perturbation Theory Methods

Andrii BOMBA[1], Sergiy LYASHKO[2] and Ihor MOROZ[1]

[1] *National University of Water and Environmental Engineering, Rivne, Ukraine*

[2] *Taras Shevchenko National University of Kyiv, Ukraine*

10.1. Introduction. Nonlinear boundary value problems of the p–i–n diodes theory

P–i–n structures (plasma diodes) are nonlinear active elements of electronic devices, which are widely used for switching the microwave electromagnetic field (Figure 10.1). The operation of such a device is based on a change in the conductivity of the diode active region (i-region) by the formation of an electron–hole plasma under the action of a control current. The charge carriers enter the active region due to the injection process from the highly doped p- and n-regions. In the active region, diffusion and drift of electrons and holes occur, their concentration changes due to generation–recombination processes. The basic characteristics of the p–i–n diode – the extinction introduced into the transmission line, the switching speed – determine the charge accumulated in the i-region and the geometry of the active region of the diode (e.g. Kwok 2002; Sze and Kwok 2006).

Computational Methods and Mathematical Modeling in Cyberphysics and Engineering Applications 1,
coordinated by Dmitri KOROLIOUK, Sergiy LYASHKO and Nikolaos LIMNIOS.

To assess the integral characteristics of p–i–n structures (conductivity, capacitance, inductance), the method of equivalent circuits is widely used (e.g. Grimalsky et al. 1994; Lebedev et al. 1998). The diffusion–drift mathematical models (e.g. Haine and Rose 1972; Adirovich et al. 1978) are more informative, and make it possible to obtain the basic characteristics of the diode based on the study (within the framework of the hydrodynamic approximation) of the process of the flow of the electron–hole current through the device, taking into account the physical properties of the semiconductor material, generation effects, warming up, etc.

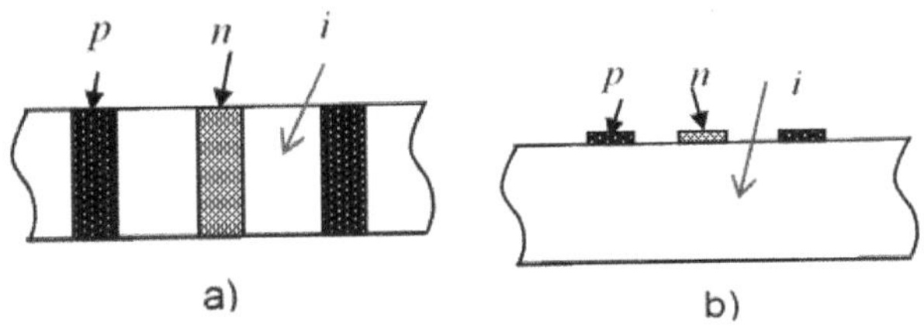

Figure 10.1. *Schematic representation of a cross-section of a diode assembly (a) and a surface-oriented structure (b)*

In the general case, the p–i–n diode mathematical model in the hydrodynamic approximation is supplied by a nonlinear system of partial differential equations, which are supplemented by the corresponding boundary conditions.

The basic mathematical model describing the distribution of the concentration of holes (n), electrons (p) and electrical potential (φ) in the active region of p–i–n diodes Ω (Figure 10.2) in the stationary mode of device operation includes the equations of continuity of holes, electrons and Poisson's equation (e.g. Adirovich et al. 1978; Bonch-Bruevich and Kalashnikov 1982; Sze and Kwok 2006):

$$\begin{cases} \Delta n = \nabla n \nabla \varphi + n \Delta \varphi + \dfrac{1}{D_n \tau_n^*} n, \\ \Delta p = -\nabla p \nabla \varphi - p \Delta \varphi + \dfrac{1}{D_p \tau_p^*} p, \\ \dfrac{\varepsilon \varepsilon_0}{e} \Delta \varphi = -(p - n + N_I), \end{cases} \qquad [10.1]$$

where τ_p^*, τ_n^* are characteristic relaxation holes and electrons lifetimes in the i-region (in the general case, they depend on the local electrons and holes concentration and are determined by a certain mechanism of charge carrier recombination); $N_l = N_a - N_d \equiv N_l(r)$ is the given function of the doping profile (and describes the difference between the concentrations of acceptors and donors in the active region); D_p and D_n are the diffusion coefficients of holes and electrons, respectively; e is the electron's charge; ε, ε_0 are the relative dielectric constant of the semiconductor and dielectric constant, respectively.

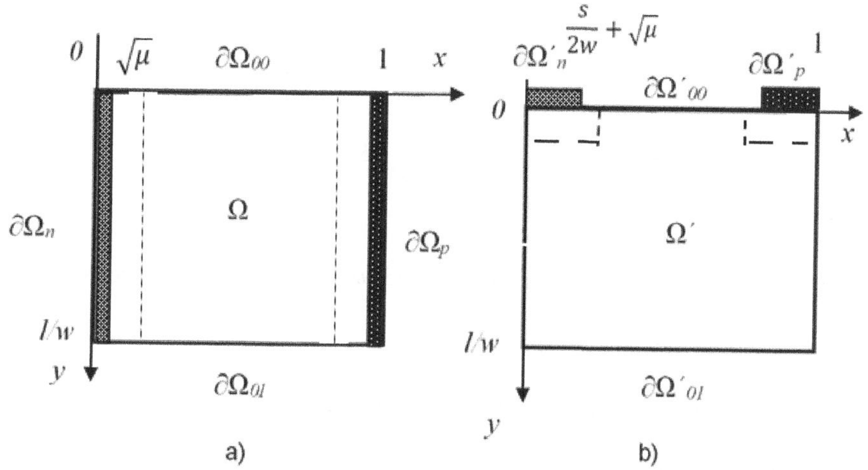

Figure 10.2. *Schematic representation of the active region elements of the diode assembly (a) and the surface-oriented structure (b)*

The boundary conditions at the injection contacts (surfaces ∂Ωn, ∂Ωp) are written, taking into account the assumption that only electrons carry current through the boundary ∂Ωn, and, respectively, through ∂Ωp – only holes (with a wide active region, most of the injected charge carriers recombine into the i region and do not have time to get into the area of opposite contact) (e.g. Adirovich et al. 1978; Kwok 2002). The surface recombination at the contacts is also taken into account:

$$\left(\vec{j}_n \cdot \vec{v}\right)\Big|_{\partial\Omega_n} = J - J_{nr}, \; \left(\vec{j}_p \cdot \vec{v}\right)\Big|_{\partial\Omega_n} = 0, \qquad [10.2]$$

$$\left(\vec{j}_p \cdot \vec{v}\right)\Big|_{\partial\Omega_p} = J - J_{pr}, \; \left(\vec{j}_n \cdot \vec{v}\right)\Big|_{\partial\Omega_p} = 0,$$

where J is the constant, which determines the injection current density (control current); $\vec{\nu}$ is the vector normal to the boundary of the region; $\vec{j}_{n,p}$ is the current density of electrons and holes, respectively; $J_{rn,rp}$ is the recombination current density.

Taking into account the ratio $\vec{j}_p = e\mu_p p\vec{E} - eD_p\vec{\nabla}p$, $\vec{j}_n = e\mu_n n\vec{E} + eD_n\vec{\nabla}n$, $J_{rp} = e\alpha_p p$, $J_{rn} = e\alpha_n n$, conditions [10.2] are given in the following form:

$$\left.\frac{\mu_n}{D_n} nE_\nu + \frac{\partial n}{\partial \nu} - 2\gamma_n n\right|_{\partial\Omega_n} = -\frac{J}{eD_n}, \quad \left.\frac{\mu_p}{D_p} pE_\nu - \frac{\partial p}{\partial \nu}\right|_{\partial\Omega_n} = 0, \qquad [10.3]$$

$$\left.-\frac{\mu_p}{D_p} pE_\nu + \frac{\partial p}{\partial \nu} - 2\gamma_p p\right|_{\partial\Omega_p} = \frac{J}{eD_p}, \quad \left.\frac{\mu_n}{D_n} nE_\nu + \frac{\partial n}{\partial \nu}\right|_{\partial\Omega_p} = 0, \qquad [10.4]$$

where $\vec{E} = -\nabla\varphi$, $\gamma_n = \frac{\alpha_n}{2D_n}$, $\gamma_p = \frac{\alpha_p}{2D_p}$ and $\alpha_{n,p}$ are the surface recombination rates of electrons and holes, respectively.

The potential function must satisfy the following boundary conditions:

$$\varphi|_{\partial\Omega_n} = 0, \quad \varphi|_{\partial\Omega_p} = U, \qquad [10.5]$$

where U is the constant, which determines the voltage on the p–i–n structure. It is assumed that the voltage is mainly applied to the high-resistance section of the p–i–n structure; the voltage drop on the injecting and ohmic contacts is neglected.

The boundary conditions for the electric potential and holes and electrons concentrations on the p–i–n diode lateral surface are determined in a similar way (the rate of surface recombination at the contacts differs from the rate of surface recombination at the lateral surface):

$$\left.\frac{\partial \varphi}{\partial \nu}\right|_{\partial\Omega_0} = 0,$$

$$\left.\frac{\mu_n}{D_n} nE_\nu + \frac{\partial n}{\partial \nu} - 2\gamma_n^* n\right|_{\partial\Omega_{00}} = 0, \quad \left.\frac{\mu_p}{D_p} pE_\nu - \frac{\partial p}{\partial \nu} - 2\gamma_p^* p\right|_{\partial\Omega_{00}} = 0, \qquad [10.6]$$

$$-\frac{\mu_n}{D_n} n E_\nu - \frac{\partial n}{\partial \nu} - 2\gamma_n^* n \bigg|_{\partial\Omega_{01}} = 0, \quad -\frac{\mu_p}{D_p} p E_\nu + \frac{\partial p}{\partial \nu} - 2\gamma_p^* p \bigg|_{\partial\Omega_{01}} = 0.$$

Note that the stationary mathematical model of the p–i–n diode [10.1]–[10.6] is one of the simplest and most widely used models, but it does not take into account the generational, thermal effects in the semiconductor volume, or the peculiarities of the injection processes at the contacts. It is obvious that the search for a solution of even a simplified system of nonlinear equations [10.1] with boundary conditions [10.3]–[10.6] is a complex mathematical problem.

The solution to problems [10.1]–[10.6] is obtained by numerical methods, or methods of linearization of the system of equations [10.1] are sought. For example, in analytical modeling, the approximation of ambipolar diffusion of charge carriers (e.g. Adirovich et al. 1978; Bonch-Bruevich and Kalashnikov 1982; Sze and Kwok 2006) is used, within which the approximate equality of the electron and hole concentrations in the diode active region is assumed. In this case, the system of equations [10.1] is reduced to a linear diffusion equation, the solution of which can be easily found in an analytical form. Obviously, the condition of equality of concentrations is violated near the injecting contacts and the result obtained ultimately does not provide the proper level of model adequacy.

It is also known that asymptotic methods have been developed and widely used to solve nonlinear equations of mathematical physics (e.g. Tikhonov 1952; Vishik and Lusternik 1957; Bomba 1982; Smith 1985; Vasil'eva et al. 1995; Bomba et al. 2017). In particular, these methods are used to simulate processes in semiconductor p–n diodes (e.g. Vasil'eva and Stel'makh 1977; Belyanin 1986). The formulation of the problems [10.1]–[10.6] differs from those illuminated in the literature by the geometry of the region (p–i–n diodes are semiconductor diodes with a wide base, and in p–n diodes there is a sharp junction between regions with different types of conductivity), boundary conditions. The authors are not aware of the application of asymptotic approaches to calculate the p–i–n diode characteristics.

10.2. Construction of an asymptotic solution of a boundary value problem for the system of the charge carrier current continuity equations and the Poisson equation

It is convenient to analyze the system of partial differential equations [10.1] in a normalized form. It is also proposed to consider a two-dimensional spatial model of a p–i–n structure element (e.g. the region $\Omega = \{(x,y): 0 < x < w, 0 < y < l\}$;

Figure 10.2(a)), since its linear dimensions along the OZ axis significantly exceed others. Let us introduce into consideration the dimensionless quantities $\tilde{x}=\frac{x}{w}$ $(0<\tilde{x}<1)$, $\tilde{y}=\frac{y}{w}$ $(0<\tilde{y}<\frac{l}{w})$, $\tilde{\varphi}=\frac{e\varphi}{kT}$, $\tilde{U}=\frac{eU}{kT}$, $\tilde{n}=\frac{n}{N_i}$ $(0\le\tilde{n}\le\frac{n_{\max}}{N_i})$, $\tilde{p}=\frac{p}{N_i}$ $(0\le\tilde{p}\le\frac{p_{\max}}{N_i})$ and $\tilde{N}_d=\frac{N_d}{N_i}$, where N_i is the constant, which determines the concentration of electrons in its own semiconductor, and depends on the selected semiconductor material; T is the constant, which determines the temperature (300 °K); k is the Boltzmann constant. We obtain a nonlinear system of equations in the following form (in order to simplify notation, the "~" sign is omitted in the following calculations):

$$\mu\Delta\varphi=-(p-n+N_l), \tag{10.7}$$

$$\Delta n=\nabla n\nabla\varphi+n\Delta\varphi+A_n n, \tag{10.8}$$

$$\Delta p=-\nabla p\nabla\varphi-p\Delta\varphi+A_p p. \tag{10.9}$$

In [10.7]–[10.9], the following notation is used: $\mu=\frac{\varepsilon\varepsilon_0 kT}{e^2w^2N_i}$, $A_n=\frac{w^2}{D_n\tau_n^*}$ and $A_p=\frac{w^2}{D_p\tau_p^*}$. Estimation of the quantity μ gives the value ~10^{-8}.

In this case, conditions [10.3]–[10.6] take the form

$$nE_v+\frac{1}{w}\frac{\partial n}{\partial v}-2\gamma_n n\bigg|_{\partial\Omega_n}=-\frac{J}{eD_nN_i},\quad pE_v-\frac{1}{w}\frac{\partial p}{\partial v}\bigg|_{\partial\Omega_n}=0, \tag{10.10}$$

$$-pE_v+\frac{1}{w}\frac{\partial p}{\partial v}-2\gamma_p p\bigg|_{\partial\Omega_p}=\frac{J}{eD_pN_i},\quad nE_v+\frac{1}{w}\frac{\partial n}{\partial v}\bigg|_{\partial\Omega_p}=0,$$

$$nE_v+\frac{1}{w}\frac{\partial n}{\partial v}-2\gamma_n^* n\bigg|_{\partial\Omega_{00}}=0,\quad pE_v-\frac{1}{w}\frac{\partial p}{\partial v}-2\gamma_p^* p\bigg|_{\partial\Omega_{00}}=0,$$

$$-nE_v-\frac{1}{w}\frac{\partial n}{\partial v}-2\gamma_n^* n\bigg|_{\partial\Omega_{01}}=0,\quad -pE_v+\frac{1}{w}\frac{\partial p}{\partial v}-2\gamma_p^* p\bigg|_{\partial\Omega_{01}}=0,$$

$$\varphi\big|_{\partial\Omega_n} = 0\,,\ \varphi\big|_{\partial\Omega_p} = U\,,\ \frac{\partial\varphi}{\partial v}\bigg|_{\partial\Omega_0} = 0\,.$$

When describing the doping profile $N_l(x,y)$, let us consider the case of a sharp boundary between the doped regions and the region of the intrinsic semiconductor:

$$N_a(x,y) = \begin{cases} 1, & (x,y)\in\Omega \\ \dfrac{N_{a0}}{N_i}, & (x,y)\in\partial\Omega_p \end{cases};\ N_d(x,y) = \begin{cases} 1, & (x,y)\in\Omega \\ \dfrac{N_{d0}}{N_i}, & (x,y)\in\partial\Omega_n \end{cases}. \qquad [10.11]$$

Let us apply the boundary functions method (e.g. Tikhonov 1952; Vishik and Lusternik 1957; Bomba 1982; Smith 1985; Vasil'eva et al. 1995; Bomba et al. 2017) to analyze problems [10.7]–[10.10]. A feature of this problem is that (in comparison with the formulations of problems (e.g. Vasil'eva and Stel'makh 1977; Belyanin 1986)) the proposed mathematical model [10.7]–[10.10] describes the flow of diffusion–drift currents of unbalanced charge carriers in semiconductor diodes with a wide base (there is no sharp boundary between the regions semiconductors of n- and p-types of conductivity). In p–i–n structures, there are two contact zones (n-i and p-i) for the different types semiconductors.

Based on the problem statement and taking into account that the structure of the solution is mainly influenced by the conditions at the contact sections ∂Ωn and ∂Ωp, through which the i-region is filled with non-basic charge carriers, we propose to look for the solution in the following asymptotic series:

$$,\ \varphi = \varphi(x,y,\mu) = \Phi_{(m)}(x,y,\mu) + \underline{\Phi}_{(m)}(\underline{\xi},\mu) + \overline{\Phi}_{(m)}(\overline{\xi},\mu) + R_{\varphi(m)}(x,y,\mu) =$$

$$= \sum_{i=0}^{m}\mu^i\varphi_i(x,y) + \sum_{i=0}^{m}\mu^i\underline{\Phi}_i(\underline{\xi}) + \sum_{i=0}^{m}\mu^i\overline{\Phi}_i(\overline{\xi}) + R_{\varphi(m)}(x,y,\mu), \qquad [10.12]$$

$$n = n(x,y,\mu) = N_{(m)}(x,y,\mu) + \underline{N}_{(m)}(\underline{\xi},\mu) + \overline{N}_{(m)}(\overline{\xi},\mu) + R_{n(m)}(x,y,\mu) =$$

$$= \sum_{i=0}^{m}\mu^i n_i(x,y) + \sum_{i=0}^{m}\mu^i\underline{N}_i(\underline{\xi}) + \sum_{i=0}^{m}\mu^i\overline{N}_i(\overline{\xi}) + R_{n(m)}(x,y,\mu),$$

$$p = p(x,y,\mu) = P_{(m)}(x,y,\mu) + \underline{P}_{(m)}(\underline{\xi},\mu) + \overline{P}_{(m)}(\overline{\xi},\mu) + R_{p(m)}(x,y,\mu) =$$

$$=\sum_{i=0}^{m}\mu^i p_i(x,y)+\sum_{i=0}^{m}\mu^i \underline{P}_i(\underline{\xi})+\sum_{i=0}^{m}\mu^i \overline{P}_i(\overline{\xi})+R_{p(m)}(x,y,\mu)$$

where $\Phi_{(m)}(x,y,\mu)$, $N_{(m)}(x,y,\mu)$, $P_{(m)}(x,y,\mu)$ are the regular part of the asymptotics; $\underline{\Phi}_{(m)}(\underline{\xi},\mu)$, $\underline{N}_{(m)}(\underline{\xi},\mu)$, $\underline{P}_{(m)}(\underline{\xi},\mu)$, $\overline{\Phi}_{(m)}(\overline{\xi},\mu)$, $\overline{N}_{(m)}(\overline{\xi},\mu)$, $\overline{P}_{(m)}(\overline{\xi},\mu)$ are the near-boundary corrections of the asymptotics, respectively, in the vicinity of the points $x=0$ and $x=l$ ($\underline{\xi}=\dfrac{x}{\sqrt{\mu}}$ and $\overline{\xi}=\dfrac{1-x}{\sqrt{\mu}}$ are the corresponding regularizing stretches); $R_{\varphi(m)}(x,y,\mu)$, $R_{n(m)}(x,y,\mu)$, $R_{p(m)}(x,y,\mu)$ are the residual members.

Substituting [10.12] into equations [10.7]–[10.9] and conditions [10.10] and applying the standard "equalization" procedure (e.g. Tikhonov 1952; Vishik and Lusternik 1957; Bomba 1982; Smith 1985; Vasil'eva et al. 1995; Bomba et al. 2017), we form the following sequence of tasks.

The principal terms of regular series [10.12] satisfy the system of equations:

$$n_0 = p_0, \qquad [10.13]$$

$$\Delta n_0-\nabla(n_0\nabla\varphi_0)-A_n n_0=0,$$

$$\Delta p_0+\nabla(p_0\nabla\varphi_0)-A_p p_0=0.$$

Zero-order boundary functions must satisfy the systems of ordinary differential equations [10.14] and [10.15] (the variable y enters the equations as a parameter):

$$\frac{\partial^2\underline{\Phi}_0}{\partial\underline{\xi}^2}=-(\underline{P}_0-\underline{N}_0), \qquad [10.14]$$

$$\frac{\partial^2\underline{N}_0}{\partial\underline{\xi}^2}-\frac{\partial}{\partial\underline{\xi}}\left(\underline{N}_0\frac{\partial\underline{\Phi}_0}{\partial\underline{\xi}}\right)=0,$$

$$\frac{\partial^2\underline{P}_0}{\partial\underline{\xi}^2}+\frac{\partial}{\partial\underline{\xi}}\left(\underline{P}_0\frac{\partial\underline{\Phi}_0}{\partial\underline{\xi}}\right)=0.$$

$$\frac{\partial^2 \overline{\Phi}_0}{\partial \overline{\xi}^2} = -\left(\overline{P}_0 - \overline{N}_0\right), \qquad [10.15]$$

$$\frac{\partial^2 \overline{N}_0}{\partial \overline{\xi}^2} - \frac{\partial}{\partial \overline{\xi}}\left(\overline{N}_0 \frac{\partial \overline{\Phi}_0}{\partial \overline{\xi}}\right) = 0,$$

$$\frac{\partial^2 \overline{P}_0}{\partial \overline{\xi}^2} + \frac{\partial}{\partial \overline{\xi}}\left(\overline{P}_0 \frac{\partial \overline{\Phi}_0}{\partial \overline{\xi}}\right) = 0.$$

In this case, the boundary conditions [10.10] for the system of equations [10.13]–[10.15] take the following form:

$$\left.\frac{\partial n_0}{\partial x} - \gamma_n w n_0 - \gamma_n w \underline{N}_0\right|_{x=0} = -\frac{J}{2eD_n}\frac{w}{N_i}, \qquad [10.16]$$

$$\left.\frac{\partial \underline{N}_0}{\partial \underline{\xi}}\right|_{\underline{\xi}=0} = 0, \quad \left.\frac{\partial \underline{P}_0}{\partial \underline{\xi}}\right|_{\underline{\xi}=0} = 0, \quad \varphi_0 + \underline{\Phi}_0\big|_{x=0} = 0,$$

$$\left.\frac{\partial n_0}{\partial x} - \gamma_p w n_0 - \gamma_p w \overline{P}_0\right|_{x=1} = \frac{J}{2eD_p}\frac{w}{N_i},$$

$$\left.\frac{\partial \overline{P}_0}{\partial \overline{\xi}}\right|_{\overline{\xi}=0} = 0, \quad \left.\frac{\partial \overline{N}_0}{\partial \overline{\xi}}\right|_{\overline{\xi}=0} = 0, \quad \varphi_0 + \overline{\Phi}_0\big|_{x=1} = U,$$

$$\left.\frac{\partial n_0}{\partial y} - \beta w n_0\right|_{y=0} = 0, \quad \left.\frac{\partial n_0}{\partial y} + \beta w n_0\right|_{y=\frac{l}{w}} = 0, \quad \left.\frac{\partial \varphi_0}{\partial y}\right|_{y=0,\, y=\frac{l}{w}} = 0 \;\; \left(\beta = \left(\gamma_p^* + \gamma_n^*\right)\right).$$

Note, that, as a result of the system of equations [10.13], we have a linear differential equation $\Delta n_0 - \frac{1}{2}\left(A_n + A_p\right) n_0 = 0$, which is also obtained within the ambipolar diffusion approximation (e.g. Bonch-Bruevich and Kalashnikov 1982) (the approximation of particle motion in a self-consistent field).

The next stages of the search for the asymptotic expansion terms are based on the sequential solution of the equation systems similar to [10.13]–[10.15] with the corresponding boundary conditions.

For example, at the second stage, we are looking for solutions to such systems of equations:

$$\Delta\varphi_0 = -(p_1 - n_1), \qquad [10.17]$$

$$\Delta n_1 - \nabla(n_1 \nabla\varphi_0) - \nabla(n_0 \nabla\varphi_1) - A_n n_1 = 0,$$

$$\Delta p_1 + \nabla(p_1 \nabla\varphi_0) + \nabla(p_0 \nabla\varphi_1) - A_p p_1 = 0;$$

$$\frac{\partial^2 \underline{\Phi}_1}{\partial \underline{\xi}^2} = -(\underline{P}_1 - \underline{N}_1), \qquad [10.18]$$

$$\frac{\partial^2 \underline{N}_1}{\partial \underline{\xi}^2} - \frac{\partial}{\partial \underline{\xi}}\left(\underline{N}_1 \frac{\partial \underline{\Phi}_0}{\partial \underline{\xi}}\right) - \frac{\partial}{\partial \underline{\xi}}\left(\underline{N}_0 \frac{\partial \underline{\Phi}_1}{\partial \underline{\xi}}\right) = 0,$$

$$\frac{\partial^2 \underline{P}_1}{\partial \underline{\xi}^2} + \frac{\partial}{\partial \underline{\xi}}\left(\underline{P}_1 \frac{\partial \underline{\Phi}_0}{\partial \underline{\xi}}\right) + \frac{\partial}{\partial \underline{\xi}}\left(\underline{P}_0 \frac{\partial \underline{\Phi}_1}{\partial \underline{\xi}}\right) = 0;$$

$$\frac{\partial^2 \overline{\Phi}_1}{\partial \overline{\xi}^2} = -(\overline{P}_1 - \overline{N}_1), \qquad [10.19]$$

$$\frac{\partial^2 \overline{N}_1}{\partial \overline{\xi}^2} - \frac{\partial}{\partial \overline{\xi}}\left(\overline{N}_1 \frac{\partial \overline{\Phi}_0}{\partial \overline{\xi}}\right) - \frac{\partial}{\partial \overline{\xi}}\left(\overline{N}_0 \frac{\partial \overline{\Phi}_1}{\partial \overline{\xi}}\right) = 0,$$

$$\frac{\partial^2 \overline{P}_0}{\partial \overline{\xi}^2} + \frac{\partial}{\partial \overline{\xi}}\left(\overline{P}_1 \frac{\partial \overline{\Phi}_0}{\partial \overline{\xi}}\right) + \frac{\partial}{\partial \overline{\xi}}\left(\overline{P}_0 \frac{\partial \overline{\Phi}_1}{\partial \overline{\xi}}\right) = 0.$$

The systems of equations [10.17]–[10.19] are supplemented by the following boundary conditions (which are similar to [10.16]):

$$\left.\frac{\partial n_1}{\partial x} - 2\gamma_n w n_1 - 2\gamma_n w \underline{N}_1\right|_{x=0} = 0, \quad \left.\frac{\partial \underline{N}_1}{\partial \underline{\xi}}\right|_{\underline{\xi}=0} = 0, \qquad [10.20]$$

$$\left.\frac{\partial p_1}{\partial x}\right|_{x=0}=0\,,\ \left.\frac{\partial \underline{P}_1}{\partial \underline{\xi}}\right|_{\underline{\xi}=0}=0\,,\ \varphi_1+\underline{\Phi}_1\big|_{x=0}=0\,,$$

$$\left.\frac{\partial p_1}{\partial x}-2\gamma_p wp_1-2\gamma_p w\overline{P}_1\right|_{x=1}=0\,,\ \left.\frac{\partial \overline{P}_1}{\partial \overline{\xi}}\right|_{\overline{\xi}=0}=0\,,$$

$$\left.\frac{\partial n_1}{\partial x}\right|_{x=1}=0\,,\ \left.\frac{\partial \overline{N}_1}{\partial \overline{\xi}}\right|_{\overline{\xi}=0}=0\,,\ \varphi_1+\overline{\Phi}_1\big|_{x=1}=0\,,$$

$$\left.\frac{\partial n_1}{\partial y}\mp\gamma_n^* wn_1\right|_{y=0,\,y=\frac{l}{w}}=0\,,\ \left.\frac{\partial p_1}{\partial y}\mp\gamma_p^* wp_1\right|_{y=0,\,y=\frac{l}{w}}=0\,,\ \left.\frac{\partial \varphi_1}{\partial y}\right|_{y=0,\,y=\frac{l}{w}}=0\,.$$

We also use additional conditions characteristic of border functions (e.g. Tikhonov 1952; Vishik and Lusternik 1957; Bomba 1982; Smith 1985; Vasil'eva et al. 1995; Bomba et al. 2017):

$$\lim_{\underline{\xi}\to\infty}\underline{N}_i\left(\underline{\xi}\right)=\lim_{\underline{\xi}\to\infty}\underline{P}_i\left(\underline{\xi}\right)=\lim_{\overline{\xi}\to\infty}\overline{N}_i\left(\overline{\xi}\right)=\lim_{\overline{\xi}\to\infty}\overline{P}_i\left(\overline{\xi}\right)=0\,,$$

$$\lim_{\underline{\xi}\to\infty}\underline{\Phi}_i\left(\underline{\xi}\right)=\lim_{\overline{\xi}\to\infty}\overline{\Phi}_i\left(\overline{\xi}\right). \qquad [10.21]$$

Using considerations similar to those presented elsewhere (e.g. Tikhonov 1952; Vishik and Lusternik 1957; Bomba 1982; Smith 1985; Vasil'eva et al. 1995; Bomba et al. 2017) on additional smoothness and consistency conditions, we arrive at estimates for the remainder of the asymptotic series:

$$R_{\varphi(m)}\left(x,y,\mu\right)=O\left(\mu^{m+1}\right),\qquad R_{n(m)}\left(x,y,\mu\right)=O\left(\mu^{m+1}\right),$$
$$R_{p(m)}\left(x,y,\mu\right)=O\left(\mu^{m+1}\right).$$

Let us find a solution to problems [10.13]–[10.16] in the one-dimensional case. This formulation of the problem is typical for bulk p–i–n structures, in which the transverse dimensions of the active region are much larger than the longitudinal ones.

By excluding the electrical potential function from the system of equations [10.13], we obtain the linear differential equation equivalent to the system [10.17]:

$$\frac{d^2 n_0}{dx^2} - \frac{1}{2}\left(A_n + A_p\right) n_0 = 0 . \qquad [10.22]$$

Equation [10.22] is also obtained within the ambipolar diffusion approximation (e.g. Adirovich et al. 1978; Bonch-Bruevich and Kalashnikov 1982; Sze and Kwok 2006) of the theory of p–i–n diodes.

The general solution of equation [10.22] is given in the form:

$$p_0(x) = n_0(x) = C_1 \exp\left(-\frac{x}{L}\right) + C_2 \exp\left(\frac{x}{L}\right), \ \frac{1}{L^2} = \frac{1}{2}\left(A_n + A_p\right) . \qquad [10.23]$$

Let us find general solutions of the system of equations [10.14] (the system of equations [10.15] is considered in a similar way). We will introduce replacements $V = \underline{P}_0 - \underline{N}_0$, $W = \underline{P}_0 + \underline{N}_0$. Then, [10.14] takes the form:

$$\begin{cases} \underline{\Phi}''_0 = V, \\ W'' + \left(V \underline{\Phi}'_0\right)' = 0, \\ V'' + \left(W \underline{\Phi}'_0\right)' = 0. \end{cases}$$

The resulting system of equations is reduced to a system of first-order equations:

$$\begin{cases} \underline{\Phi}'_0 = \Pi, \\ \Pi' = V, \\ W' + \left(V\Pi\right) = C_W, \\ V' + \left(W\Pi\right) = C_V, \end{cases} \qquad [10.24]$$

where C_W and C_V are the constants of integration.

The system of equations [10.24] is supplemented by the boundary conditions at $\underline{\xi} = 0$ and $\underline{\xi} \to \infty$, which we obtain from relations [10.16] and [10.21] in the form

$$\left.\frac{\partial V}{\partial \underline{\xi}}\right|_{\underline{\xi}=0} = 0, \ \left.\frac{\partial W}{\partial \underline{\xi}}\right|_{\underline{\xi}=0} = 0, \ \lim_{\underline{\xi}\to\infty} V\left(\underline{\xi}\right) = \lim_{\underline{\xi}\to\infty} W\left(\underline{\xi}\right) = 0 . \qquad [10.25]$$

To establish the nature of the behavior of function $\Pi(\underline{\xi})$, let us use its physical content. According to [10.24], this is a function of the strength of the electric field created by boundary charges. At the boundary of the region ($x=0$), the value of the potential is constant; therefore, $\left.\frac{\partial \Phi_0}{\partial \underline{\xi}}\right|_{\underline{\xi}=0} = \Pi|_{\underline{\xi}=0} = 0$. In the depth of the active region (at $\underline{\xi} \to \infty$), the field strength is constant ($\lim_{\underline{\xi}\to\infty} \Pi = \Pi_\infty = const$).

After substituting V from the second equation of system [10.24] into the third, performing integration and using the boundary conditions [10.25], we obtain:

$$W = -\frac{\Pi^2}{2} + \frac{\Pi_\infty^{\ 2}}{2}, \ C_W = 0. \qquad [10.26]$$

From the third and fourth equations of system [10.24], it follows the nature of the relationship between the functions $V(\underline{\xi})$, $W(\underline{\xi})$: $V^2 = W^2 \Rightarrow V = \pm W$. From the physical considerations and taking into account the method of introducing these functions, the sign "–" is chosen in the last expression. Finally:

$$V = \frac{\Pi^2}{2} - \frac{\Pi_\infty^{\ 2}}{2}. \qquad [10.27]$$

Substituting [10.27] into the second equation of system [10.24] makes it possible to obtain a differential equation for determining the function $\Pi(\underline{\xi})$:

$$\Pi' = \frac{\Pi^2}{2} - \frac{\Pi_\infty^{\ 2}}{2}. \qquad [10.28]$$

As a result of integration [10.28], we obtain one of the unknown functions of system [10.24] in the form

$$\Pi(\underline{\xi}) = -\Pi_\infty \frac{sh\left(\frac{\Pi_\infty \underline{\xi}}{2}\right)}{ch\left(\frac{\Pi_\infty \underline{\xi}}{2}\right)}. \qquad [10.29]$$

Thus, by using the near-boundary concentration of charge carriers from relations [10.29], [10.26] and [10.27] and $V = \underline{P}_0 - \underline{N}_0$, $W = \underline{P}_0 + \underline{N}_0$, we obtain:

$$\underline{P}_0 = 0, \qquad [10.30]$$

$$\underline{N}_0\left(\underline{\xi}\right) = \Pi_\infty^{\ 2}\left(1 - th^2\left(\frac{\Pi_\infty \underline{\xi}}{2}\right)\right).$$

We find solutions to problems [10.15] and [10.16] in a similar way. As a result, we obtain boundary functions for $x = 1$:

$$\overline{P}_0\left(\overline{\xi}\right) = \Pi_\infty^{\ 2}\left(1 - th^2\left(\frac{\Pi_\infty \overline{\xi}}{2}\right)\right), \qquad [10.31]$$

$$\overline{N}_0 = 0.$$

Note that functions [10.30] and [10.31] contain the same constant Π_∞, since in the depth of the active region (at $\underline{\xi} \to \infty$ and $\overline{\xi} \to \infty$), the values of the corresponding electric field potentials coincide.

To construct particular solutions to the problem, it is necessary to find the unknown constants from the boundary conditions [10.16]. As a result, we obtain the following system of equations for the unknown constants C_1 and C_2, Π_∞ :

$$\left(-\frac{1}{L}C_1 + \frac{1}{L}C_2\right) - \gamma_n w\left(C_1 + C_2\right) - \gamma_n w \Pi_\infty^{\ 2} = -\frac{J}{eD_n}\frac{w}{N_i}, \qquad [10.32]$$

$$\frac{1}{L}\left(-C_1 \exp\left(-\frac{1}{L}\right) + C_2 \exp\left(\frac{1}{L}\right)\right) - \gamma_p w\left(C_1 \exp\left(-\frac{1}{L}\right) + C_2 \exp\left(\frac{1}{L}\right)\right) - \gamma_p w \Pi_\infty^{\ 2} =$$
$$= \frac{J}{eD_p}\frac{w}{N_i},$$

$$\int_0^1 \left(th\left(\frac{\Pi_\infty x}{2\sqrt{\mu}}\right) + th\left(\frac{\Pi_\infty (1-x)}{2\sqrt{\mu}}\right)\right) dx = \frac{U}{\Pi_\infty}.$$

The approximate solution to the nonlinear problems [10.13]–[10.16] is the distribution function of the charge carriers concentration in the i-region of a p–i–n diode of the form (contains the leading terms of asymptotic series [10.10]):

$$p(x)+n(x)=2C_1\exp\left(-\frac{x}{L}\right)+2C_2\exp\left(\frac{x}{L}\right)+$$
$$+\Pi_\infty{}^2\left(\left(1-th^2\left(\frac{\Pi_\infty x}{2\sqrt{\mu}}\right)\right)+\left(1-th^2\left(\frac{\Pi_\infty(1-x)}{2\sqrt{\mu}}\right)\right)\right),$$

$$C_2=\frac{B(1+\gamma_n wL)-A(1+\gamma_p wL)e^{-\frac{1}{L}}}{(1+\gamma_n wL)(1+\gamma_p wL)e^{\frac{1}{L}}-(1-\gamma_n wL)(1+\gamma_p wL)e^{-\frac{1}{L}}}, \qquad [10.33]$$

$$C_1=\frac{-A(1+\gamma_n wL)+(1-\gamma_n wL)C_2}{(1+\gamma_n wL)},$$

$$\ln ch\left(\frac{\Pi_\infty}{2\sqrt{\mu}}\right)=\frac{U}{4\sqrt{\mu}},$$

$$A=-\frac{J}{eD_n}\frac{wL}{N_i}+\gamma_n wL\Pi^2{}_\infty,\quad B=\frac{J}{eD_p}\frac{wL}{N_i}+\gamma_p wL\Pi^2{}_\infty.$$

The distribution function of the charge carriers concentration obtained by solving the ambipolar diffusion equation has the following form:

$$p(x)+n(x)=2\tilde{C}_1\exp\left(-\frac{x}{L}\right)+2\tilde{C}_2\exp\left(\frac{x}{L}\right), \qquad [10.34]$$

where

$$\tilde{C}_2=\frac{B(1+\gamma_n wL)-A(1+\gamma_p wL)e^{-\frac{1}{L}}}{(1+\gamma_n wL)(1+\gamma_p wL)e^{\frac{1}{L}}-(1-\gamma_n wL)(1+\gamma_p wL)e^{-\frac{1}{L}}},$$

$$\tilde{C}_1=\frac{-A(1+\gamma_n wL)+(1-\gamma_n wL)C_2}{(1+\gamma_n wL)},$$

$$A=-\frac{J}{eD_n}\frac{wL}{N_i},\quad B=\frac{J}{eD_p}\frac{wL}{N_i}.$$

Note that as a result of applying the asymptotic method, the traditional solution naturally appears at one of the stages of solving the problem as one of the main members of the asymptotics. The found boundary functions refine the traditional model (e.g. Sze and Kwok 2006). In addition, the difference between [10.33] and [10.34] provides different types of boundary conditions. In the proposed model of the plasma concentration distribution [10.33], there is an additional parameter – the voltage at the contacts U, which also to some extent provides an increase in the level of adequacy of the constructed model.

The results of numerical simulation are shown in the graphs of the dependence of the electron and hole concentration in the active region of the p–i–n diode (Figures 10.3–10.5). The following values of the constants were used in the experiments: $N_i=10^{16}$ cm^{-3}, $T=300$ °K, $k=1{,}38\cdot10^{23}$ J/K, $J=0.4$ mA/cm^2, $U=5$ mV, $H=30$ mkm, $D_n=30$ cm^2/s, $D_p=6$ cm^2/s, $\tau^*_n=20$ mks and $\tau^*_p=25$ mks.

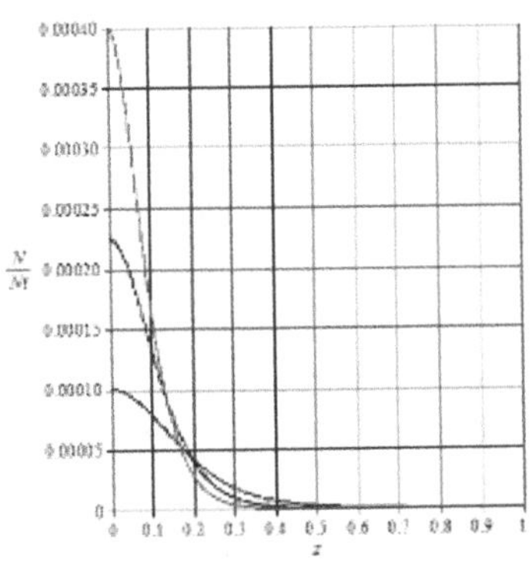

Figure 10.3. *Dependence of the boundary electron concentration on the coordinate at various relative voltages at the contacts. For a color version of this figure, see www.iste.co.uk/korolliouk/computational.zip*

Figure 10.4 shows the distribution of the charge carrier concentration in the i-region. The thin curve reflects the results of modeling by asymptotic methods, the thickened one – the results of modeling in the framework of the ambipolar diffusion approximation. Figure 10.5 shows similar dependences (without taking into account the contribution of the boundary functions): the thin line shows the distribution of the charge carriers concentration, which is supplied by the main terms of the regular part of the asymptotics; the thick line shows the result of solving the ambipolar diffusion equation. The values of the accumulated charge in the active region at the above-mentioned current density and voltage at the contacts, estimated using the improved model and within the ambipolar diffusion approximation, differ within 3%. With an increase in the value of the current density and the applied voltage, the difference in the estimates of the value of the accumulated charge also grows. This behavior is explained by an increase in the charge carriers' concentration in the contact zones, which cannot be tracked using other model approaches.

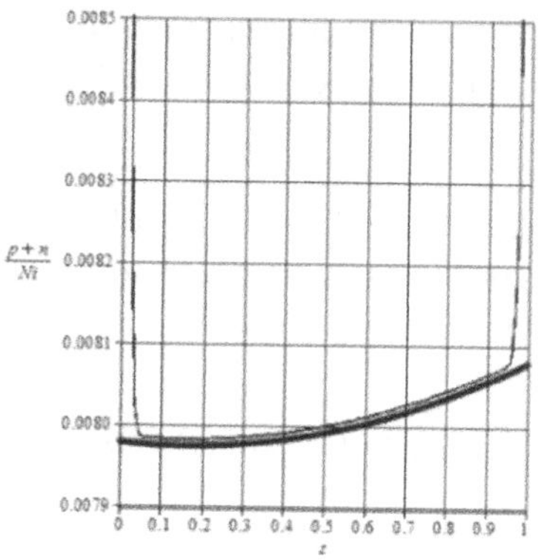

Figure 10.4. *Dependence of the total concentration of electrons and holes (thickened line – ambipolar diffusion model). For a color version of this figure, see www.iste.co.uk/koroliouk/computational.zip*

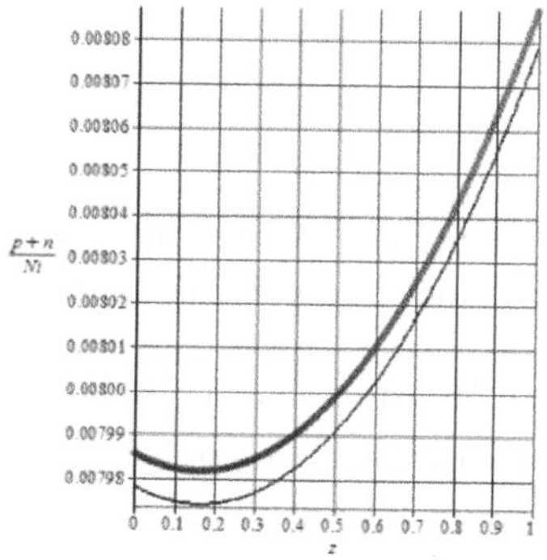

Figure 10.5. *Dependence of the concentration of electrons and holes (regular members of the schedule) on the coordinate. The thin line is a model of ambipolar diffusion. For a color version of this figure, see www.iste.co.uk/koroliouk/computational.zip*

10.3. Simulation of the charge carriers' stationary distribution in the electron–hole plasma of the p–i–n diode assembly elements

The main disadvantage of the p–i–n diode one-dimensional mathematical model is that such a model does not allow the influence of surface phenomena (e.g. surface recombination phenomena (e.g. Adirovich et al. 1978; Bonch-Bruevich and Kalashnikov 1982; Sze and Kwok 2006)) on the formation of electron–hole plasma in the active region of the diode to be studied. Let us find solutions to problems [10.7]–[10.10] in the domain Ω (Figure 10.2(a). That is, the case is considered when the element of the p–i–n diode assembly has finite transverse dimensions. In this case, we will use the algorithm proposed above for solving the problem.

The leading terms of asymptotics [10.12] are found by solving problems [10.13]–[10.16]. The solution of the system of equations [10.13] with the corresponding boundary conditions is proposed to be sought by the method of separation of variables in the following form:

$$n_0(x,y)=\sum_{i=1}^{K}\Big[\big(C_{1i}\exp(\eta_i x)+C_{2i}\exp(-\eta_i x)\big)\sin a_i y+\big(D_{1i}\exp(\eta_i x)+$$

$$+D_{2i}\exp(-\eta_i x)\big)\cos a_i y\Big], \qquad [10.35]$$

where $\eta_i^2=a_i^2+\dfrac{1}{L^2}$. The constants a_i must satisfy a transcendental equation of the form

$$tga_i\frac{l}{w}=\frac{2\beta w a_i}{a_i^2-(\beta w)^2}, \qquad [10.36]$$

which is obtained by substituting [10.35] into the boundary conditions [10.16] at $y=0$ and $y=l/w$ and performing the corresponding transformations.

Boundary conditions [10.16] at $x=0$ and $x=1$, after applying the projection procedure to the system of basis functions of the problem ($\sin(a_i y),\cos(a_i y)$), make it possible to obtain a system of linear algebraic equations for the unknowns C_{1i}, C_{2i}, D_{1i} and D_{2i}:

$$\sum_{i=1}^{K}\Big[\big((\eta_i+\gamma_n w)C_{1i}-(\eta_i-\gamma_n w)C_{2i}\big)I_{ssij}+\big((\eta_i+\gamma_n w)D_{1i}-(\eta_i-\gamma_n w)D_{2i}\big)I_{csij}\Big]=$$

$$=\left(\frac{Jw}{2eD_nN_i}-\gamma_n w\Pi_\infty^2\right)I_{sj}, \qquad [10.37]$$

$$\sum_{i=1}^{K}\left[\left((\eta_i+\gamma_n w)C_{1i}-(\eta_i-\gamma_n w)C_{2i}\right)I_{scij}+\left((\eta_i+\gamma_n w)D_{1i}-(\eta_i-\gamma_n w)D_{2i}\right)I_{ccij}\right]=$$

$$=\left(\frac{Jw}{2eD_nN_i}-\gamma_n w\Pi_\infty^2\right)I_{cj},$$

$$\sum_{i=1}^{K}\left[\left((\eta_i-\gamma_p w)\exp(\eta_i)C_{1i}-(\eta_i+\gamma_p w)\exp(-\eta_i)C_{2i}\right)I_{ssij}+\right.$$

$$\left.+\left((\eta_i-\gamma_p w)\exp(\eta_i)D_{1i}-(\eta_i+\gamma_p w)\exp(-\eta_i)D_{2i}\right)I_{csij}\right]=\left(-\frac{Jw}{2eD_pN_i}+\gamma_p w\Pi_\infty^2\right)I_{sj},$$

$$\sum_{i=1}^{K}\left[\left((\eta_i-\gamma_p w)\exp(\eta_i)C_{1i}-(\eta_i+\gamma_p w)\exp(-\eta_i)C_{2i}\right)I_{scij}+\right.$$

$$\left.+\left((\eta_i-\gamma_p w)\exp(\eta_i)D_{1i}-(\eta_i+\gamma_p w)\exp(-\eta_i)D_{2i}\right)I_{ccij}\right]=\left(-\frac{Jw}{2eD_pN_i}+\gamma_p w\Pi_\infty^2\right)I_{cj},$$

where $1\le j\le K$, $I_{si}=\int_0^{l/w}\sin(a_iy)\,dy=-\frac{1}{a_i}\left(\cos a_i\frac{l}{w}-1\right)$,

$$I_{ci}=\int_0^{l/w}\cos(a_iy)\,dy=\frac{1}{a_i}\sin a_i\frac{l}{w},$$

$$I_{ssij}=\int_0^{l/w}\sin(a_iy)\sin(a_jy)\,dy=\frac{(a_j-a_i)\sin\frac{(a_j+a_i)l}{w}-(a_j+a_i)\sin\frac{(a_j-a_i)l}{w}}{2\left(a_j^2-a_i^2\right)},$$

$$I_{ccij}=\int_0^{l/w}\cos(a_iy)\cos(a_jy)\,dy=\frac{(a_j-a_i)\sin\frac{(a_j+a_i)l}{w}+(a_j+a_i)\sin\frac{(a_j-a_i)l}{w}}{2\left(a_j^2-a_i^2\right)},$$

$$I_{scij} = \int_0^{l/w} \sin(a_i y)\cos(a_j y)dy = -\frac{(a_j - a_i)\cos\frac{(a_j + a_i)l}{w} - (a_j + a_i)\cos\frac{(a_j - a_i)l}{w} + 2a_i}{2(a_j^2 - a_i^2)}.$$

The value $K = \frac{M}{4}$ (M-dimension of system [10.37]) is chosen so as to ensure the required accuracy of the problem solutions [10.13] and [10.16].

The boundary corrections are found from the systems of equations [10.14] and [10.15], supplemented by the corresponding boundary conditions [10.16]. The problem statements for finding the boundary functions are identical to the analogous problems that were considered for the one-dimensional case. Therefore, in this case, the boundary corrections also take the form [10.30] and [10.31].

Thus, the mathematical model of the stationary distribution of the electron–hole plasma concentration in the active region of the p–i–n structure element takes the following form:

$$n_0(x,y) = \sum_{i=1}^{K}\Big[\big(C_{1i}\exp(\eta_i x) + C_{2i}\exp(-\eta_i x)\big)\sin a_i y + \big(D_{1i}\exp(\eta_i x) +$$

$$+D_{2i}\exp(-\eta_i x)\big)\cos a_i y\Big] + \Pi_\infty^{\ 2}\left(\left(1 - th^2\left(\frac{\Pi_\infty x}{2\sqrt{\mu}}\right)\right) + \left(1 - th^2\left(\frac{\Pi_\infty(1-x)}{2\sqrt{\mu}}\right)\right)\right). \quad [10.38]$$

10.4. Modeling the charge carriers stationary distribution in the active region of the integrated surface-oriented p–i–n structures

Figures 10.1 and 10.2 schematically show the elements of p–i–n structures, which differ in the geometry of the arrangement of the injecting n–i and p–i contacts. The advantage of the surface arrangement of contacts (Figure 10.1b)) is due to the possibility of using the technology of manufacturing microcircuits in the industrial production of p–i–n structures. This raises a reasonable question: can a p–i–n diode structure with surface arrangement of elements provide the proper functionality of the diode? The answer to this question can be obtained by comparing the concentration distribution of the p–i–n structure of the electron–hole plasma formed in the active region (i-region) with the data typical for the bulk diodes.

Let us apply the approach described above to model the distribution of the electron–hole plasma concentration in the active region of surface-oriented p–i–n structures. The leading terms of asymptotics [10.12] are found by solving the system of equations [10.13]–[10.15] in the domain $\Omega' = \left\{(x,y): 0<x<1, 0<y<\frac{l}{w}\right\}$ (Figure 10.2(b)). The sought functions must satisfy the boundary conditions of the form:

$$\left.\frac{\partial n_0}{\partial y} - \gamma_n w n_0 - \gamma_n w \underline{N}_0\right|_{y=0} = -\frac{J}{2eD_n}\frac{w}{N_i} \text{ at } 0 \le x \le \frac{s}{2w}, \qquad [10.39]$$

$$\left.\frac{\partial n_0}{\partial y} - \beta w n_0\right|_{y=0} = 0 \text{ at } \frac{s}{2w} \le x \le 1 - \frac{s}{2w},$$

$$\left.\frac{\partial n_0}{\partial y} + \gamma_p w n_0 + \gamma_p w \overline{P}_0\right|_{y=0} = -\frac{J}{2eD_p}\frac{w}{N_i} \text{ at } 1 - \frac{s}{2w} \le x \le 1,$$

$$\left.\frac{\partial n_0}{\partial y} + \beta w n_0\right|_{y=\frac{l}{w}} = 0, \; n_0(0,y) = n_0(x \pm 2i, y),$$

$$\left.\frac{\partial \underline{N}_0}{\partial \underline{\xi}}\right|_{\underline{\xi}=0} = 0, \; \left.\frac{\partial \underline{P}_0}{\partial \underline{\xi}}\right|_{\underline{\xi}=0} = 0, \; \left.\frac{\partial \overline{P}_0}{\partial \overline{\xi}}\right|_{\overline{\xi}=0} = 0, \; \left.\frac{\partial \overline{N}_0}{\partial \overline{\xi}}\right|_{\overline{\xi}=0} = 0,$$

$$\varphi_0 + \underline{\Phi}_0\Big|_{\substack{y=0 \\ 0 \le x \le \frac{s}{2w}}} = 0, \; \varphi_0 + \overline{\Phi}_0\Big|_{\substack{y=0 \\ 1-\frac{s}{2w} \le x \le 1}} = U, \; \left.\frac{\partial \varphi_0}{\partial y}\right|_{y=0,\, y=\frac{l}{w}} = 0.$$

The general solution of the system of equations [10.13] with the corresponding boundary conditions [10.39] is given in the following form:

$$n_0(x,y) = \sum_{i=0}^{M-1}\left(C_{1i}\exp(-\eta_i y) + C_{2i}\exp(\eta_i y)\right)\cos\alpha_i x, \qquad [10.40]$$

where $\alpha_i = \pi i$.

To determine the constants C_{1i} and C_{2i}, conditions [10.39] are used on the boundaries of the domain Ω' at $y=0$ and $y=l/w$, which, after being projected onto the system of basis functions $\cos\alpha_j x$, take the form:

$$\int_0^1 \sum_i (-\eta_i C_{1i} + \eta_i C_{2i}) \cos\alpha_i x \cos\alpha_j x dx = \int_0^{\frac{s}{2w}} \left(-\frac{J}{2eD_n}\frac{w}{N_i} + \gamma_n w \tilde{\Pi}_\infty^2 \right) \cos\alpha_j x dx +$$

$$+\gamma_n w \int_0^{\frac{s}{2w}} \sum_i (C_{1i} + C_{2i}) \cos\alpha_i x \cos\alpha_j x dx + \beta w \int_{\frac{s}{2w}}^{1-\frac{s}{2w}} \sum_i (C_{1i} + C_{2i}) \cos\alpha_i x \cos\alpha_j x dx -$$

$$-\int_{1-\frac{s}{2w}}^1 \left(\frac{J}{2eD_p}\frac{w}{N_i} + \gamma_p w \tilde{\Pi}_\infty^2 \right) \cos\alpha_j x dx - \gamma_p w \int_{1-\frac{s}{2w}}^1 \sum_i (C_{1i} + C_{2i}) \cos\alpha_i x \cos\alpha_j x dx,$$

$$C_{2i} = \frac{(\eta_i - \beta w)}{(\eta_i + \beta w)} \exp(-2\eta_i) C_{1i}.$$

The system of linear algebraic equations obtained in this way for the unknowns C_{1i} and C_{2i} can be rewritten in the following form:

$$-\eta_0 C_{10} + \eta_0 C_{20} = \left(-\frac{J}{2eD_n}\frac{w}{N_i} + \gamma_n w \tilde{\Pi}_\infty^2 \right) \frac{S}{2w} + \gamma_n w \sum_{i=0}^{M-1} (C_{1i} + C_{2i}) I_{1i} + \quad [10.41]$$

$$+\beta w \sum_{i=0}^{M-1} (C_{1i} + C_{2i}) I_{3i} - \left(\frac{J}{2eD_p}\frac{w}{N_i} + \gamma_p w \tilde{\Pi}_\infty^2 \right) \frac{S}{2w} - \gamma_p w \sum_{i=0}^{M-1} (C_{1i} + C_{2i}) I_{2i},$$

$$(-\eta_j C_{1j} + \eta_j C_{2j}) \frac{1}{2} = \left(-\frac{J}{2eD_n}\frac{w}{N_i} + \gamma_n w \tilde{\Pi}_\infty^2 \right) I_{1j} + \gamma_n w \sum_{i=0}^{M-1} (C_{1i} + C_{2i}) I_{2ij} +$$

$$+\beta w \sum_{i=0}^{M-1} (C_{1i} + C_{2i}) I_{4ij} - \left(\frac{J}{2eD_p}\frac{w}{N_i} + \gamma_p w \tilde{\Pi}_\infty^2 \right) I_{2j} - \gamma_p w \sum_{i=0}^{M-1} (C_{1i} + C_{2i}) I_{3ij},$$

$$C_{2i} = \frac{(\eta_i - \beta w)}{(\eta_i + \beta w)} \exp(-2\eta_i) C_{1i},$$

where M is the dimension of the system:

$$I_{1i} = \frac{1}{\pi i}\sin\left(\frac{\pi is}{2w}\right),\ I_{2i} = \frac{\cos(\pi i)}{\pi i}\sin\left(\frac{\pi is}{2w}\right),\ I_{3i} = \frac{(-\cos(\pi i)-1)}{\pi i}\sin\left(\frac{\pi is}{2w}\right),$$

$$I_{1j} = I_{1i},\ I_{2j} = I_{2i},$$

$$I_{2ij} = \frac{(i-j)\sin\left(\frac{(i+j)\pi s}{2w}\right) + (i+j)\sin\left(\frac{(i-j)\pi s}{2w}\right)}{2\pi\left(i^2 - j^2\right)},$$

$$I_{3ij} = -\frac{(i-j)\sin\left(\frac{(i+j)\pi(2w-s)}{2w}\right) + (i+j)\sin\left(\frac{(i-j)\pi(2w-s)}{2w}\right)}{2\pi\left(i^2 - j^2\right)},$$

$$I_{4ij} = \frac{(i-j)\left[\sin\left((i+j)\pi\left(1-\frac{s}{2w}\right)\right) - \sin\left((i+j)\pi\frac{s}{2w}\right)\right]}{2\pi\left(i^2 - j^2\right)} +$$

$$+\frac{(i+j)\left[\sin\left((i-j)\pi\left(1-\frac{s}{2w}\right)\right) - \sin\left((i-j)\pi\frac{s}{2w}\right)\right]}{2\pi\left(i^2 - j^2\right)}.$$

The boundary corrections in the domain Ω' for this formulation of the problem are also described by relations of the form [10.30] and [10.31]. This conclusion follows from similar formulations of problems for determining boundary corrections (systems of equations [10.14] and [10.15] with boundary conditions [10.39] are considered). The difference is that the following equality is used to determine the constant of integration $\tilde{\Pi}_\infty$:

$$\int_{\frac{s}{2w}}^{1-\frac{s}{2w}} \left(th\left(\frac{\tilde{\Pi}_\infty t}{2\sqrt{\mu}}\right) + th\left(\frac{\tilde{\Pi}_\infty (1-t)}{2\sqrt{\mu}}\right)\right) dt = \frac{U}{\tilde{\Pi}_\infty}.$$

Thus, the distribution of the concentration of the electron–hole plasma in the active region of the integrated surface-oriented p–i–n structure is determined by a function of the form:

$$n(x,y)=\sum_{i=0}^{M-1}\left(C_{1i}\exp(-\eta_i y)+C_{2i}\exp(\eta_i y)\right)\sin\alpha_i x+$$

$$+\begin{cases}\tilde{\Pi}_{\infty}^{2}\left(1-th^{2}\left(\dfrac{\tilde{\Pi}_{\infty}y}{2\sqrt{\mu}}\right)\right), & 0\le x\le\dfrac{s}{2w},1-\dfrac{s}{2w}\le x\le 1\\ 0, & \dfrac{s}{2w}<x<1-\dfrac{s}{2w}\end{cases}. \qquad [10.42]$$

Using considerations similar to those presented elsewhere (e.g. Tikhonov 1952; Vishik and Lusternik 1957; Bomba 1982; Smith 1985; Vasil'eva et al. 1995; Bomba et al. 2017), for additional conditions of smoothness and their consistency, we arrive at estimates for the remainder of the asymptotic series:

$$R_{\varphi(m)}(x,y,\mu)=O\left(\mu^{m+1}\right), \qquad R_{n(m)}(x,y,\mu)=O\left(\mu^{m+1}\right),$$

$$R_{p(m)}(x,y,\mu)=O\left(\mu^{m+1}\right).$$

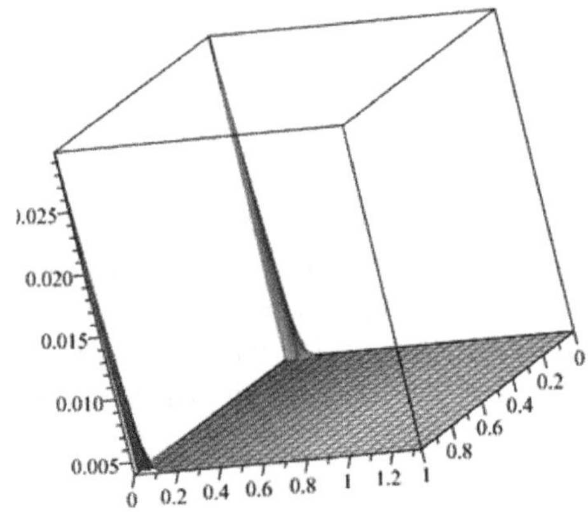

Figure 10.6. *The distribution of the electron–hole plasma concentration in the active region according to the refined model, which takes into account the boundary corrections*

The results of numerical simulation are shown in the graphs of the stationary distribution of concentration of the electrons and holes in the active region of the element of the surface-oriented p–i–n structure (Figures 10.6 and 10.7). The following values of the constants were used in the experiments: $N_i = 10^{16}$ cm^{-3}, $T = 300$ °K, $k = 1{,}38 \cdot 10^{-23}$ J/K, $J = 0.4$ mA/cm^2, $U = 5$ mV, $w = 30$ mkm, s=10 mkm, $D_n = 30$ cm^2/s, $D_p = 6$ cm^2/s, $\tau^*_n = 20$ mks and $\tau^*_p = 25$ mks.

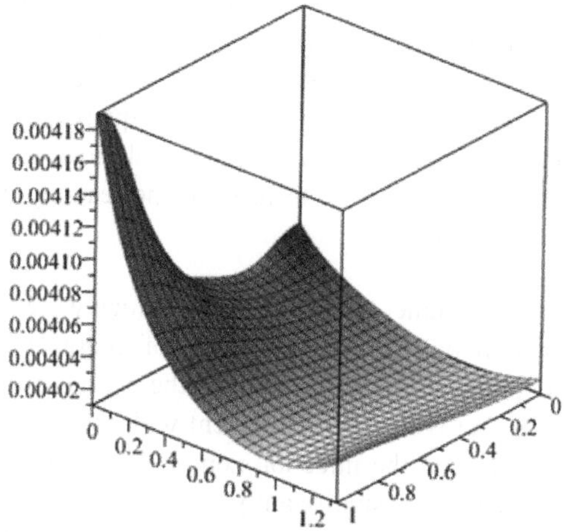

Figure 10.7. *The distribution of the charge carriers' concentration in the active region according to the ambipolar diffusion model*

Figure 10.6 illustrates the stationary distribution of the charge carriers in the i-region of the p–i–n diode according to the improved mathematical model [10.42]. A sharp increase of the electrons' concentration in the region near the n–i junction and holes near the p–i junction is observed. An increased value of the concentration of charge carriers (~ 10 times for the specified set of input data) is observed at a distance $\sim\sqrt{\mu}$ from the boundary of the n–i and p–i junctions. Such a character of the stationary distribution of the plasma concentration is seen as natural, since objectively, in the injection zone of a certain type charge carrier, their concentration should significantly exceed the concentration of charges of the opposite type due to the action of injection processes. This feature of the behavior of the plasma concentration distribution in the diode cannot be displayed using the classical model of ambipolar diffusion. For comparison, the simulation results within the framework of the classical approach are shown in Figure 10.7. Note that according to the forecast for the improved mathematical model of the p–i–n diode, the accumulated charge in the i-region is 3–5% higher than the value of the same indicator, obtained by traditional methods. In this case, a redistribution of the plasma concentration is observed – in the middle of the diode active region, the plasma concentration decreases slightly.

10.5. Final considerations

As part of research on this topic, mathematical models of the stationary distribution of electron–hole plasma in the active region of the various types p–i–n structures have been constructed in the form of nonlinear singularly perturbed boundary value problems for a system of nonlinear equations of continuity of the charge carriers current and Poisson. Approximate solutions to the corresponding boundary value problems are found in the form of the leading terms of asymptotic series in powers of a small parameter.

According to the data of numerous experiments, which were carried out using an improved mathematical model, the concentration of charge carriers in the near-contact regions of the diode significantly exceeds the charge carriers' concentration in the inner region. This fact is in good agreement with the physical concepts of the operation of a p–i–n diode – in the injection zones of a certain type charge carrier, the concentration of these carriers significantly exceeds the concentration of charges of the opposite sign, here the conditions for equality of the concentrations of holes and electrons are not provided, within which the classical approximation of ambipolar diffusion works. Thus, the improved mathematical model of the stationary distribution of electron–hole plasma solves the contradiction inherent in the traditional mathematical model of the theory of p–i–n diodes.

A preliminary analysis of the results obtained indicates that the proposed approach is promising. In addition to the fact that the classical formulations of the problems of modeling the characteristics of p–i–n structures are automatically included in the framework of the proposed scheme for finding solutions to the problem, the presented method allows us to make significant amendments to the solution. This not only increases the level of modeling adequacy, but also provides an understanding of the features of a number of physical processes (diffusion-drift, recombinant, injection) in the near-contact regions of the p–i–n diodes active region.

We consider the proposed approach a promising tool for studying nonlinear thermal, diffusion–drift, generation–recombination stationary and non-stationary processes occurring in bulk and integrated p–i–n structures under the action of external microwave radiation, and predicting new physical effects in the systems under study, for example, due to the influence of local surface and bulk defects (described by "cap" type functions) on the characteristics of p–i–n structures.

10.6. References

Adirovich, E.I., Karageorgii-Alkalaev, P.M., Leiderman, A.Iu. (1978). Toki dvoinoi inzhektsii v poluprovodnikakh [Currents double injection in semiconductors]. *Moscow: Sov. Radio*, 320.

Belyanin, M.P. (1986). Ob asimptoticheskom reshenii odnoi modeli p-n- perehoda [On the asymptotic solution of one p-n-junction model]. *U.S.S.R. Comput. Math. Math. Phys.*, 26(2), 306–311 [Online]. Available at: http://www.mathnet.ru/links/f91b663c991258d0733ea2d02cf7f68a/zvmmf9248.pdf.

Bomba, A. (1982). Pro asymptotychnyi metod nablygennogo rozv'yazannya odnijei zadachi masoperenosu pry fil'tratsii v porystomu seredovyschi [On the approximate solution asymptotic method of one problem of mass transfer during filtration in a porous medium]. *Ukr. Math. J.*, 34(4), 37–40.

Bomba, A., Prysyagnuk, I., Prysyagnuk, O. (2017). Metody teorii zburen' prognozuvannya protsesiv teplomasoperenesennya v porystyh ta microporystyh seredovyschah [Perturbation theory methods for predicting heat and mass transfer processes in porous and microporous media]. *O.Zen'*, Rivne, 291.

Bonch-Bruevich, V.L. and Kalashnikov, S.G. (1982). *Physics of Semiconductors*. VEB, Berlin.

Grimalsky, V.V., Kishenko, Y.I., Koshevaya, S.V., Moroz, I.P. (1994). The interaction of powerful electromagnetic waves with integrated p-i-n-structures. *Doc. of Int. Symp. "Physics and Engineering of Mm and Submm Waves"*, June 7-10, Kharkov, 1, 238–239.

Haine, M.E. and Rose, F.W.G. (1972). On the conductance of p-i-n junctions at high microwave fields. *Solid-State Electron.*, 15(6), 687–705.

Kwok, K. (2002). *Complete Guide to Semiconductor Devices*. Wiley-Interscience, New York [Online]. Available at: https://ieeexplore.ieee.org/book/5271197.

Lebedev, I.V., Shnitnikov, A.S., Dyakov, I.V., Borisova, N.A. (1998). Impedance properties of high-frequency PIN diodes. *Solid-State Electron.*, 41(1), 121–128.

Smith, D.R. (1985). *Singular-Perturbation Theory. An Introduction with Applications*. Cambridge University Press, Cambridge [Online]. Available at: https://books.google.com.ua/books?id=cEszbdam0zwC&printsec=frontcover&hl=ru#v=onepage&q&f=false.

Sze, S. and Kwok, K. (2006). *Physics of Semiconductor Devices*. Wiley-Interscience, New York. doi: 10.1002/0470068329.

Tikhonov, A.N. (1952). Sistemy differentsyal'nyh uravnenij, sodergaschie malye parametry [Systems of differential equations containing small parameters in the derivatives]. *Mat. Sb.*, 31(73), 3, 575–586 [Online]. Available at: http://www.mathnet.ru/links/7bf5e98e10e302c132f30df4fdad6b9b/sm5548.pdf.

Vasil'eva, A.B. and Stel'makh, V.G. (1977). Singulyarno vozmuschennye sistemy teorii poluprovodnikovyh priborov [Singularly perturbed systems in the theory of semiconductors]. *U.S.S.R. Comput. Math. Math. Phys.*, 17(2), 339–348 [Online]. Available at: http://www.mathnet.ru/links/5d86b56f4673e4843abf1810d61ba4e8/zvmmf6023.pdf.

Vasil'eva, A.B., Butusov, V.F., Kalachev, L.V. (1995). *The Boundary Function Method for Singular Perturbation Problems*. SIAM, Philadelphia. doi: 10.1137/1.9781611970784.

Vishik, M. and Lusternik, L.A. (1957). Regulyarnoe vyrogdenie i pogranichnyi sloi dlya lineinyh differentsyal'nyh uravnenii s malym parametrom [Regular degeneration and boundary layer for linear differential equations with small parameter]. *Usp. Mat. Nauk.*, 12(5), 3–122. [Online]. Available at: http://www.mathnet.ru/links/3ad5badbcecb855d7a6ed5e79bf964d1/rm7705.pdf.

11

Diffusion Perturbations in Models of the Dynamics of Infectious Diseases Taking into Account the Concentrated Effects

Serhii BARANOVSKY[1], Andrii BOMBA[1], Sergiy LYASHKO[2] and Oksana PRYSHCHEPA[1]

[1] *National University of Water and Environmental Engineering, Rivne, Ukraine*
[2] *Taras Shevchenko National University of Kyiv, Ukraine*

11.1. Introduction

Today, in the time of the Covid-19 pandemic, despite the advance of medicinal science, infectious diseases are still a problem and a danger to the community. An already established approach to studying different types of infectious diseases is to consider them from a unique position as processes of interaction of the immune system with pathogens. This approach is based on the formulated on the clonal selection theory of Burnett (1971) fundamental mechanisms of immune defense. Among other things, this approach also contributes to the widespread use of mathematical modeling methods that allow finding the characteristics of the

Computational Methods and Mathematical Modeling in Cyberphysics and Engineering Applications 1,
coordinated by Dmitri KOROLIOUK, Sergiy LYASHKO and Nikolaos LIMNIOS.

organism immune system in viral disease to analyze various model treatment programs based on the control of the immune system.

In the papers by Marchuk (1980, 1991, 1997), the so-called simplest infectious disease model is presented. It is built on the principle of balance for each active factors that determine the basic mechanisms of the humoral immune response to detected viruses in the organism. The model is formulated by a system of nonlinear differential equations with time delays that determine change speeds of the number of antigens, plasma cells, antibodies and measure of target organ damage

$$\begin{aligned} V_t' &= (\beta - \gamma F)V, \\ C_t' &= \xi(m)\alpha V(t-\tau)F(t-\tau) - \mu_C\left(C - C^*\right), \\ F_t' &= \rho C - \left(\mu_f + \eta\gamma V\right)F, \\ m_t' &= \sigma V - \mu_m m \end{aligned} \qquad [11.1]$$

for conditions as

$$\begin{aligned} &C(t_0) = C^0,\ m(t_0) = m^0,\ V(\tilde{t})(\tilde{t}) = V^0(\tilde{t}), \\ &F(\tilde{t}) = F^0(\tilde{t}),\ t_0 - \tau \le \tilde{t} \le t_0 \end{aligned} \qquad [11.2]$$

for conditions as where $V(t)$, $C(t)$, $F(t)$ and $m(t)$ are the antigens, plasma cells, antibodies concentrations, and the value of the damage degree to the target organ, at time t, respectively; C^* is the plasma cell concentrations in a healthy organism; β is the antigens reproduction rate; γ is the coefficient that takes into account the result of the antigens interaction with antibodies; τ is the period of time delay required to form a plasma cells cascade; μ_C is the value inverse to the plasma cells lifespan; α is the immune system stimulation coefficient; ρ is the rate of antibodies production by one plasma cell; μ_f is the value inverse of the antibodies duration; η is the cost of antibodies to neutralize one antigen; σ is the rate of the target organ cells damage; μ_m is the rate of the target organ recovery. The function

$$\xi(m) = \begin{cases} 1, & 0 \le m < m^*, \\ (m-1)/\left(m^* - 1\right), & m^* \le m < 1, \end{cases}$$

takes into account the effect of reducing the antibody production productivity in significant damage to the target organ. If m^*, the maximum value of the degree of the target organ damage, at which the immune system normal functioning is still possible, then on the interval $0 \le m \le m^*$ value $\xi(m)$ is equal to one, regardless of the lesion, the immunological organs function fully. For $m^* \le m < 1$, the organism efficiency declines rapidly. A small number of variables of the simplest (basic) infectious disease model make it possible to establish and to explain the properties of stationary solutions, in particular, their stability.

In the paper by Marchuk (1991), it is shown that the stationary solution $V = 0,\ C = C^*,\ \ F = F^* = \rho C^* / \mu_f,\ m = 0$ describes the state of a healthy organism. The solution in condition of $\beta < \gamma F^*$ is asymptotically stable and maintains such resistance when a healthy organism is infected be dose of antigen V^0 that does not exceed some level V^* of the immunological barrier:

$$V^0 < V^* = \mu_f \left(\gamma F^* - \beta\right) / (\beta \eta \gamma) \qquad [11.3]$$

The other stationary solution (Marchuk 1991, 1997)

$$\bar{V} = \frac{\mu_c \mu_f \left(\beta - \gamma F^*\right)}{\beta(\alpha\rho - \mu_c \eta \gamma)},\ \bar{C} = \frac{\alpha\beta\mu_f - \eta\mu_c\gamma^2 C^*}{\gamma(\alpha\rho - \mu_c\eta\gamma)},$$

$$\bar{F} = \frac{\beta}{\gamma},\ \bar{m} = \frac{\sigma\bar{V}}{\mu_m} = \frac{\sigma\mu_c\mu_f\left(\beta - \gamma F^*\right)}{\mu_m\beta(\alpha\rho - \mu_c\eta\gamma)},$$

that is obtained for $V > 0$ and $\xi(m) \equiv 1$ and describes the so-called chronic disease process, is also stable.

The approaches used in the basic infectious disease model are developed and expanded in antiviral and antibacterial immune response models (Marchuk 1991, 1997). In contrast to the simplest model, in addition to the humoral immune response with the antibodies production, the cellular type of immunity with the cytotoxic *T*-lymphocyte effectors accumulation is taken into account. As mentioned in Marchuk (1991, 1997), antibodies are able to neutralize viral antigens that circulate freely in the blood or lymph, but cannot penetrate into infected cells and neutralize viruses that multiply in them. Detection and destruction of infected cells is carried out by accumulated cytotoxic *T*-killer lymphocytes. The antiviral immune response model, as well as the basic model, is represented by a nonlinear differential equations system with time delay:

$$\begin{aligned}
&V'_{f\,t} = h_1 C_V + h_2 C_V E - h_3 F V_f - h_4 V_f - h_5\left(C^* - C_V - m\right)V_f,\\
&M'_{V\,t} = h_6 V_f - h_7 M_V,\\
&H'_{E\,t} = h_8 \xi(m) M_V(t-\tau_1) H_E(t-\tau_1) - h_9 M_V H_E - h_{10} M_V H_E E +\\
&+h_{11}\left(H_E^* - H_E\right),\\
&H'_{B\,t} = h_{12} \xi(m) M_V(t-\tau_2) H_B(t-\tau_2) - h_{13} M_V H_B - h_{14} M_V H_B B +\\
&+h_{15}\left(H_B^* - H_B\right),\\
&E'_t = h_{16} \xi(m) M_V(t-\tau_3) H_E(t-\tau_3) E(t-\tau_3) - h_{17} M_V H_E E - h_{18} C_V E +\\
&+h_{19}\left(E^* - E\right),\\
&B'_t = h_{20} \xi(m) M_V(t-\tau_4) H_B(t-\tau_4) B(t-\tau_4) - h_{21} M_V H_B B + h_{22}\left(B^* - B\right),\\
&P'_t = h_{23} \xi(m) M_V(t-\tau_5) H_B(t-\tau_5) B(t-\tau_5) + h_{24}\left(P^* - P\right),\\
&F'_t = h_{25} P - h_{26} V_f F - h_{27} F,\\
&C'_{V\,t} = h_{28} V_f \left(C^* - C_V - m\right) - h_{29} C_V E - h_{30} C_V,\\
&m'_t = h_{31} C_V E + h_{32} C_V - h_{33} m,
\end{aligned} \qquad [11.4]$$

under conditions

$$V_f(0) = V_f^0,\ M_V(0) = 0,\ H_E(0) = H_E^0,\ H_B(0) = H_B^0,\ E(0) = E^0,$$

$$B(0) = B^0,\ P(0) = P^0,\ F(0) = F^0,\ C_V(0) = 0,\ m(0) = 0,$$

$$M_V(\tilde{t}) H_E(\tilde{t}) = 0,\ -\tau_1 \le \tilde{t} < 0,\ M_V(\tilde{t}) H_B(\tilde{t}) = 0,\ -\tau_2 \le \tilde{t} < 0, \qquad [11.5]$$

$$M_V(\tilde{t}) H_E(\tilde{t}) E(\tilde{t}) = 0, \qquad -\tau_3 \le \tilde{t} < 0, \qquad M_V(\tilde{t}) H_B(\tilde{t}) B(\tilde{t}) = 0,$$
$$-\tilde{\tau} \le \tilde{t} < 0,$$

$$\tilde{\tau} = \max\{\tau_4, \tau_5\},$$

which describe the change rate of free circulating in the body viruses V_f, antibodies F, infected with viruses of target organ cells C_V, T-killer lymphocytes E, T-helper lymphocytes of cellular immunity H_E, T-lymphocytes helpers of humoral immunity H_B, B-lymphocytes B, plasma P cells, stimulated M_V macrophages and the nonfunctioning part of the virus-affected organ m. Here, $h_1, h_2, ..., h_{33}$ are model parameters.

The infectious disease basic model, the antiviral immune response model and other immunology models (Nowak and May 2000; Wodarz 2007; Keeling and Rohani 2008) are built under the assumption that the "organism" is a homogeneous environment, in which all the process components are instantly mixed and end up evenly distributed. On the contrary, antigen detection and the launch of the immune system of appropriate mechanisms of response to them do not occur immediately after the antigens penetration into the body. That is, some of the antigens that were not immediately neutralized by the immune defense can penetrate into the cells, and multiply in them. As a result, the infection foci with higher antigen concentrations form around the affected cells. The antigens generated in the organism will eventually be redistributed from the initial infection foci to the surrounding uninfected areas, increasing the affected area and decreasing the antigen concentrations values in the respective infection epicenters. An approach for taking into account small spatially distributed diffusion effects on infectious disease dynamics is presented in Bomba et al. (2020a, 2020b, 2020c), Baranovsky and Bomba (2020) and Bomba and Baranovsky (2020).

11.2. Model problem of infectious disease dynamics taking into account diffusion perturbation and asymptotics of the solution

We assume that small-scale spatially distributed influences on the development of the infectious disease process have a diffusive character. Let us describe the corresponding spatio-temporal dynamics of infectious disease factors in the set $G_Z = \{(x,t): -\infty < x < +\infty;\ t_0 < t < +\infty\}$ by the generalized system of nondimensional differential equations with time delay, according to equations [11.1] and [11.2]:

$$
\begin{aligned}
&v'_t(x,t) = (h_1 - h_2 f(x,t))v(x,t) + \varepsilon h_9 v''_{xx}(x,t),\\
&s'_t(x,t) = \xi(m)h_3 f(x,t-\tau)v(x,t-\tau) - h_5(s(x,t)-1) + \varepsilon^2 h_{10} s''_{xx}(x,t),\\
&f'_t(x,t) = h_4\left(s(x,t) - f(x,t)\right) - h_8 v(x,t) f(x,t) + \varepsilon h_{11} f''_{xx}(x,t),\\
&m'_t(x,t) = h_6 v(x,t) - h_7 m(x,t) + \varepsilon^2 h_{12} m''_{xx}(x,t),
\end{aligned}
\qquad [11.6]
$$

under conditions

$$s(x,t_0) = s^0(x),\ m(x,t_0) = m^0(x),\ v(x,\tilde{t}) = v^0(x,\tilde{t}),$$
$$f(x,\tilde{t}) = f^0(x,\tilde{t}),\ t_0 - \tau \le \tilde{t} \le t_0, \quad [11.7]$$

where $v(x,t) = V(x,t)/V_m$, $s(x,t) = C(x,t)/C^*$, $f(x,t) = F(x,t)/F^*$; V_m is some scale factor for the antigens concentration, for example, the biologically acceptable antigens concentration in the body; C^* and F^* are the plasma cells and antibodies concentration in a healthy organism; $h_1 = \beta$, $h_2 = \gamma F^*$, $h_3 = \alpha V_m F^*/C^*$, $h_4 = \mu_f$, $h_5 = \mu_C$, $h_6 = \sigma V_m$, $h_7 = \mu_m$, $h_8 = \eta\gamma V_m$; εh_9, εh_{11} and $\varepsilon^2 h_{12}$ are diffusion redistribution coefficients of antigens, antibodies, plasma and affected cells, respectively, ε is a small parameter that characterizes the respective components small impact compared to the dominant components of the process.

At the beginning, we consider the case when the level of damage to the target organ by antigens remains such that it does not lead to a decrease in the productivity of antibody production, that is, $\xi(m) = 1$. Then, the solution of problem with time delay [11.6]–[11.7] is reduced to a sequence of problems without time delay (Elsholz and Norkin 1971):

$$\begin{cases} v'_{0t}(x,t) = \left(h_1 - h_2 f_0(x,t)\right) v_0(x,t) + \varepsilon h_9 v''_{0xx}(x,t), \\ s'_{0t}(x,t) = h_3 v^0(x,t-\tau) f^0(x,t-\tau) - h_5\left(s_0(x,t) - 1\right) + \varepsilon^2 h_{10} s''_{0xx}(x,t), \\ f'_{0t}(x,t) = h_4\left(s_0(x,t) - f_0(x,t)\right) - h_8 v_0(x,t) f_0(x,t) + \varepsilon h_{11} f''_{0xx}(x,t), \\ m'_{0t}(x,t) = h_6 v_0(x,t) - h_7 m_0(x,t) + \varepsilon^2 h_{12} m''_{0xx}(x,t), \\ s_0(x,t_0) = s^0(x),\ m_0(x,t_0) = m^0(x), \\ v_0(x,t_0) = v^0(x,t_0),\ f_0(x,t_0) = f^0(x,t_0),\ t_0 \le t \le t_0 + \tau; \end{cases} \quad [11.8]$$

$$\begin{cases} v'_{rt}(x,t) = \left(h_1 - h_2 f_r(x,t)\right) v_r(x,t) + \varepsilon h_9 v''_{r\,xx}(x,t), \\ s'_{rt}(x,t) = h_3 v_{r-1}(x,t-\tau) f_{r-1}(x,t-\tau) - h_5\left(s_r(x,t) - 1\right) + \varepsilon^2 h_{10} s''_{r\,xx}(x,t), \\ f'_{rt}(x,t) = h_4\left(s_r(x,t) - f_r(x,t)\right) - h_8 v_r(x,t) f_r(x,t) + \varepsilon h_{11} f''_{r\,xx}(x,t), \\ m'_{rt}(x,t) = h_6 v_r(x,t) - h_7 m_r(x,t) + \varepsilon^2 h_{12} m''_{r\,xx}(x,t), \\ s_r(x,t_0+r\tau) = s_{r-1}(x,t_0+r\tau),\ m_r(x,t_0+r\tau) = m_{r-1}(x,t_0+r\tau), \\ v_r(x,t_0+r\tau) = v_{r-1}(x,t_0+r\tau),\ F_r(x,t_0+r\tau) = F_{r-1}(x,t_0+r\tau), \\ t_0 + r\tau \le t \le t_0 + (r+1)\tau, \quad r = 1,2,\ldots . \end{cases} \qquad [11.9]$$

To ensure sufficient smoothness of the corresponding solutions at $t = t_0 + \tau$, $t = t_0 + 2\tau$, ..., $t = t_0 + r\tau$, ... in addition to the traditional smoothness conditions with respect to the initial conditions functions in the infectious disease model, it is necessary to impose conditions of their consistency at $t = t_0 - \tau$, $t = t_0$, ... (Bomba 1982; Bomba et al. 2008). In particular, the condition must be met:

$$s'_{0t}(x,t_0) = h_3\left(f^0(x,t_0-\tau) + k^0(x,t_0-\tau)\right) v^0(x,t_0-\tau) - \\ -h_5\left(s_0(x,t_0) - 1\right) + \varepsilon^2 h_{14} s''_{0xx}(x,t_0).$$

We also assume that conditions are given and are in agreement with equation [11.7] at infinity (Bomba 1982; Bomba et al. 2008) (in order to simplify the building of asymptotic series of solutions, we do not pay attention to these conditions).

Given that the diffusion redistributions of active factors are small compared to other components of the infectious disease process, we use the asymptotic method to solve the corresponding singularly perturbed model problems [11.8]–[11.9] (Bomba 1982; Bomba et al. 2008). In particular, the solutions of problems [11.8]–[11.9] are formally represented as asymptotic series $v_j(x,t) = \sum_{i=0}^{n} \varepsilon^i v_{ij}(x,t) + R^v_{nj}(x,t,\varepsilon)$, $s_j(x,t) = \sum_{i=0}^{n} \varepsilon^i s_{ij}(x,t) + R^s_{nj}(x,t,\varepsilon)$, $f_j(x,t) = \sum_{i=0}^{n} \varepsilon^i f_{ij}(x,t) + R^f_{nj}(x,t,\varepsilon)$, $m_j(x,t) = \sum_{i=0}^{n} \varepsilon^i m_{ij}(x,t) + R^m_{nj}(x,t,\varepsilon)$ as perturbations of the corresponding degenerate problems solution (Bomba 1982), where $j = 0,\ldots,r,\ldots$, $v_{ij}(x,t)$, $s_{ij}(x,t)$, $f_{ij}(x,t)$, $m_{ij}(x,t)$ are the required functions (members of the asymptotics), $R^v_{nj}(x,t,\varepsilon)$, $R^s_{nj}(x,t,\varepsilon)$, $R^f_{nj}(x,t,\varepsilon)$, $R^m_{nj}(x,t,\varepsilon)$ are relevant residual members. After substituting the asymptotic series and performing the standard procedure of "equalization", we obtain the following problems to determine the functions $v_{ij}(x,t)$, $s_{ij}(x,t)$, $f_{ij}(x,t)$, $m_{ij}(x,t)$ $(i = 0,1,\ldots,n,\ j = 0,1,\ldots,r,\ldots)$:

$$\begin{cases} v'_{0,0\,t}(x,t) = \left(h_1 - h_2 f_{0,0}(x,t)\right) v_{0,0}(x,t), \\ s'_{0,0\,t}(x,t) = h_3 v^0(x,t-\tau) f^0(x,t-\tau) - h_5\left(s_{0,0}(x,t) - 1\right), \\ f'_{0,0\,t}(x,t) = h_4\left(s_{0,0}(x,t) - f_{0,0}(x,t)\right) - h_8 v_{0,0}(x,t) f_{0,0}(x,t), \\ m'_{0,0\,t}(x,t) = h_6 v_{0,0}(x,t) - h_7 m_{0,0}(x,t), \\ s_{0,0}(x,t_0) = s^0(x),\ m_{0,0}(x,t_0) = m^0(x), \\ v_{0,0}(x,t_0) = v^0(x,t_0),\ f_{0,0}(x,t_0) = f^0(x,t_0),\ t_0 \le t \le t_0 + \tau, \end{cases} \quad [11.10]$$

$$\begin{cases} v'_{1,0\,t}(x,t) = h_1 v_{1,0}(x,t) - h_2\left(a_{0,0}(x,t) f_{1,0}(x,t) + b_{0,0}(x,t) v_{1,0}(x,t)\right) + \\ +\Phi_{v\,1,0}(x,t), \\ s'_{1,0\,t}(x,t) = h_3\left(a_{0,0}(x,t-\tau) f_{1,0}(x,t-\tau) + b_{0,0}(x,t-\tau) f_{1,0}(x,t-\tau)\right) - \\ +h_5 s_{1,0}(x,t), \\ f'_{1,0\,t}(x,t) = h_4\left(s_{1,0}(x,t) - f_{1,0}(x,t)\right) - h_8\left(a_{0,0}(x,t) f_{1,0}(x,t) + \right. \\ \left. +b_{0,0}(x,t) v_{1,0}(x,t)\right) + \Phi_{f\,1,0}(x,t), \\ m'_{1,0\,t}(x,t) = h_6 v_{1,0}(x,t) - h_7 m_{1,0}(x,t), \\ s_{1,0}(x,t_0) = 0,\ m_{1,0}(x,t_0) = 0, \\ v_{1,0}(x,t_0) = 0,\ f_{1,0}(x,t_0) = 0,\ t_0 \le t \le t_0 + \tau, \end{cases} \quad [11.11]$$

. . . .

$$\begin{cases} v'_{i,0\,t}(x,t) = h_1 v_{i,0}(x,t) - h_2\left(a_{0,0}(x,t) f_{i,0}(x,t) + b_{0,0}(x,t) v_{i,0}(x,t)\right) + \\ +\Phi_{V\,i,0}(x,t), \\ s'_{i,0\,t}(x,t) = h_3\left(a_{0,0}(x,t-\tau) f_{i,0}(x,t-\tau) + b_{0,0}(x,t-\tau) v_{i,0}(x,t-\tau)\right) - \\ -h_5 s_{i,0}(x,t) + \Phi_{s\,i,0}(x,t), \\ f'_{i,0\,t}(x,t) = h_4\left(s_{i,0}(x,t) - f_{i,0}(x,t)\right) - h_8\left(a_{0,0}(x,t) f_{i,0}(x,t) + \right. \\ \left. +b_{0,0}(x,t) v_{i,0}(x,t)\right) + \Phi_{f\,i,0}(x,t), \\ m'_{i,0\,t}(x,t) = h_6 v_{i,0}(x,t) - h_7 m_{i,0}(x,t) + \Phi_{m\,i,0}(x,t), \\ s_{i,0}(x,t_0) = 0,\ m_{i,0}(x,t_0) = 0, \\ v_{i,0}(x,t_0) = 0,\ f_{i,0}(x,t_0) = 0,\ t_0 \le t \le t_0 + \tau, \end{cases} \quad [11.12]$$

$$\begin{cases} v'_{0,r\,t}(x,t) = \left(h_1 - h_2 f_{0,r}(x,t)\right) v_{0,r}(x,t), \\ s'_{0,r\,t}(x,t) = h_3 v_{0,r-1}(x,t-\tau) f_{0,r-1}(x,t-\tau) - h_5\left(s_{0,r}(x,t) - 1\right), \\ f'_{0,r\,t}(x,t) = h_4\left(s_{0,r}(x,t) - f_{0,r}(x,t)\right) - h_8 v_{0,r}(x,t) f_{0,r}(x,t), \\ m'_{0,r\,t}(x,t) = h_6 v_{0,r}(x,t) - h_7 m_{0,r}(x,t), \\ s_{0,r}(x,t_0 + r\tau) = s_{0,r-1}(x,t_0 + r\tau),\ m_{0,r}(x,t_0 + r\tau) = m_{0,r-1}(x,t_0 + r\tau), \\ v_{0,r}(x,t_0 + r\tau) = v_{0,r-1}(x,t_0 + r\tau),\ f_{0,r}(x,t_0 + r\tau) = f_{0,r-1}(x,t_0 + r\tau), \\ t_0 + r\tau \le t \le t_0 + (r+1)\tau, \end{cases} \quad [11.13]$$

$$\begin{cases} v'_{1,r\,t}(x,t) = h_1 v_{1,r}(x,t) - h_2\left(a_{0,r}(x,t) f_{1,r}(x,t) + b_{0,r}(x,t) v_{1,r}(x,t)\right) + \\ +\Phi_{v1,r}(x,t), \\ s'_{1,r\,t}(x,t) = h_3\left(a_{0,r}(x,t-\tau) f_{1,r-1}(x,t-\tau) + b_{0,r}(x,t-\tau) v_{1,r-1}(x,t-\tau)\right) - \\ -h_5 s_{1,r}(x,t), \\ f'_{1,r\,t}(x,t) = h_4\left(s_{1,r}(x,t) - f_{1,r}(x,t)\right) - h_8\left(a_{0,r}(x,t) f_{1,r}(x,t) + \right. \\ \left. +b_{0,r}(x,t) v_{1,r}(x,t)\right) + \Phi_{f1,r}(x,t), \\ m'_{1,r\,t}(x,t) = h_6 V_{1,r}(x,t) - h_7 m_{1,r}(x,t), \\ s_{1,r}(x,t_0 + r\tau) = s_{1,r-1}(x,t_0 + r\tau),\ m_{1,r}(x,t_0 + r\tau) = m_{1,r-1}(x,t_0 + r\tau), \\ v_{1,r}(x,t_0 + r\tau) = v_{1,r-1}(x,t_0 + r\tau),\ f_{1,r}(x,t_0 + r\tau) = f_{1,r-1}(x,t_0 + r\tau), \\ t_0 + r\tau \le t \le t_0 + (r+1)\tau, \end{cases} \quad [11.14]$$

....

$$\begin{cases} v'_{i,r\,t}(x,t) = h_1 v_{i,r}(x,t) - h_2\left(a_{0,r}(x,t) f_{i,r}(x,t) + b_{0,r}(x,t) v_{i,r}(x,t)\right) + \\ +\Phi_{vi,r}(x,t), \\ s'_{i,r\,t}(x,t) = h_3\left(a_{0,r}(x,t-\tau) f_{i,r-1}(x,t-\tau) + b_{0,r}(x,t-\tau) v_{i,r-1}(x,t-\tau)\right) - \\ -h_5 s_{i,r}(x,t) + \Phi_{s\,i,r}(x,t), \\ f'_{i,r\,t}(x,t) = h_4\left(s_{i,r}(x,t) - f_{i,r}(x,t)\right) - h_8\left(a_{0,r}(x,t) f_{i,r}(x,t) + \right. \\ \left. +b_{0,r}(x,t) v_{i,r}(x,t)\right) + \Phi_{f\,i,r}(x,t), \\ m'_{i,r\,t}(x,t) = h_6 v_{i,r}(x,t) - h_7 m_{i,r}(x,t) + \Phi_{m\,i,r}(x,t), \\ s_{i,r}(x,t_0 + r\tau) = s_{i,r-1}(x,t_0 + r\tau),\ m_{i,r}(x,t_0 + r\tau) = m_{i,r-1}(x,t_0 + r\tau), \\ v_{i,r}(x,t_0 + r\tau) = v_{i,r-1}(x,t_0 + r\tau),\ f_{i,r}(x,t_0 + r\tau) = f_{i,r-1}(x,t_0 + r\tau), \\ t_0 + r\tau \le t \le t_0 + (r+1)\tau, \end{cases} \quad [11.15]$$

where $a_{0,j}(x,t) = v_{0,j}(x,t)$, $b_{0,j}(x,t) = f_{0,j}(x,t)$

$$\Phi_{v1,j}(x,t) = h_9 v''_{0,j\,xx}(x,t),\ \Phi_{f1,j}(x,t) = h_{11} f''_{0,j\,xx}(x,t);$$

$$\Phi_{vi,j}(x,t) = -h_2 \sum_{l=1}^{i-1} v_{l,j}(x,t) f_{i-l,j}(x,t) + h_9 v''_{i-1,j\,xx}(x,t),$$

$$\Phi_{s\,i,j}(x,t) = \sum_{l=1}^{i-1} h_3 v_{l,j}(x,t-\tau) f_{i-l,j}(x,t-\tau) + h_{10} s''_{i-2,j\,xx}(x,t),$$

$$\Phi_{fi,j}(x,t) = -h_8 \sum_{l=1}^{i-1} v_{l,j}(x,t) f_{i-l,j}(x,t) + h_{11} f''_{i-1,j\,xx}(x,t),$$

$$\Phi_{m\,i,j}(x,t) = h_{12} m''_{i-2,j\,xx}(x,t),\ i = 2,3,...,n,\ j = 0,1,...,r,... .$$

In the iteration process, the solution of each of these problems is found by numerical methods (e.g. Runge–Kutta method), using obtained solutions of problems that were obtained on previous stage.

Note that the proposed approach is easy to "transfer" to other, in particular, finite domains G_Z. In this case, of course, instead of those described above, more complex schedules should be used (see, for example, Bomba (1982) and Bomba et al. (2008)). Estimation of residual terms $R^v_{nj}(x,t,\varepsilon)$, $R^s_{nj}(x,t,\varepsilon)$, $R^f_{nj}(x,t,\varepsilon)$ and $R^m_{nj}(x,t,\varepsilon)$ and establishment of spatio-temporal intervals of convergence for forecasting of concrete processes are done on the basis of the principle of type of a maximum similarly to Bomba (1982) and Bomba et al. (2008).

11.3. Modeling of diffusion perturbations of infectious disease process taking into account the concentrated effects and immunotherapy

The simplest infectious disease model, the immune response model and their modifications after estimation of parameters according to clinical data make it possible to forecast the nature and result of the disease, identify and study the most common patterns of external influences on the process dynamics, model the control of immune response, build, analyze and estimate various treatment programs.

In particular, in the paper by Bolodurina et al. (2009), the modification of the infectious disease model with an additional control function $u_2(t)$ and the dynamics of the function that are described by the differential equation $F_t' = \rho C - (\mu_f + \eta\gamma V)F + u_2$ are proposed to study the influence of the introduction procedure of the ready-to-use immunoglobulin or donor antibodies on the immune system. In Rusakov and Chirkov (2012), it proposes to describe the dynamics of native and donor antibodies in the organism separately and to introduce an additional function of the donor antibodies concentration $K(t)$.

At the modification, the basic infectious disease model in order to take into account immunotherapy procedures, it is proposed to consider the concentration of native and donor antibodies in the organism separately and to introduce into the model an additional variable $K(t)$ that describes the concentration of donor antibodies in the organism.

In the paper by Bomba et al. (2020b), a modification of the basic mathematical model for taking into account diffusion perturbations on the infectious disease dynamics under conditions of immunotherapy is proposed. We generalize this kind of the basic model modification by introducing additional terms describing the influences that are concentrated, close to the impulse character.

Theoretical studies of similar mathematical models with concentrated influences related to the existence and uniqueness of generalized solutions were studied in Lyashko et al. (2010), Sandrakov et al. (2019) and Klyushin et al. (2020). In Sandrakov et al. (2019), the optimization of similar models in the classes of generalized influences of finite order and in Lyashko and Mankovskiui (1983) and Lyashko and Semenov (2001) the problem of their control were studied.

The spatio-temporal dynamics of the model components of the infectious disease process, taking into account the concentrated influences in the domain $G_Z = \{(x,t) : -\infty < x < +\infty;\ t_0 < t < +\infty\}$, will be described by such a singularly perturbed dimensionless nonlinear differential equations system (Baranovsky et al. 2021):

$$\begin{aligned}
&v_t'(x,t) = h_{16}u_V(x,t) + \left(h_1 - h_2 f(x,t) - h_{13}k(x,t)\right)v(x,t) + \varepsilon h_9 v_{xx}''(x,t),\\
&s_t'(x,t) = \xi(m)h_3\left(f(x,t-\tau) + k(x,t-\tau)\right)v(x,t-\tau) - h_5\left(s(x,t)-1\right) +\\
&+\varepsilon^2 h_{10}s_{xx}''(x,t),\\
&f_t'(x,t) = h_4\left(s(x,t) - f(x,t)\right) - h_8 v(x,t) f(x,t) + \varepsilon h_{11} f_{xx}''(x,t),\\
&k_t'(x,t) = h_{16}u_K(x,t) - \left(h_{14} + h_{15}v(x,t)\right)k(x,t) + \varepsilon h_{17}k_{xx}''(x,t),\\
&m_t'(x,t) = h_6 v(x,t) - h_7 m(x,t) + \varepsilon^2 h_{12} m_{xx}''(x,t),
\end{aligned} \qquad [11.16]$$

under conditions

$$\begin{aligned}
&s(x,t_0) = s^0(x),\ m(x,t_0) = m^0(x),\ v(x,\tilde{t}) = v^0(x,\tilde{t}),\\
&f(x,\tilde{t}) = f^0(x,\tilde{t}),\ k(x,\tilde{t}) = k^0(x,\tilde{t}),\ t_0 - \tau \le \tilde{t} \le t_0,
\end{aligned} \qquad [11.17]$$

where $k(x,t) = K(x,t)/F^*$, $K(x,t)$ is the concentration of donor antibodies at the point x at time t; $h_{13} = \gamma_1 F^*$, $h_{14} = \mu_k$, $h_{15} = \eta_1\gamma_1 V_m$ and $h_{16} = 1/F^*$; γ_1 is the coefficient that takes into account the result of the antigens interaction with donor antibodies; η_1 is the cost of donor antibodies to neutralize one antigen; εh_{17} is the diffusion redistribution coefficient of donor antibodies.

The introduction of donor antibodies into the organism can be carried out in various ways, in particular by injecting a solution for injection through a syringe. In this case, the change in the concentration of donor antibodies is also described as close to the pulse-point function of the source $u_K(x,t)$ with the maximum values at points x_{Kj} at times t_{Kj}:

$$u_K(x,t) = \sum_{j=1}^{n_K} A_{Kj} e^{-\alpha_{Kj}(x-x_{Kj})^2} e^{-\beta_{Kj}(t-t_{Kj})^2} \qquad [11.18]$$

where A_{Kj}, α_{Kj} and β_{Kj} are parameters that determine the intensity, "duration" and the spatial length of the injection. In the case, when the injection solution with donor antibodies is step-by-step injected through a dropper during intervals of time $(\overline{t}_j, \overline{\overline{t}}_j)$ at points $\hat{x}_{Kj}$ (or on the intervals $(\overline{x}_j, \overline{\overline{x}}_j)$), we describe the change of the donor antibodies concentration by the function of the source $u_K(x,t)$ in the forms of

$$u_K(x,t) = \sum_{j=1}^{n_K} A_{Kj} e^{-\alpha_{Kj}(x-\hat{x}_{Kj})^2} \left[\frac{e^{\beta_{Kj}(t-\bar{t}_{Kj})}}{e^{\beta_{Kj}(t-\bar{t}_{Kj})}+1} - \frac{e^{\beta_{Kj}(t-\bar{\bar{t}}_{Kj})}}{e^{\beta_{Kj}(t-\bar{\bar{t}}_{Kj})}+1} \right] \quad [11.19]$$

or

$$u_K(x,t) = \sum_{j=1}^{n_K} A_{Kj} e^{-\alpha_{Kj}(x-\hat{x}_{Kj})^2} \left[\frac{e^{\beta_{Kj}(t-\bar{t}_{Kj})}}{e^{\beta_{Kj}(t-\bar{t}_{Kj})}+1} - \frac{e^{\beta_{Kj}(t-\bar{\bar{t}}_{Kj})}}{e^{\beta_{Kj}(t-\bar{\bar{t}}_{Kj})}+1} \right] \quad [11.20]$$

Virus replication leads to the death of infected cells. As a result, a significant number of new-formed viral particles are released in the places of a destroyed cell. Such close to the pulse change concentration of the antigens at some points x_{Vj} at time t_{Vj}, we describe by the function $u_V(x,t)$ as

$$u_V(x,t) = \sum_{j=1}^{n_V} A_{Vj} e^{-\alpha_{Vj}(x-x_{Vj})^2} e^{-\beta_{Vj}(t-t_{Vj})^2} \quad [11.21]$$

where A_{Vj}, α_{Vj} and β_{Vj} are parameters that determine the intensity and "duration" of the change in antigen concentration. The generalization allows the choice as the initial conditions and, in particular, the conditions that characterize the stationary solution, which corresponds to the healthy organism state, that is

$$s(x,t_0) = 1,\ m(x,t_0) = 0,\ v(x,\tilde{t}) = 0,\ f(x,\tilde{t}) = 1,\ k(x,\tilde{t}) = 0,$$
$$t_0 - \tau \le \tilde{t} \le t_0 \quad [11.22]$$

As before, we will consider development of the infection disease process under the condition $\xi(m) = 1$ and reduce the solution of problems [11.16]–[11.17] with time delay to sequence of problems without time delay. In particular, at the first time stage, we will have the following task:

$$
\begin{cases}
v'_{0t}(x,t) = h_{16}u_V(x,t) + \left(h_1 - h_2 f_0(x,t) - h_{13}k_0(x,t)\right)v_0(x,t) + \\
+\varepsilon h_9 v''_{0xx}(x,t), \\
s'_{0t}(x,t) = h_3\left(f^0(x,t-\tau) + k^0(x,t-\tau)\right)v^0(x,t-\tau) - h_5\left(s_0(x,t)-1\right) + \\
+\varepsilon^2 h_{10} s''_{0xx}(x,t), \\
f'_{0t}(x,t) = h_4\left(s_0(x,t) - f_0(x,t)\right) - h_8 v_0(x,t) f_0(x,t) + \varepsilon h_{11} f''_{0xx}(x,t), \\
k'_{0t}(x,t) = h_{16}u_K(x,t) - \left(h_{14} + h_{15}v_0(x,t)\right)k_0(x,t) + \varepsilon h_{17}k''_{0xx}(x,t), \\
m'_{0t}(x,t) = h_6 v_0(x,t) - h_7 m_0(x,t) + \varepsilon^2 h_{12} m''_{0xx}(x,t), \\
s_0(x,t_0) = s^0(x),\ m_0(x,t_0) = m^0(x),\ v_0(x,t_0) = v^0(x,t_0), \\
f_0(x,t_0) = f^0(x,t_0),\ k_0(x,t_0) = k^0(x,t_0),\ t_0 \le t \le t_0 + \tau;
\end{cases}
\quad [11.23]
$$

As in the previous case, we consider small-scare diffusion redistributions of the active components and present the solutions of the obtained problems without time delay in the form of asymptotic series:

$$
\begin{aligned}
&v_j(x,t) = \sum_{i=o}^{n} \varepsilon^i v_{ij}(x,t) + R_{nj}^v(x,t,\varepsilon),\ s_j(x,t) = \sum_{i=o}^{n} \varepsilon^i s_{ij}(x,t) + R_{nj}^s(x,t,\varepsilon), \\
&f_j(x,t) = \sum_{i=o}^{n} \varepsilon^i f_{ij}(x,t) + R_{nj}^f(x,t,\varepsilon),\ k_j(x,t) = \sum_{i=o}^{n} \varepsilon^i k_{ij}(x,t) + R_{nj}^k(x,t,\varepsilon), \quad [11.24] \\
&m_j(x,t) = \sum_{i=o}^{n} \varepsilon^i m_{ij}(x,t) + R_{nj}^m(x,t,\varepsilon).
\end{aligned}
$$

We will obtain the problems for getting the functions $v_{ij}(x,t)$, $s_{ij}(x,t)$, $f_{ij}(x,t)$, $k_{ij}(x,t)$, $m_{ij}(x,t)$ $(i = 0,1,\dots,n,\ j = 0,1,\dots,r,\dots)$, similarly to the previous one, after performance of procedure "equalization". In particular, for the first time stage:

$$\begin{cases} v'_{0,0\,t}(x,t) = h_{16}u_V(x,t) + \left(h_1 - h_2 f_{0,0}(x,t) - h_{13}k_{0,0}(x,t)\right)v_{0,0}(x,t), \\ s'_{0,0\,t}(x,t) = h_3\left(f^0(x,t-\tau) + k^0(x,t-\tau)\right)v^0(x,t-\tau) - h_5\left(s_{0,0}(x,t) - 1\right), \\ f'_{0,0\,t}(x,t) = h_4\left(s_{0,0}(x,t) - f_{0,0}(x,t)\right) - h_8 v_{0,0}(x,t) f_{0,0}(x,t), \\ k'_{0,0\,t}(x,t) = h_{16}u_K(x,t) - \left(h_{14} + h_{15}v_{0,0}(x,t)\right)k_{0,0}(x,t), \\ m'_{0,0\,t}(x,t) = h_6 v_{0,0}(x,t) - h_7 m_{0,0}(x,t), \\ s_{0,0}(x,t_0) = s^0(x),\ m_{0,0}(x,t_0) = m^0(x),\ v_{0,0}(x,t_0) = v^0(x,t_0), \\ f_{0,0}(x,t_0) = f^0(x,t_0),\ k_{0,0}(x,t_0) = k^0(x,t_0),\ t_0 \le t \le t_0 + \tau, \end{cases} \quad [11.25]$$

. . . .

$$\begin{cases} v'_{l,0\,t}(x,t) = h_1 v_{l,0}(x,t) - h_2\left(a_{0,0}(x,t) f_{l,0}(x,t) + b_{0,0}(x,t) v_{l,0}(x,t)\right) - \\ -h_{13}\left(a_{0,0}(x,t) k_{l,0}(x,t) + c_{0,0}(x,t) v_{l,0}(x,t)\right) + \Phi_{v\,l,0}(x,t), \\ s'_{l,0\,t}(x,t) = h_3\left(a_{0,0}(x,t-\tau)\left(f_{l,0}(x,t-\tau) + k_{l,0}(x,t-\tau)\right) + \left(b_{0,0}(x,t-\tau) + \right.\right. \\ \left.\left. + c_{0,0}(x,t-\tau)\right) v_{l,0}(x,t-\tau)\right) - h_5 s_{l,0}(x,t) + \Phi_{s\,l,0}(x,t), \\ f'_{l,0\,t}(x,t) = h_4\left(s_{l,0}(x,t) - f_{l,0}(x,t)\right) - h_8\left(a_{0,0}(x,t) f_{l,0}(x,t) + b_{0,0}(x,t) \times \right. \\ \left. \times v_{l,0}(x,t)\right) + \Phi_{f\,l,0}(x,t), \\ k'_{l,0\,t}(x,t) = -h_{14}k_{l,0}(x,t) - h_{15}\left(a_{0,0}(x,t) k_{l,0}(x,t) + c_{0,0}(x,t) v_{l,0}(x,t)\right) + \\ + \Phi_{k\,l,0}(x,t), \\ m'_{l,0\,t}(x,t) = h_6 v_{l,0}(x,t) - h_7 m_{l,0}(x,t) + \Phi_{m\,l,0}(x,t), \\ s_{l,0}(x,t_0) = 0,\ m_{l,0}(x,t_0) = 0,\ v_{l,0}(x,t_0) = 0, \\ f_{l,0}(x,t_0) = 0,\ k_{l,0}(x,t_0) = 0,\ t_0 \le t \le t_0 + \tau,\ l = 1,2,\ldots,n, \end{cases} \quad [11.26]$$

where $a_{0,0}(x,t) = v_{0,0}(x,t),\ b_{0,0}(x,t) = f_{0,0}(x,t),\ c_{0,0}(x,t) = k_{0,0}(x,t)$;

$$\Phi_{v1,0}(x,t) = h_9 v''_{0,0\,xx}(x,t),\ \Phi_{s1,0}(x,t) = 0,\ \Phi_{m1,0}(x,t) = 0,$$

$$\Phi_{f1,0}(x,t) = h_{11} f''_{0,0xx}(x,t),\ \Phi_{k1,0}(x,t) = h_{17} k''_{0,0xx}(x,t);$$

$$\Phi_{v\,l,0}(x,t) = -h_2 \sum_{q=1}^{l-1} v_{q,0}(x,t) f_{l-q,0}(x,t) + h_9 v''_{l-1,0\,xx}(x,t),$$

$$\Phi_{s\,l,0}(x,t)=h_3\sum_{q=1}^{l-1}v_{q,0}(x,t-\tau)f_{l-q,0}(x,t-\tau)+h_{10}s''_{l-2,0xx}(x,t),$$

$$\Phi_{f\,l,0}(x,t)=-h_8\sum_{q=1}^{l-1}v_{q,0}(x,t)f_{l-q,0}(x,t)+h_{11}f''_{l-1,0xx}(x,t),$$

$$\Phi_{k\,l,0}(x,t)=-h_{15}\sum_{q=1}^{l-1}v_{q,0}(x,t)k_{l-q,0}(x,t)+h_{17}k''_{l-1,0xx}(x,t),$$

$$\Phi_{m\,l,0}(x,t)=h_{12}m''_{l-2,0xx}(x,t),\ \ l=2,3,\dots,n\,.$$

Note that the implementation of immunotherapy procedures does not limit all of the methods of treatment of infectious diseases. To reduce the level of pathogenic antigens in the organism, other drugs different to immunobiological can be used. Using that may, in particular, suppress the natural response of the immune system to pathogens of the disease. In Bomba and Baranovsky (2020), the modification of the basic infectious disease model for the account of diffusion perturbations in the conditions of type of pharmacotherapy is presented.

11.4. Modeling the influence of diffusion perturbations on development of infectious diseases under convection

Together with diffusion redistribution, convection transfer (e.g. antigens and antibodies by lymph from tissues at first to the lymphatic and then to the circulatory system, etc.) must also be taken into account. Below we note the generalization of the basic disease model taking into account diffusion perturbations in the convection under immunotherapy and concentrated effects.

The spatio-temporal dynamics of the infection disease process taking into account diffusion perturbations and convection in the conditions of immunotherapy in the set $G_Z=\{(x,t):\ 0<x<l,\ t_0<t<+\infty\}$ is described by the following singularly perturbed system of nonlinear differential equations (with time delay):

$$v'_t(x,t)=h_{16}u_V(x,t)+\left(h_1-h_2f(x,t)-h_{13}k(x,t)\right)v(x,t)+\varepsilon h_9v''_{xx}(x,t)-$$
$$-\upsilon_Vv'_x,$$

$$s'_t(x,t) = \xi(m)h_3\left(f(x,t-\tau)+k(x,t-\tau)\right)v(x,t-\tau) - h_5\left(s(x,t)-1\right)+$$
$$+\varepsilon h_{10}s''_{xx}(x,t) - \upsilon_C s'_x,$$
$$f'_t(x,t) = h_4\left(s(x,t)-f(x,t)\right) - h_8 v(x,t)f(x,t) + \varepsilon h_{11}f''_{xx}(x,t) - \upsilon_F f'_x, \qquad [11.27]$$
$$k'_t(x,t) = h_{16}u_K(x,t) - \left(h_{14}+h_{15}v(x,t)\right)k(x,t) + \varepsilon h_{17}k''_{xx}(x,t) - \upsilon_K k'_x,$$
$$m'_t(x,t) = h_6 v(x,t) - h_7 m(x,t) + \varepsilon h_{12}m''_{xx}(x,t) - \upsilon_m m'_x,$$

under conditions

$$s(x,t_0) = s^0(x),\ m(x,t_0) = m^0(x),\ v(x,\tilde{t}) = v^0(x,\tilde{t}),\ f(x,\tilde{t}) = f^0(x,\tilde{t}),$$
$$k(x,\tilde{t}) = k^0(x,\tilde{t}),\ t_0 - \tau \le \tilde{t} \le t_0, \qquad [11.28]$$

$$L_{*s}s(0,t) = s_*(t),\ L_{*m}m(0,t) = m_*(t),\ L_{*v}v(0,t) = v_*(t),\ L_{*f}f(0,t) = f_*(t),$$
$$L_{*k}k(0,t) = k_*(t),\ \overset{*}{L}_s s(l,t) = s^*(t),\ \overset{*}{L}_m m(l,t) = m^*(t),\ \overset{*}{L}_v v(l,t) = v^*(t), \qquad [11.29]$$
$$\overset{*}{L}_f f(l,t) = f^*(t),\ \overset{*}{L}_k k(l,t) = k^*(t),$$

where $\upsilon_V = \delta_V \upsilon$, $\upsilon_F = \delta_F \upsilon$, $\upsilon_K = \delta_K \upsilon$, $\upsilon_C = \delta_C \upsilon$ and $\upsilon_m = \delta_m \upsilon$ are convection transfer rates of the corresponding active factors of the process (υ is the move rate of the environment) $0 < \delta_V, \delta_C, \delta_F, \delta_K, \delta_m \le 1$; L_{*s}, L_{*m}, L_{*v}, L_{*f}, L_{*k}, $\overset{*}{L}_s$, $\overset{*}{L}_m$, $\overset{*}{L}_v$, $\overset{*}{L}_f$ and $\overset{*}{L}_k$ are operators of boundary conditions (of 1st, 2nd, 3rd kind, periodicity, etc.).

As before, we reduce the solution of problems [11.27]–[11.29] with time delay to a sequence of solutions of problems without time delay. In particular, in the case $\xi(m) = 1$ and under boundary conditions of the 1st kind on intervals $r\tau < t \le (r+1)\tau$ ($r = 1,2,\ldots$), we have

$$v'_{rt}(x,t) = h_{16}u_V(x,t) + \left(h_1 - h_2 f_r(x,t) - h_{13}k_r(x,t)\right)v_r(x,t)+$$
$$+\varepsilon h_9 v''_{r\,xx}(x,t) - \upsilon_V v'_{r\,x}(x,t),$$
$$s'_{rt}(x,t) = h_3\left(f_{r-1}(x,t-\tau) + k_{r-1}(x,t-\tau)\right)v_{r-1}(x,t-\tau) -$$
$$-h_5\left(s_r(x,t)-1\right) + \varepsilon h_{10}s''_{r\,x}(x,t) - \upsilon_C s'_{r\,xx}(x,t), \qquad [11.30]$$
$$f'_{rt}(x,t) = h_4\left(s_r(x,t) - f_r(x,t)\right) - h_8 v_r(x,t)f_r(x,t) + \varepsilon h_{11}f''_{r\,xx}(x,t) -$$
$$-\upsilon_F f'_{r\,x}(x,t),$$

$$k'_{rt}(x,t) = h_{16}u_K(x,t) - (h_{14} + h_{15}v_r(x,t))k_r(x,t) + \varepsilon h_{17}k''_{r\,xx}(x,t) -$$
$$-\upsilon_K k'_{r\,x}(x,t),$$
$$m'_{rt}(x,t) = h_6 v_r(x,t) - h_7 m_r(x,t) + \varepsilon h_{12} m''_{r\,xx}(x,t) - \upsilon_m m'_{r\,x}(x,t),$$

$$\begin{gathered} s_r(x,r\tau) = s_{r-1}(x,r\tau),\ m_r(x,r\tau) = m_{r-1}(x,r\tau),\ v_r(x,r\tau) = v_{r-1}(x,r\tau), \\ f_r(x,r\tau) = f_{r-1}(x,r\tau),\ k_r(x,r\tau) = k_{r-1}(x,r\tau),\ s_r(0,t) = s_*(t), \\ m_r(0,t) = m_*(t),\ v_r(0,t) = v_*(t),\ f_r(0,t) = f_*(t),\ k_r(0,t) = k_*(t), \\ s_r(l,t) = s^*(t),\ m_r(l,t) = m^*(t),\ v_r(l,t) = v^*(t),\ f_r(l,t) = f^*(t), \\ k_r(l,t) = k^*(t),\ r\tau < t \le (r+1)\tau. \end{gathered} \qquad [11.31]$$

The solutions of such problems are presented in the form of asymptotic series:

$$\begin{gathered} v_j(x,t) = \sum_{i=0}^{n} \varepsilon^i v_{ij}(x,t) + \sum_{i=0}^{n} \varepsilon^i \Pi_{v\,ij}(\xi,t) + R_{nj}^{v}(x,t,\varepsilon), \\ s_j(x,t) = \sum_{i=0}^{n} \varepsilon^i s_{ij}(x,t) + \sum_{i=0}^{n} \varepsilon^i \Pi_{s\,ij}(\xi,t) + R_{nj}^{s}(x,t,\varepsilon), \\ f_j(x,t) = \sum_{i=0}^{n} \varepsilon^i f_{ij}(x,t) + \sum_{i=0}^{n} \varepsilon^i \Pi_{f\,ij}(\xi,t) + R_{nj}^{f}(x,t,\varepsilon), \\ k_j(x,t) = \sum_{i=0}^{n} \varepsilon^i k_{ij}(x,t) + \sum_{i=0}^{n} \varepsilon^i \Pi_{k\,ij}(\xi,t) + R_{nj}^{k}(x,t,\varepsilon), \\ m_j(x,t) = \sum_{i=0}^{n} \varepsilon^i m_{ij}(x,t) + \sum_{i=0}^{n} \varepsilon^i \Pi_{m\,ij}(\xi,t) + R_{nj}^{m}(x,t,\varepsilon), \end{gathered} \qquad [11.32]$$

where $\Pi_{vij}(\xi,t)$, $\Pi_{sij}(\xi,t)$, $\Pi_{f\,ij}(\xi,t)$, $\Pi_{k\,ij}(\xi,t)$ and $\Pi_{mij}(\xi,t)$ are functions of boundary layer type (Bomba et al. 2008) in neighborhood $x = l$, and $\xi = (l - x)\cdot\varepsilon^{-1}$ is the regularizing transformation (tension variable). After performing the standard procedure of "equalization" (Bomba 1982; Bomba et al. 2008), we obtain the problems to determine these functions. In particular, for $i = 0$, $j = r > 0$ ($r\tau \le t \le (r+1)\tau$), we have

$$v'_{0,rt}(x,t)+\upsilon_V v'_{0,rx}(x,t)=h_{16}u_V(x,t)+\left(h_1-h_2 f_{0,r}(x,t)-h_{13}k_{0,r}(x,t)\right)\times$$
$$\times v_{0,r}(x,t),$$
$$s'_{0,rt}(x,t)+\upsilon_C s'_{0,r\,xx}(x,t)=-h_5\left(s_{0,r}(x,t)-1\right)+\Phi_{s\,0,r},$$

$$f'_{0,rt}(x,t)+\upsilon_F f'_{0,rx}(x,t)=h_4\left(s_{0,r}(x,t)-f_{0,r}(x,t)\right)-h_8 v_{0,r}(x,t)\times$$
$$\times f_{0,r}(x,t),$$
$$k'_{0,rt}(x,t)+\upsilon_K k'_{0,rx}(x,t)=h_{16}u_K(x,t)-\left(h_{14}+h_{15}v_{0,r}(x,t)\right)k_{0,r}(x,t), \qquad [11.33]$$
$$m'_{0,rt}+\upsilon_m m'_{0,rx}(x,t)(x,t)=h_6 v_{0,r}(x,t)-h_7 m_{0,r}(x,t),$$

$$s_{0,r}(x,r\tau)=s_{0,r-1}(x,r\tau),\ m_{0,r}(x,r\tau)=m_{0,r-1}(x,r\tau),$$
$$v_{0,r}(x,r\tau)=v_{0,r-1}(x,r\tau),\ f_{0,r}(x,r\tau)=f_{0,r-1}(x,r\tau),$$
$$k_{0,r}(x,r\tau)=k_{0,r-1}(x,r\tau),\ s_{0,r}(0,t)=s_*(t),\ m_{0,r}(0,t)=m_*(t), \qquad [11.34]$$
$$v_{0,r}(0,t)=v_*(t),\ f_{0,r}(0,t)=f_*(t),\ k_{0,r}(0,t)=k_*(t),$$

where $\Phi_{s0,r}=h_3\left(f_{0,r-1}(x,t-\tau)+k_{0,r-1}(x,t-\tau)\right)v_{0,r-1}(x,t-\tau)$;

$$\begin{cases} h_9\Pi_{v0,r}{}''_{\xi\xi}+\upsilon_V\Pi_{v0,r}{}'_{\xi}=0, \\ \Pi_{v0,r}(0,t)=v^*(t)-v_{0,r}(l,t),\quad \Pi_{v0,r}(\xi,t)\xrightarrow[\xi\to\infty]{}0, \end{cases}$$
$$\begin{cases} h_{10}\Pi_{s0,r}{}''_{\xi\xi}+\upsilon_C\Pi_{s0,r}{}'_{\xi}=0, \\ \Pi_{c0,r}(0,t)=s^*(t)-s_{0,r}(l,t),\quad \Pi_{s0,r}(\xi,t)\xrightarrow[\xi\to\infty]{}0, \end{cases}$$
$$\begin{cases} h_{11}\Pi_{f0,r}{}''_{\xi\xi}+\upsilon_F\Pi_{f0,r}{}'_{\xi}=0, \\ \Pi_{f0,r}(0,t)=f^*(t)-f_{0,r}(l,t),\quad \Pi_{f0,r}(\xi,t)\xrightarrow[\xi\to\infty]{}0, \end{cases} \qquad [11.35]$$
$$\begin{cases} h_{17}\Pi_{k0,r}{}''_{\xi\xi}+\upsilon_K\Pi_{k0,r}{}'_{\xi}=0, \\ \Pi_{k0,r}(0,t)=k^*(t)-k_{0,r}(l,t),\quad \Pi_{k0,r}(\xi,t)\xrightarrow[\xi\to\infty]{}0, \end{cases}$$
$$\begin{cases} h_{12}\Pi_{m0,r}{}''_{\xi\xi}+\upsilon_m\Pi_{m0,r}{}'_{\xi}=0, \\ \Pi_{m0,r}(0,t)=m^*(t)-m_{0,r}(l,t),\quad \Pi_{m0,r}(\xi,t)\xrightarrow[\xi\to\infty]{}0. \end{cases}$$

In the process of iterations, the solutions of each problems [11.33]–[11.34] are found numerically using the ideas of the method of characteristics (Bomba 1982; Bomba et al. 2008) and data from the solutions of the problems of the previous stages. And the solutions of problems [11.35], for example, under conditions of 1st kind at $x = l$, can be obtained analytically in the form:

$$\Pi_{v0,r}(\xi,t) = (\mathrm{v}^*(t) - v_{0,r}(l,t)) \cdot e^{-\frac{\upsilon_V}{h_9}\cdot\xi},$$

$$\Pi_{s0,r}(\xi,t) = (s^*(t) - s_{0,r}(l,t)) \cdot e^{-\frac{\upsilon_s}{h_{10}}\cdot\xi},$$

$$\Pi_{f0,r}(\xi,t) = (f^*(t) - f_{0,r}(l,t)) \cdot e^{-\frac{\upsilon_F}{h_{11}}\cdot\xi},$$

$$\Pi_{k0,r}(\xi,t) = (k^*(t) - k_{0,r}(l,t)) \cdot e^{-\frac{\upsilon_K}{h_{17}}\cdot\xi},$$

$$\Pi_{m0,r}(\xi,t) = (m^*(t) - m_{0,r}(l,t)) \cdot e^{-\frac{\upsilon_m}{h_{12}}\cdot\xi}, \; \ldots .$$

Note that the proposed approach can be easily to "transfer", in particular, to the set $G_Z = \{(x,t)\colon\ 0 < x < l,\ 0 < y < d,\ t_0 < t < +\infty\}$. At the same time, we introduce additional functions of the boundary layer type in the neighborhoods $y = 0$ and $y = d$, and the angular functions (Bomba et al. 2008) in the case when the consistency of the initial and boundary conditions is insufficient.

11.5. Numerical experiment results

Figure 11.1 presents the spatial–temporal dynamics of the level of concentration of pathogenic antigens according to models [11.6]–[11.7], under the conditions in the space of: $v(x,\tilde{t}) = v^0(\tilde{t}),\ t_0 - \tau \le \tilde{t} \le t_0$. The small-scale spatially distributed diffuse influences in this case do not cause changes in the dynamics of the infectious disease process. In the case when the distribution of pathogenic antigens at the initial time t_0 is spatial nonuniform $v(x,\tilde{t}) = v^0(x,\tilde{t}),\ t_0 - \tau \le \tilde{t} \le t_0$, taking into account the small-scale spatial distribution of diffuse influences in [11.6]–[11.7] leads to changes in the corresponding model dynamics of infectious disease processes. Figure 11.2 presents the spatial–temporal dynamics of the concentration of pathogenic antigens in the organism in the case of a separate locus of infection when the small-scale spatially distributed diffuse influences are absent (Figure 11.2a) and present (Figure 11.2b). The results of numerical experiments show a decrease in the maximum level of concentration of pathogenic antigens in the infection locus in the case of diffuse influences. It is the result of the "diffusion" spread of antigens from the places with their high concentration to places with lower concentration.

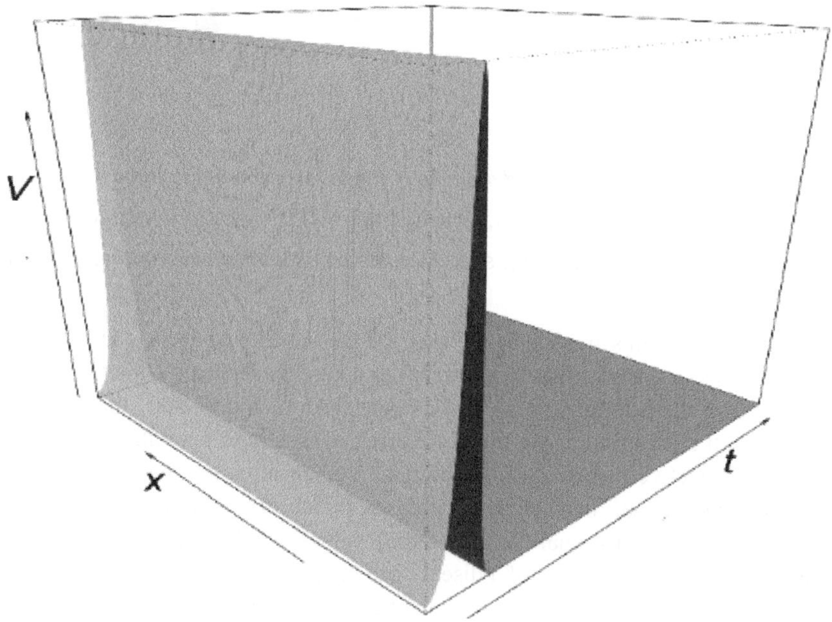

Figure 11.1. *Spatio-temporal dynamics of the level of the concentration of pathogenic antigens under condition* $v(x,\tilde{t}) = v^0(\tilde{t}),\ t_0 - \tau \le \tilde{t} \le t_0$. *For a color version of this figure, see www.iste.co.uk/koroliouk/computational.zip*

As it is known, the infectious disease can take various forms that depend on the intensity of the immune system's response to antigens. In the papers by Marchuk (1980, 1991, 1997), four possible types of solutions of the simplest model are described, which correspond to four forms of disease dynamics: subclinical, acute, chronic and lethal.

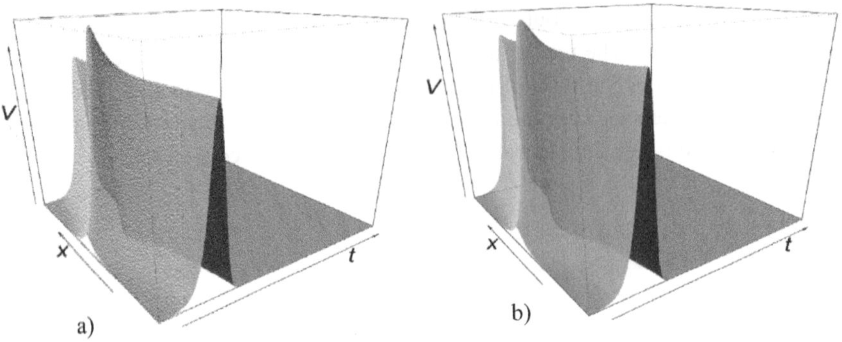

Figure 11.2. *Spatio-temporal dynamics of the level of the pathogenic antigens concentration under condition* $v^0(x,t_0) = \delta/(1+(x-\lambda)^2)$ *: (a)* $\varepsilon = 0$ *and (b)* $\varepsilon = 0.1$. *For a color version of this figure, see www.iste.co.uk/koroliouk/computational.zip*

In particular, in this model, in the case of the subclinical form, the number of special antigens in the organism is sufficient to neutralize antigens without triggering the mechanism of their formation. In the acute form of the disease, the amount of antibodies in the organism at the initial time is insufficient to neutralize all antigens, but with an effective response of the immune system after starting the mechanism of antibody production, their number increases, which causes a gradual decrease of antigen concentration to small (almost zero) values. Under the weak reaction of immune system, the acute form of a disease can pass to chronic when the antigens are not completely neutralized and over time their concentration in an organism become at a certain stationary level. It is clear that the choice of a treatment program should also be consistent with the predicted form of the viral disease, because in cases of subclinical and acute forms, the immune system is able to overcome the disease without external treatment. Therefore, the numerical experiments focused on the study features of the course of infectious disease in various situational conditions under diffusion perturbations, taking into account close to pulse influences.

The model dynamics of antigen concentration in the infection locus under conditions of acute infectious disease at different intensities of diffusion "redistribution" is illustrated in Figure 11.3a. At the initial time t_0, the antigen concentration is uneven in space: $v(x,\tilde{t}) = v^0(x,\tilde{t}) = \lambda e^{-\delta(x-\theta)^2} e^{-\delta(\tilde{t}-t_0)^2}$, $t_0 - \tau \le \tilde{t} \le t_0$, that is, there is an initial infection of the organism in separate center with the maximum concentration of antigens at the point $x = \theta$. Under conditions without diffusion perturbation (ε = 0), as in the base model, first, the concentration of antigens increases, then decreases and is established almost zero over time. Under conditions of diffusion perturbation, the maximum value of the model antigen concentration decreases. As the intensity of diffusion "redistribution" increases, starting from the certain level, the concentration of antigens in the infection locus will never increase. In the case of the corresponding diffusion "redistribution" of antigens in the center of infection, their concentration in the locus decreases to values at which the accepted level of immune protection in the model becomes sufficient to prevent further increase in the model concentration of antigens and to decrease to almost zero without "exacerbations" over time. Figure 11.3b presents the change in time of the concentration of antigens in the infection locus when, under the same conditions and at the initial time, the immune system antigens in the organism have not yet been detected (i.e. $v(x,\tilde{t}) = v^0(x,\tilde{t}) = 0,\ t_0 - \tau \le \tilde{t} \le t_0$). But due to the destruction of previously infected cell at the moment $t_1 > t_0$, there is a sharp increase in their concentration at the point $x = \theta$: $u_V(x,t) = \lambda e^{-\delta(x-\theta)^2} e^{-\delta(t-t_1)^2}$.

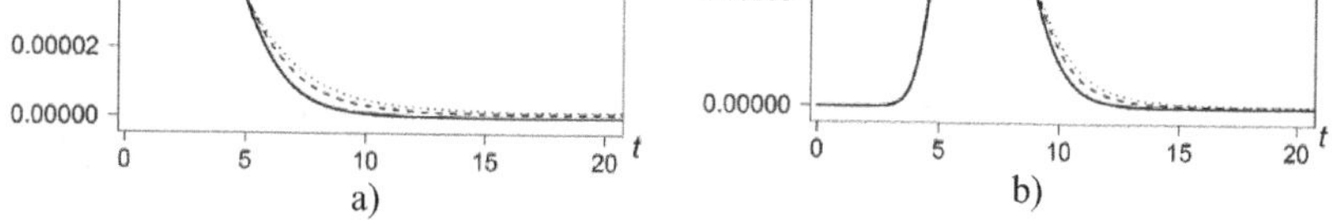

Figure 11.3. *Dynamics of the antigen concentration in the disease locus in the acute form at different levels of diffusion "redistribution" intensity under the conditions: a) presence of the initial infection distribution* $v^0(x,\tilde{t}) = \lambda e^{-\delta(x-\theta)^2} e^{-\delta(\tilde{t}-t_0)^2}$, $t_0 - \tau \le \tilde{t} \le t_0$ *and b) presence of the pulse source of antigens* $u_V(x,t) = \lambda e^{-\delta(x-\theta)^2} e^{-\delta(t-t_1)^2}$, $t_1 > t_0$

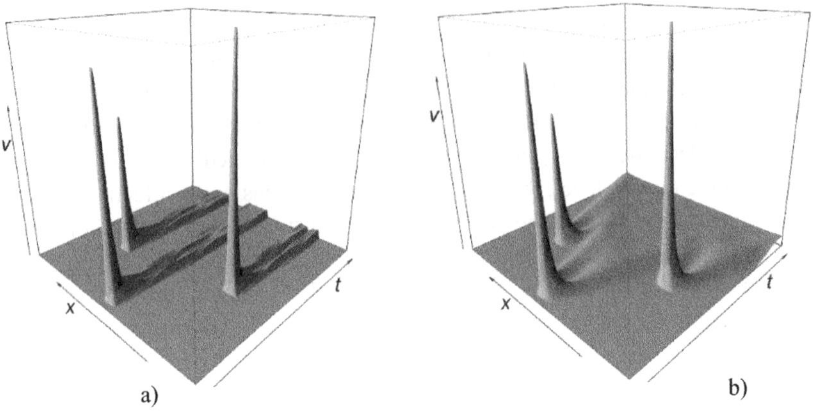

Figure 11.4. *Spatial–temporal dynamics of antigen concentration in the presence of some sources of pulse type in the case of chronic infectious disease. For a color version of this figure, see www.iste.co.uk/korolіouk/computational.zip*

Figure 11.4 shows the model of spatio-temporal dynamics of antigen concentrations with the development of viral disease process in the chronic form in cases without taking into account (Figure 11.4a) and taking into account (Figure 11.4b) diffusion perturbations under conditions when the initial values of active process factors correspond to the values of the stationary solution that characterizes the state of a healthy organism. At different times ($t_1 < t_2 < t_3$) and at different points ($x_3 < x_1 < x_2$) (in different foci of infection), there are sharp, close to pulse changes in antigen concentrations. The numerical experiments results illustrate a certain "localization" in some area of the model process of infectious disease in the chronic form without taking into account the diffusion "redistribution". The influence of diffusion "redistribution" of the excessive number of antigens in the infection locus smoothes this "localization" of the model process development and "equalizes" the spatial distribution of antigen concentration over time.

In particular, the weak immune response to viral elements that is associated with insufficient concentration and antibody production rate can lead to the chronic form development of infectious diseases. An effective decision on the choice of treatment in this case can be the immunotherapy (Bomba and Baranovsky 2020; Bomba et al. 2020b), which involves the injection of donor antibodies. The injection of these additional antibodies can be carried out in various ways, in particular, by injecting using a syringe or by using a dropper. In this model, the injection of donor antibodies can be described as close to the impulse functions of type

[11.18]–[11.19]. Figure 11.5 illustrates the spatio-temporal changes of the model concentrations of antigens with the development of the viral disease process in the chronic form in the presence of the pulse source of antigens and two pulse sources of donor antibodies with maximum values at the point $x_V = x_{K1} = x_{K2}$ in the case of the injection with a syringe at the time $t_V < t_{K1} < t_{K2}$ (Figure 11.5a) and with a dropper at the intervals $(\bar{t}_{K1}, \bar{\bar{t}}_{K1})$ and $(\bar{t}_{K2}, \bar{\bar{t}}_{K2})$, $\bar{\bar{t}}_{K1} < \bar{t}_{K2}$ (Figure 11.5b). The results show the decrease in the antigen concentration of in the areas around the points of the injection of donor antibodies. Moreover, in the case of intravenous injections, the choice of different duration of injection of the drug solution makes it possible to control over the duration and effectivity of the effect of the injected donor antibodies on the process development.

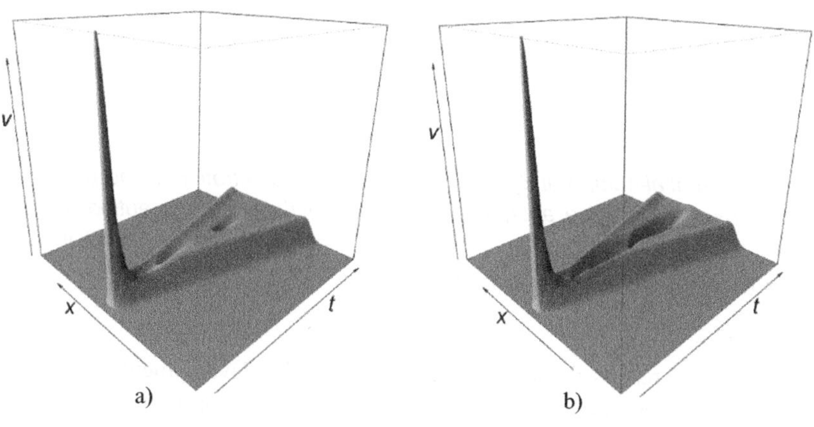

Figure 11.5. *Spatial–temporal dynamics of antigen concentration in the presence of sources of donor antibodies that model cases of injection using: a) syringe and b) dropper. For a color version of this figure, see www.iste.co.uk/korolliouk/computational.zip*

The effectivity of immunotherapy depends on both the number of injected donor antibodies and the time of their injection. Figure 11.6 presents the spatial–temporal dynamics of antigen concentration with the development of chronic viral disease in situations when injections of donor antibodies are carried out at different intervals according to the time of detection by the immune system of the infection locus. The concentration of antibodies in the injections is the same in both cases. These results illustrate the increase, in particular, in the duration of the "therapeutic effect" at the operative injection of donor antibodies into an organism (Figure 11.6b). At later injections of donor antibodies (Figure 11.6a), after the detection of the locus of viral infection, the duration of their action is shorter due to the higher rate of their using for neutralization of the increased number of antigens. Rational choice of time for the injections can help to monitor the effectivity of immunotherapy procedures.

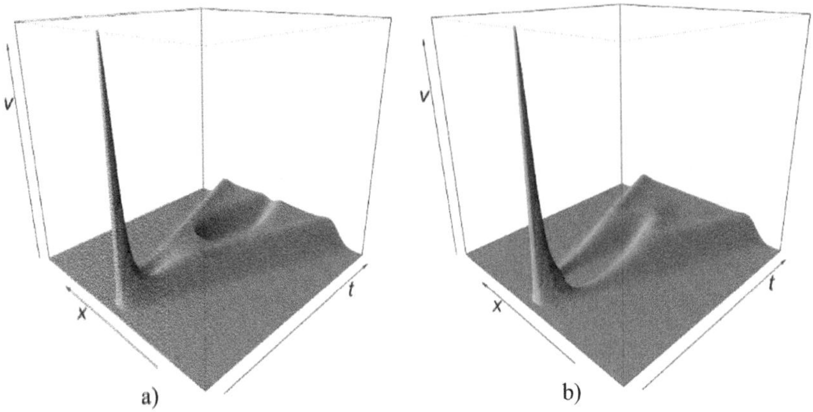

Figure 11.6. *Spatial–temporal dynamics of antigen concentration in the presence of a source close to the pulse antigen and the introduction of donor antibodies at the time: a)* $t_{K1} = 75$ *and b)* $t_{K1} = 20$. *For a color version of this figure, see www.iste.co.uk/koroliouk/computational.zip*

The proposed results demonstrate the effect on the development of the model process of infectious disease taking into account the diffusion perturbations of not only antigens but also antibodies (both native and donor). The intensity and duration of action of donor antibodies in the area of their injection taking into account diffusion "redistribution" decrease compared to the model situation without taking into account such "redistribution", caused by diffusion "erosion" of donor antibodies from this area over time. It is possible to increase the effectivity of donor antibody actions, in particular, by increasing the area of injection of the donor antibodies. To describe the source of antibodies, in this case, the proposed model can be used as the "hat" function [11.20]. Figure 11.7 presents the spatio-temporal dynamics of viral antigen concentrations with the development of the viral disease process in the chronic form when injections of the donor antibodies are carried out by drip on the interval $(\bar{t}_{K1}, \bar{\bar{t}}_{K1})$ at some point (Figure 11.7a) and at some area $(\bar{x}_{K1}, \bar{\bar{x}}_{K1})$ (Figure 11.7b). The obtained results illustrate a significant increase in the effectivity of donor antibodies activity with increasing area of their injection that can also help to monitor the effectivity of immunotherapy procedures.

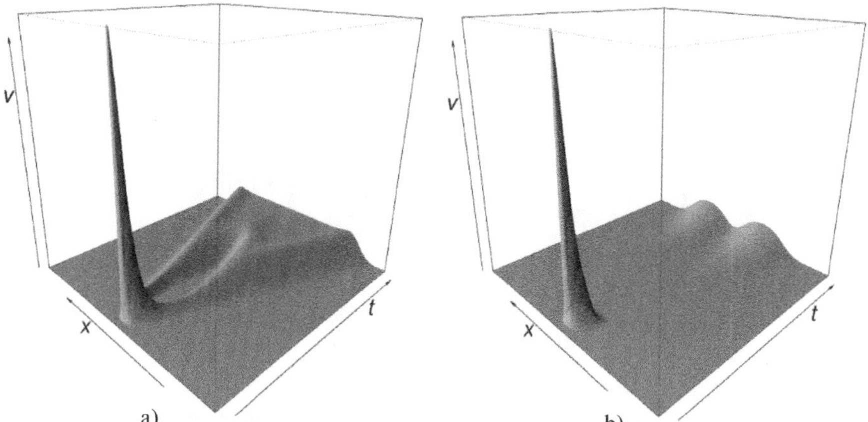

Figure 11.7. *Spatial–temporal dynamics of antigen concentration in the presence of sources of donor antibodies: a) pulse type and b) type "hat". For a color version of this figure, see www.iste.co.uk/koroliouk/computational.zip*

Figure 11.8 presents the model spatio-temporal dynamics of antigen concentration $v(x,t)$ with the development of viral disease in chronic form in cases without taking into account (Figure 11.8a) and taking into account (Figure 11.8b) diffusion perturbations and convection under conditions when at the initial time antigens in the organism have not yet been detected by the immune system (i.e. $v^0(x,\tilde{t}) = 0, \ -\tau \leq \tilde{t} \leq 0$). But due to the destruction of previously infected cells at different times ($t_{v1} < t_{v2}$) and different points ($x_{v1} > x_{v2}$), there are sharp, close to pulse changes in antigen concentrations: $u_V(x,t) = \sum_{j=1}^{2} A_{V\,j} e^{-\alpha_{V\,j}(x-x_{V\,j})^2} e^{-\beta_{V\,j}(t-t_{V\,j})^2}$.

Figure 11.9a and b presents the spatial–temporal dynamics of the antigen concentration with the development of infectious disease in cases when different time injections of donor antibodies are given ($u_K(x,t) = A_{K1} e^{-\alpha_{K1}(x-x_{K1})^2}$ $e^{-\beta_{K1}(t-t_{K1})^2}$). These results illustrate the higher "therapeutic effect" of early injection of donor antibodies (Figure 11.9b). The highest effect is achieved with the injection of donor antibodies with a constant intensity u_K=0.0125 (Figure 11.9c). Figure 11.9 also shows the change over time the location of the areas with the maximum values of antigen concentrations that is caused due the influence of convection.

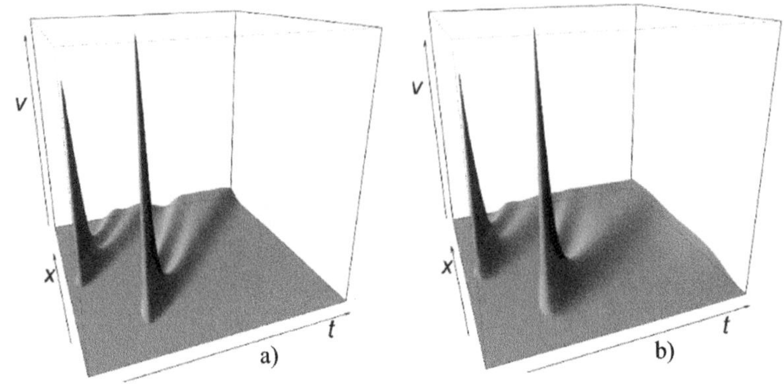

Figure 11.8. *Spatial–temporal dynamics of antigen concentration in the presence some of their concentrated sources in the chronic form of infectious disease under a)* $\varepsilon = 0{,}0$ *and b)* $\varepsilon = 0{,}1$. *For a color version of this figure, see www.iste.co.uk/koroliouk/computational.zip*

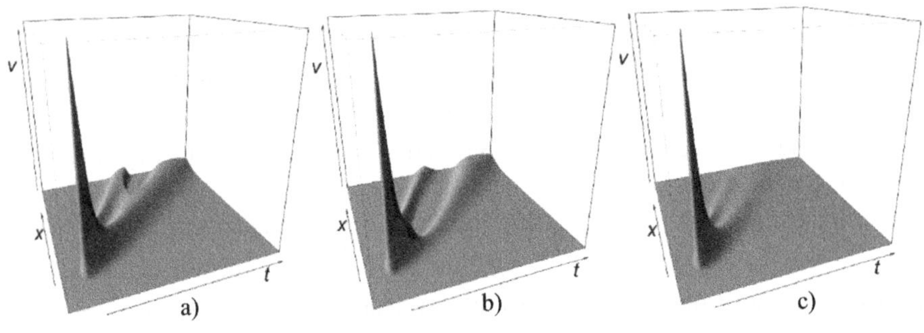

Figure 11.9. *Spatial–temporal dynamics of antigen concentration in: the presence of a concentrated source of antigens and injection of a donor antibodies at time a)* $t_{K1} = 50$ *and b)* $t_{K1} = 20$. *c) The introduction of donor antibodies with a constant rate* u_K=0.0125. *For a color version of this figure, see www.iste.co.uk/koroliouk/computational.zip*

11.6. Conclusion

Based on the generalization of the known modification of the simplest infectious disease model, an approach is presented to take into account spatially distributed diffusion perturbations on the development of infectious diseases under the condition that close to impulse influences are present (Bomba 1982; Bolodurina et al. 2009). The corresponding model problem with time delay is reduced to a sequence of problems without time delay, for which representations of the required functions in the form of asymptotic series as perturbation of solutions of the corresponding degenerate problems are constructed.

The numerical experiments results illustrate the decrease in the maximum value of the antigens concentration in the infection loci due to their diffusion "redistribution", including for different situational conditions. It has been shown that even when the initial antigen concentration or the intensity of the pulsed antigen source in some area of the infection zone exceeds a certain critical value, the diffusion "redistribution" over a period of time can reduce the critical antigen concentration to a level below the critical the reduction can be provided by the available level of antibodies, as well as a more economical mode of administration of donor antibodies by injection. That is, under this model, the "severity" of the infectious disease in such cases will decrease, so to speak, at low cost.

The developed computational procedure can be an element of designing specialized expert systems for making a wide range of decisions such as: can we in this case, according to the values of relevant input data, in particular (on the intensity of diffusion "redistribution", on the level of immune protection available in the organism, or, otherwise) carry out external therapeutic effects. And in a situation where it is decided that such an effect should be exercised, in particular by means of injections of donor antibodies, to establish the most rational frequency of their introduction and an acceptable concentration of antibodies for each injection.

It is also promising to take into account such impulse effects in spatially distributed diffusion perturbations in the study of viral and bacterial diseases based on more general models, in particular antiviral and antibacterial immune response models by Marchuk (1991).

11.7. References

Baranovsky, S.V. and Bomba, A.Ya. (2020). Generalization of anti-virus immune response mathematical model by Marchuk and Petrov taking into account the influence of small spatially distributed particles. *Mathematical and Computer Modeling. Series: Technical Sciences*, 21, 5–24.

Baranovsky, S.V., Bomba, A.Ya., Lyashko, S.I. (2021). Decision-making in modeling the dynamics of an infectious disease taking into account diffusion perturbations and concentrated effects. *Problems of Management and Informatics*, 3, 115–129.

Bolodurina, I.P., Lugovskaya, Y.P., Zakirov, D.A. (2009). Management of the dynamic interaction of the human immune system with infectious diseases. *Bull. TVSU Seri. Appl. Math.*, 14, 101–117.

Bomba, A.Ya. (1982). On the asymptotic method for the solution of one problem of mass transfer in the course of filtration in a porous medium [in Ukrainian]. *Ukr. Mat. Zhurn.*, 4(4), 493–496.

Bomba, A.Ya. and Baranovsky, S.V. (2020). Modeling of small spatially distributed influences on the dynamics of an infectious disease in the conditions of the type of pharmacotherapy. *J. Comput. Appl. Math.*, 1(133), 5–17.

Bomba, A.Ya., Baranovskyi, S.V., Prisyajnyuk, I.M. (2008). *Neliniini synhuliarno zbureni zadachi typu "konvektsiia-dyfuziia"* [in Ukrainian]. NUVGP, Rivne.

Bomba, A.,Ya., Baranovsky, S.V., Pasichnyk, M.S., Pryshchepa, O.V. (2020a). Modeling small-scale spatial distributed influences on the development of infectious disease process. *Mathe. Model. Comput.*, 7(2), 310–321.

Bomba, A.Ya., Baranovskii, S.V., Pryshchepa, O.V. (2020b). Modeling influence of small-scale diffusion perturbations on the development of infectious diseases under immunotherapy. In *Proceedings of the International Scientific and Practical Conference "Modeling, Control and Information Technologies"*, National University of Water and Environmental Engineering, Rivne, 15–18.

Bomba, A., Baranovskii, S., Pasichnyk, M., Malash, K. (2020c). Modeling of infectious disease dynamics under the conditions of spatial perturbations and taking into account impulse effects. In *International Conference on Informatics & Data-Driven Medicine (IDDM 2020)*, 119–128.

Burnet, F.M. (1971). *Cellular Immunology: Trans. With English.* Mir, Moscow.

Elsholz, L.E. and Norkin, S.B. (1971). *Vvedenie v teoriiu differentsial'nykh uravenii s otkloniaiushchimsia argumentom* [in Russian]. Nauka, Moscow.

Keeling, M.J. and Rohani, P. (2008). *Modeling Infectious Diseases in Humans and Animals.* Princeton University Press, New Jersey.

Klyushin, D.A., Lyashko, S.I., Lyashko, N.I., Bondar, O.S., Tymoshenko, A.A. (2020). Generalized optimization of processes of drug transport in tumors. *Cybern. Syst. Anal.*, 56(5), 758–765.

Lyashko, S.I. and Mankovskiui, A.A. (1983). Simultaneous optimization of impulse and intensities in control problems for parabolic equations. *Cybernet. Syst. Anal.*, 19(5), 687–690.

Lyashko, S.I. and Semenov, V.V. (2001). On the controllability of linear distributed systems in classes of generalized actions. *Cybernet. Syst. Anal.*, 37(1), 13–32.

Lyashko, S.I., Klyushin, D.A., Pavlychko, V.V. (2010). Model-based analysis of biological tissue heating by point ultrasound sources. *J. Autom. In. Sci.*, 42(2), 44–50.

Marchuk, G.I. (1980). *Mathematical Models in Immunology* [in Russian]. Nauka, Moscow.

Marchuk, G.I. (1991). *Mathematical Models in Immunology. Computational Methods and Experiments* [in Russian]. Nauka, Moscow.

Marchuk, G.I. (1997). *Mathematical Models of Immune Response in Infectious Diseases.* Kluwer Press, Dordrecht.

Nowak, M.A. and May, R.M. (2000). *Virus Dynamics. Mathematical Principles of Immunology and Virology.* Oxford University Press, New York.

Rusakov, S.V. and Chirkov, M.B. (2012). Mathematical model of the influence of immunotherapy on the dynamics of the immune response. *Management Problems*, 6, 45–50.

Sandrakov, G.V., Lyashko, S.I., Bondar, E.S., Lyashko, N.I. (2019). Modeling and optimization of microneedle systems. *J. Autom. In. Sci.*, 51(6), 1–11.

Wodarz, D. (2007). *Killer Cell Dynamics: Mathematical and Computational Approaches to Immunology.* Springer-Verlag, New York.

12

Solitary Waves in "Shallow Water" Environments

Yurii TURBAL, Mariana TURBAL and Andrii BOMBA
National University of Water and Environmental Engineering, Rivne, Ukraine

12.1. Introduction

Localized soliton-like perturbations are considered in various aspects. A number of works are devoted to solutions of the famous Corteweg de Vries equation in particular, which describe solitons (Lyashko et al. 2021). Solitary waves are considered to be base constructive elements of dark matter (Mielke 2019; Ureña-López 2019; Asselmeyer-Maluga and Kró 2020). These waves are considered in plasmas and in the atmosphere (Petviashvili and Pokhotelov 2016). In the study of seismic processes (Turbal et al. 2014; Bomba and Turbal 2015), a separate wave can be considered to be a generator of the seismic shock if it passes through the zone of accumulation of seismic energy. A similar effect can occur when several waves hit an area at the same time, because then there may be cracks in the Earth's crust and, accordingly, generate seismic shocks. Such an effect occurred, for example, during the earthquake in Fukushima Prefecture, Japan, in 2011 (Turbal et al. 2014). Information about the trajectories of isolated waves and the nature of their movement is of great practical importance in cases where the isolated waves themselves pose a danger, such as with a tsunami. If the soliton occurs on

Computational Methods and Mathematical Modeling in Cyberphysics and Engineering Applications 1,
coordinated by Dmitri KOROLIOUK, Sergiy LYASHKO and Nikolaos LIMNIOS.

the surface of the water, it can also be dangerous, for example, in the case of the so-called killer wave, studied in the MaxWave project (Chabchoub et al. 2012). Therefore, the study of the trajectories of localized waves is an important task. It is important to identify the soliton and predict its trajectory.

The problem of modeling the processes of the origin and propagation of isolated waves has several aspects, in particular, an analytical description of the wave profile and the presence of a number of the soliton-specific properties (Bazeia et al. 2018; Basak et al. 2021), which are guaranteed by some methods at the stage of their application and do not require further study. The presence of such methods determines their complexity. Mathematical models that describe the propagation of isolated waves are divided into two classes: integrated and non-integrated, using the method of the inverse scattering problem (Antonova 1983; Basak et al. 2021); in collisions, they retain their characteristics and there is only a phase shift. Many papers have shown that in integrated models, localized waves behave like particles. This fact has been confirmed by numerous experiments in which plasma waves, liquids with gas bubbles, stratified liquids and electromagnetic waves have been studied. At the same time, for systems that do not integrate using the inverse scattering problem, a number of other effects occur, including the radiation of nonlinear wave trains, splitting of individual waves and formation of new soliton-like waves, merging of several waves and formation of a new series of elastic reflections.

At the same time, the wave profile may change over time, then such a wave will no longer be a soliton in the classical sense, but may have soliton properties that require further study. Soliton solutions of equations are considered in the framework of the structural–phenomenological approach; for example, they are isolated partial cases of solutions and do not make it possible to solve the problem of formation of a separate wave from an arbitrary initial perturbation. The use of δ-solitons in the asymptotic case allows us to ignore the shape of the wave profile. However, general functions as models of real physical phenomena have a number of disadvantages, including infinity, specific mathematical content of derivatives and others.

The analytical description of isolated waves in multidimensional cases is especially problematic now. Even for a well-studied system of equations such as "shallow water" (Antonova 1983; Bila et al. 2006), periodic analytical solutions have only recently been found and the author does not know the methods of analytical study of the isolated waves.

12.2. T-forms for the solitary wave approximation

A number of approaches to finding soliton solutions of the corresponding differential equations are known today. The method of the inverse scattering problem plays a fundamental role. There is an infinite number of differential equations that can be differentiated using this method. However, the method does not exhaust all the equations that the soliton solutions can have. In addition, there are other methods that allow us to find both exact and approximate soliton solutions of the differential equations. However, the corresponding methods are often special in nature and are effective only for a certain class of equations. Most of them relate to the one-dimensional case.

In recent years, in order to constructively study the trajectories of localized soliton-like perturbations, a slightly different approach has been proposed, called the T-representation method. This method has approximate variants when the solutions are Gaussian and can easily be extended to multidimensional cases.

In Turbal et al. 2015, the method for finding solutions of motion equations is presented in the form

$$(u^{(1)},u^{(2)},...,u^{(m)})^T = (\psi_1(t),\psi_2(t),...,\psi_m(t))^T W(x_1,x_2,...,x_n,t)\,, \qquad [12.1]$$

where $u^{(1)},u^{(2)},...,u^{(m)}$ are the displacements in the Cartesian coordinate system;

$$\begin{aligned} &W(x_1,x_2,...,x_n,\tilde{x}_1(t),\tilde{x}_2(t),...,\tilde{x}_n(t),t) = \\ &= \exp(-\mu((x_1,x_2,...,x_n),(\tilde{x}_1(t),\tilde{x}_2(t),...,\tilde{x}_n(t)))/\varepsilon) \end{aligned};$$

$\mu(a,b)$ is a measure defined on a set of intervals $\{[a.b],a,b\in R^n\}$, ($\mu(\alpha,\beta)=0$ if $\alpha=\beta$), $\psi_1(t)$, $\psi_2(t)$, ..., $\psi_m(t)$ are functions that determine the components of the amplitude of the corresponding perturbations, ε is a parameter that determines the localization of the perturbation, $\tilde{x}_1(t)$, $\tilde{x}_2(t)$,..., $\tilde{x}_n(t)$ are functions that determine the trajectory of the point of maximum perturbation.

The solution in form [12.1] defines a localized wave, the maximum of which is at a point $(\tilde{x}_1(t),\tilde{x}_2(t),...,\tilde{x}_n(t))$. Note that in the case of an arbitrary measure μ, in form [12.1], we can represent an arbitrary solution that has the character of the unimodal isolated wave, the shape of which does not change with time. The measure μ determines the form of perturbation. We can consider a more general case when the functions $\tilde{x}_i(.)$ depend not only on time but also on spatial coordinates $\tilde{x}_i(t) = \tilde{x}_i(x_1, x_2, ..., x_n, t)$.

Shallow water equations are of great practical importance because they describe large-scale atmospheric and ocean currents, particularly in mid-latitudes. In Cartesian coordinates, they have the form (Bila et al. 2006):

$$\begin{aligned} &u_t + uu_x + vu_y + gh_x = fv, \\ &v_t + uv_x + vv_y + gh_y = -fu, \\ &h_t + (uh)_x + (vh)_y = 0, \end{aligned} \qquad [12.2]$$

where u, v are components of the velocity vector, h is the depth of the liquid layer above the flat bottom, f is the Coriolis parameter (rotation frequency) and g is the acceleration of free fall. The approximate model [12.2] is obtained from Euler's equations of an ideal fluid under the following assumptions: the ratio of the characteristic vertical scale of the flow to the characteristic horizontal scales is much less than 1; the pressure in the liquid is hydrostatic at depth; the axis of rotation coincides with the vertical axis z.

Note that the shallow water equations are usually solved by numerical methods due to significant difficulties in obtaining analytical solutions. In recent years, group theory methods have been used to analyze shallow water-type equations (Chesnokov 2008;). In particular, with the help of group analysis methods, the Lie algebra in the absence of rotation is constructed. In Bila (2006) and Chesnokov (2008), a system of subalgebras containing 179 representatives is constructed. On the basis of the corresponding algebra, periodic exact solutions are constructed, which can be interpreted as pulsations of the liquid under the action of gravity or Coriolis. Numerical methods for solving shallow water equations show that there are solutions that simulate isolated waves, such as the circular tsunami waves.

The system of shallow water equations can easily be generalized to the case of polytropic gas flow, which is considered in the barotropic approximation. The motion of the compressible viscous gas can be conveniently studied in the polar coordinate system.

Such problems include a number of applied engineering and geophysical problems, including the study of atmospheric phenomena on a planetary scale, the study of astrophysical objects associated with the motion of matter around the center of gravity, including the gas disks of galaxies. The shallow water-type equations in the polar coordinate system (Morozov et al. 2005) will look like this:

$$\frac{\partial\sigma}{\partial t}+\sigma(\frac{\partial u}{\partial r}+\frac{u}{r}+\frac{\partial v}{r\partial\phi})+u\frac{\partial\sigma}{\partial r}+\frac{v\partial\sigma}{r\partial\phi}=0,$$

$$\frac{\partial u}{\partial t}+u\frac{\partial u}{\partial r}+\frac{v}{r}\frac{\partial u}{\partial\phi}-\frac{v^2}{r}=-\frac{\partial\Phi}{\partial r}-B\gamma_s\sigma^{\gamma_s-2}\frac{\partial\sigma}{\partial r}, \qquad [12.3]$$

$$\frac{\partial v}{\partial t}+u\frac{\partial v}{\partial r}+\frac{v}{r}\frac{\partial v}{\partial\phi}+\frac{uv}{r}=-\frac{1}{r}\frac{\partial\Phi}{\partial\phi}-\sigma^{\gamma_s-2}\frac{B\gamma_s}{r}\frac{\partial\sigma}{\partial\phi},$$

where $u(r,\phi,t)$, $v(r,\phi,t)$ are radial and azimuthal components of the gas velocity, respectively, $\sigma(r,\phi,t)$ is the surface density of the gas disk, $\Phi(r,\phi,t)$ is the potential that describes the action of an external force and B,D are positive constants. Note that system [12.3] is quasilinear.

We will look for the solution of system [12.3] in the form of the sum of equilibrium quantities and some perturbations:

$$\sigma(r,\phi,t)=\sigma_0(r,\phi,t)+\sigma_1(r,\phi,t),$$

$$\Phi(r,\phi,t)=\Phi_0(r)+\Phi_1(r,\phi,t), \qquad [12.4]$$

$$u(r,\phi,t)=u_1(r,\phi,t),$$

$$v(r,\phi,t)=v_0(r,\phi,t)+v_1(r,\phi,t).$$

The gravitational potential satisfies the Poisson equation:

$$\frac{\partial}{r\partial r}(r\frac{\partial\Phi_1}{\partial r})+\frac{\partial^2\Phi_1}{r^2\partial\phi^2}+\frac{\partial^2\Phi_1}{\partial z^2}=4\pi G\sigma_1\delta(z). \qquad [12.5]$$

We will substitute relations [12.4] into system [12.3]. As a result, we obtain a system for perturbations

$$\begin{cases} \frac{\partial \sigma_1}{\partial t} + \frac{v_0}{r}\frac{\partial \sigma_1}{\partial \phi} + \sigma_0(\frac{\partial u_1}{\partial r} + \frac{u_1}{r} + \frac{\partial v_1}{r\partial \phi}) + u_1\frac{\partial \sigma_0}{\partial r} + \frac{v_1}{r}\frac{\partial \sigma_0}{\partial \phi} + \\ + \frac{1}{r}(\sigma_1 u_1 + u_1 r\frac{\partial \sigma_1}{\partial r} + \sigma_1 r\frac{\partial u_1}{\partial r} + \sigma_1\frac{\partial v_1}{\partial \phi} + v_1\frac{\partial \sigma_1}{\partial \phi}) = 0, \\ \frac{\partial u_1}{\partial t} + \frac{v_0}{r}\frac{\partial u_1}{\partial \phi} - \frac{2v_0 v_1}{r} = -\frac{\partial \Phi_1}{\partial r} - B\gamma_s(\gamma_s - 2)\sigma_1\sigma_0^{\gamma_s - 3}\frac{\partial \sigma_0}{\partial r} - B\gamma_s {\sigma_0}^{\gamma_s - 2}\frac{\partial \sigma_1}{\partial r} - \\ -B\gamma_s(\gamma_s - 2)\sigma_1\sigma_0^{\gamma_s - 3}\frac{\partial \sigma_1}{\partial r} + \frac{3}{2r}BD(\gamma_s - 1){\sigma_0}^{\gamma_s - 2}\sigma_1 + \frac{{v_1}^2}{r} - u_1\frac{\partial u_1}{\partial r} - \frac{v_1}{r}\frac{\partial u_1}{\partial \phi}, \\ \frac{\partial v_1}{\partial t} + u_1\frac{\partial v_0}{\partial r} + \frac{v_0}{r}\frac{\partial v_1}{\partial \phi} + \frac{u_1 v_0}{r} = -\frac{1}{r}\frac{\partial \Phi_1}{\partial \phi} - \frac{B\gamma_s}{r}(\gamma_s - 2)\sigma_1\sigma_0^{\gamma_s - 3}\frac{\partial \sigma_0}{\partial \phi} - \\ -\frac{B\gamma_s}{r}{\sigma_0}^{\gamma_s - 2}\frac{\partial \sigma_1}{\partial \phi} - \frac{B\gamma_s}{r}(\gamma_s - 2)\sigma_0^{\gamma_s - 3}\sigma_1\frac{\partial \sigma_1}{\partial \phi} - u_1\frac{\partial v_1}{\partial r} - \frac{v_1}{r}\frac{\partial v_1}{\partial \phi} - \frac{v_1}{r}\frac{\partial v_0}{\partial \phi} - \\ -\frac{u_1 v_1}{r}. \end{cases} \quad [12.6]$$

Based on the above-described method of T-representations, the vector-function of the perturbations $f_1(r,\varphi,t) = (u_1(r,\varphi,t), v_1(r,\varphi,t), \sigma_1(r,\varphi,t))$ will be found in the form:

$$f_1(r,\phi,t) = \varepsilon_0 \psi_f(t)\omega(r,\phi,t,\varepsilon), \quad [12.7]$$

where $\psi_f(t) = (\psi_u(t), \psi_v(t), \psi_\sigma(t))$, $\psi_u(t)$, $\psi_v(t)$, $\psi_\sigma(t)$, $\tilde{r}(t)$, $\tilde{\phi}(t)$ are any functions, $\tilde{r}(t) \geq 0$, $\psi_\sigma(t) \geq 0$, ε is a small parameter, ε_0 is constant, $\omega(r,\phi,t,\varepsilon) = \exp\{-\frac{g(r-\tilde{r}(t)) + g(\phi - \tilde{\phi}(t))}{\varepsilon}\}$.

$g(x)$ is a function that has properties:

1) $g(x) \geq 0$, $x \in (-\infty, +\infty)$,

2) $g(0) = 0$,

3) $g(-x) = g(x)$,

4) there exists constants $\alpha_1, \tilde{\alpha}_1$, $\alpha_2, \tilde{\alpha}_2$, $c_1 > 0, c_2 > 0$, $\varepsilon > 0$ those that,

$$\tilde{c}_1 g^{\tilde{\alpha}_1}(x) \leq |g'(x)| \leq c_1 g^{\alpha_1}(x), |g''(x)| \leq c_2 g^{\alpha_2}(x) \text{ in region } \{x : g(x) \leq -\varepsilon \ln \varepsilon\}.$$

In this case, the perturbation is a single wave, and the point of maximum perturbation moves along the trajectory described in the polar coordinate system by the functions $\tilde{r}(t)$ and $\tilde{\varphi}(t)$. We introduce a vector function of perturbations $f_1(r,\phi,t) = (u_1(r,\phi,t), v_1(r,\phi,t), \sigma_1(r,\phi,t))$. We see that in this case the perturbation is the single wave and the point of maximum perturbation moves along the trajectory which is described in the polar coordinate system by the functions $\tilde{r}(t)$ and $\tilde{\varphi}(t)$. Differentiate relationship [12.7]. The result is

$$\frac{\partial f_1}{\partial r} = -\varepsilon_0 \frac{g'(r - \tilde{r}(t))}{\varepsilon} \psi_f(t)\omega(r,\varphi,t,\varepsilon), \quad [12.8]$$

$$\frac{\partial f_1}{\partial \varphi} = -\varepsilon_0 \frac{g'(\varphi - \tilde{\varphi}(t))}{\varepsilon} \psi_f(t)\omega(r,\varphi,t,\varepsilon),$$

$$\frac{\partial f_1}{\partial t} = \varepsilon_0 \psi_f'(t)\omega(r,\varphi,t,\varepsilon) + \frac{1}{\varepsilon}\varepsilon_0 \psi_f(t)\omega(r,\varphi,t,\varepsilon)[\tilde{\varphi}'(t)g'(\varphi - \tilde{\varphi}(t)) + \quad [12.9]$$

$$+\tilde{r}'(t)g'(r - \tilde{r}(t)).$$

We will substitute representations [12.8]–[12.9] in system [12.6]. We have

$$\varepsilon_0 \omega(r,\varphi,t,\varepsilon)(\psi_\sigma'(t) + \sigma_0 \frac{\psi_u(t)}{r} + \psi_u(t)\frac{\partial \sigma_0}{\partial r} + \frac{\psi_v(t)}{r}\frac{\partial \sigma_0}{\partial \varphi} + \frac{1}{r}\varepsilon_0 \psi_\sigma(t)\psi_u(t) \times$$
$$\times \omega(r,\varphi,t,\varepsilon)) + \varepsilon_0 \omega(r,\varphi,t,\varepsilon)(\frac{1}{\varepsilon}\psi_\sigma(t)[\tilde{\varphi}'(t)g'(\varphi - \varphi(t)) + \tilde{r}'(t)g'(r - \tilde{r}(t))] -$$
$$-\frac{v_0}{r}\frac{g'(\varphi - \tilde{\varphi}(t))}{\varepsilon}\psi_\sigma(t) + \sigma_0(-\frac{g'(r - \tilde{r}(t))}{\varepsilon}\psi_u(t) - \frac{g'(\varphi - \tilde{\varphi}(t))}{\varepsilon r}\psi_v(t)) + \quad [12.10]$$
$$+\frac{1}{r}(-\varepsilon_0 \psi_u(t) r \frac{g'(r - \tilde{r}(t))}{\varepsilon}\psi_\sigma(t)\omega(r,\varphi,t,\varepsilon) - \psi_\sigma(t) r \varepsilon_0 \frac{g'(r - \tilde{r}(t))}{\varepsilon}\psi_u(t) \times$$
$$\times \omega(r,\varphi,t,\varepsilon) - \varepsilon_0 \psi_\sigma(t)\frac{g'(\varphi - \tilde{\varphi}(t))}{\varepsilon}\psi_v(t)\omega(r,\varphi,t,\varepsilon) -$$
$$-\varepsilon_0 \psi_v(t)\frac{g'(\varphi - \tilde{\varphi}(t))}{\varepsilon}\psi_\sigma(t)\omega(r,\varphi,t,\varepsilon))) = 0.$$

$$\varepsilon_0\psi_u{}'(t)\omega(r,\varphi,t,\varepsilon)+\frac{1}{\varepsilon}\varepsilon_0\psi_u(t)\omega(r,\varphi,t,\varepsilon)[\tilde{\varphi}'(t)g'(\varphi-\varphi(t))+$$
$$+\tilde{r}'(t)g'(r-\tilde{r}(t))]-\frac{v_0}{r}\varepsilon_0\frac{g'(\varphi-\tilde{\varphi}(t))}{\varepsilon}\psi_u(t)\omega(r,\varphi,t,\varepsilon)-$$
$$-\frac{2v_0\varepsilon_0\psi_v(t)\omega(r,\varphi,t,\varepsilon)}{r}=-\frac{\partial\Phi_1}{\partial r}-B\gamma_s(\gamma_s-2)\varepsilon_0\psi_\sigma(t)\omega(r,\varphi,t,\varepsilon)\sigma_0^{\gamma_s-3}\frac{\partial\sigma_0}{\partial r}+$$
$$+B\gamma_s\sigma_0{}^{\gamma_s-2}\varepsilon_0\frac{g'(r-\tilde{r}(t))}{\varepsilon}\psi_\sigma(t)\omega(r,\varphi,t,\varepsilon)+$$
$$+B\gamma_s(\gamma_s-2)\varepsilon_0\psi_\sigma(t)\omega(r,\varphi,t,\varepsilon)\sigma_0^{\gamma_s-3}\varepsilon_0\frac{g'(r-\tilde{r}(t))}{\varepsilon}\psi_\sigma(t)\omega(r,\varphi,t,\varepsilon)+$$
$$+\frac{3}{2r}BD(\gamma_s-1)\sigma_0{}^{\gamma_s-2}\varepsilon_0\psi_\sigma(t)\omega(r,\varphi,t,\varepsilon)+\varepsilon_0\psi_u(t)\omega(r,\varphi,t,\varepsilon)\varepsilon_0\frac{g'(r-\tilde{r}(t))}{\varepsilon}\times \qquad [12.11]$$
$$\times\psi_u(t)\omega(r,\varphi,t,\varepsilon)+\frac{\varepsilon_0\psi_v(t)\omega(r,\varphi,t,\varepsilon)}{r}\varepsilon_0\frac{g'(\varphi-\tilde{\varphi}(t))}{\varepsilon}\psi_u(t)\omega(r,\varphi,t,\varepsilon)+$$
$$+\frac{(\varepsilon_0\psi_v(t)\omega(r,\varphi,t,\varepsilon))^2}{r}.$$

$$\varepsilon_0\psi_v{}'(t)\omega(r,\varphi,t,\varepsilon)+\frac{1}{\varepsilon}\varepsilon_0\psi_v(t)\omega(r,\varphi,t,\varepsilon)[\tilde{\varphi}'(t)g'(\varphi-\varphi(t))+$$
$$+\tilde{r}'(t)g'(r-\tilde{r}(t))]+\varepsilon_0\psi_u(t)\omega(r,\varphi,t,\varepsilon)\frac{\partial v_0}{\partial r}+$$
$$-\frac{v_0}{r}\varepsilon_0\frac{g'(\varphi-\tilde{\varphi}(t))}{\varepsilon}\psi_v(t)\omega(r,\varphi,t,\varepsilon)+\frac{\varepsilon_0\psi_u(t)\omega(r,\varphi,t,\varepsilon)v_0}{r}=$$
$$-\frac{1}{r}\frac{\partial\Phi_1}{\partial\varphi}-\frac{B\gamma_s}{r}(\gamma_s-2)\sigma_0^{\gamma_s-3}\frac{\partial\sigma_0}{\partial\varphi}\varepsilon_0\psi_\sigma(t)\omega(r,\varphi,t,\varepsilon)+$$
$$+\frac{B\gamma_s}{r}\sigma_0{}^{\gamma_s-2}\varepsilon_0\frac{g'(\varphi-\tilde{\varphi}(t))}{\varepsilon}\psi_\sigma(t)\omega(r,\varphi,t,\varepsilon)+ \qquad [12.12]$$
$$+\frac{B\gamma_s}{r}(\gamma_s-2)\sigma_0^{\gamma_s-3}\varepsilon_0\psi_\sigma(t)\omega(r,\varphi,t,\varepsilon)\varepsilon_0\frac{g'(\varphi-\tilde{\varphi}(t))}{\varepsilon}\psi_\sigma(t)\omega(r,\varphi,t,\varepsilon)+$$
$$+\varepsilon_0\psi_u(t)\omega(r,\varphi,t,\varepsilon)\varepsilon_0\frac{g'(r-\tilde{r}(t))}{\varepsilon}\psi_v(t)\omega(r,\varphi,t,\varepsilon)-\frac{\varepsilon_0\psi_v(t)\omega(r,\varphi,t,\varepsilon)}{r}\frac{\partial v_0}{\partial\varphi}+$$
$$+\frac{\varepsilon_0\psi_v(t)\omega(r,\varphi,t,\varepsilon)}{r}\varepsilon_0\frac{g'(\varphi-\tilde{\varphi}(t))}{\varepsilon}\psi_v(t)\omega(r,\varphi,t,\varepsilon)-$$
$$-\frac{\varepsilon_0\psi_u(t)\omega(r,\varphi,t,\varepsilon)\varepsilon_0\psi_v(t)\omega(r,\varphi,t,\varepsilon)}{r}.$$

Further analysis will be divided into two parts – the results that relate to the existence of partial solutions in the form of isolated waves and those related to the nature of the trajectories.

We will consider the following areas:

$$G_\varepsilon^1 = \{(r,\varphi,t) : g(r-\tilde{r}(t)) + g(\varphi - \tilde{\varphi}(t)) \geq -\varepsilon \ln \varepsilon, t \in [0,T], r \leq R\},$$

$$G_\varepsilon^2 = \{(r,\varphi,t) : g(r-\tilde{r}(t)) + g(\varphi - \tilde{\varphi}(t)) \leq -\varepsilon \ln \varepsilon, t \in [0,T], r \leq R\}.$$

12.3. Existence of the solution of the gas dynamics equations in the form of solitary waves

In this section, the main results will be formulated in the form of the following propositions.

PROPOSITION 12.1.– Let $\varepsilon_0 = \varepsilon^{1-\alpha_1}\vartheta(\varepsilon)$, $\vartheta(\varepsilon)$ be any function, such that $\lim_{\varepsilon \to 0} \vartheta(\varepsilon) = 0$, $\sigma_0(r)$, $v_0(r)$, $\frac{d\sigma_0(r)}{dr}$, $\frac{dv_0(r)}{dr}$ are limited functions. Then, there are the functions $\tilde{r}(t) \geq 0$, $\tilde{\varphi}(t)$ and constant $T > 0$, such that the partial solution of system [12.6] can be found in form [12.7] with precision:

$$\delta(\varepsilon) = \begin{cases} O(\varepsilon_0(-\varepsilon \ln \varepsilon)^{\alpha_1}), & (r,\varphi,t) \in G_\varepsilon^1, \\ O(\vartheta(\varepsilon)), & (r,\varphi,t) \in G_\varepsilon^2. \end{cases}$$

In this solution, $\psi_u(t) \equiv 0$, $\psi_v(t) \equiv 0$, $\psi_\sigma(t) = \psi_\sigma \geq 0$.

PROPOSITION 12.2.– Let the region $G = \{(r,\phi) : r_1 \leq r \leq r_2\}$, the surface density has the form:

$$\sigma_0(r) = Cr^D - \frac{\varepsilon_0 \psi_\sigma}{2}, \qquad [12.13]$$

where C is such constant that $\sigma_0(r) \geq 0$. Then, there are the functions $\tilde{r}(t) \geq 0$, $\tilde{\varphi}(t)$ and constant $T > 0$, such that in the interval $[0,T]$, the partial solution of system [12.6] can be found in form [12.7] with precision:

$$\delta(\varepsilon) = \begin{cases} O(\varepsilon_0(-\varepsilon \ln \varepsilon)^{\alpha_1}), & (r,\varphi,t) \in G_\varepsilon^1, \\ O(\varepsilon_0 \varepsilon^{(1+\alpha)\alpha_1 - 1}), & (r,\varphi,t) \in G_\varepsilon^\alpha. \end{cases} \qquad [12.14a]$$

In this solution, $\psi_u(t)\equiv 0$, $\psi_v(t)\equiv 0$, $\psi_\sigma(t)\equiv\psi_\sigma>0$, ε_0 and α are positive constants.

PROPOSITION 12.3.– Let $\sigma_0(r)\geq 0, v_0(r)$ be continuously differentiable and limited in some area $G=\{(r,\phi): r_1\leq r\leq r_2\}$ functions. Then, there are such values $T>0$ and limited differentiation functions $\psi_u(t),\psi_v(t),\psi_\sigma(t)\geq 0$,$\tilde{r}(t)\geq 0$,$\tilde{\varphi}(t)$ such that in the interval $[0,T]$ the solution of system [12.6] can be found in form [12.7] with accuracy [12.14]. In this case, $\psi_\sigma'(t)\geq 0$, the constant $\varepsilon_0>0$ must be such that the function $\frac{\partial\sigma_0(\tilde{r}(t))}{\partial r}+\frac{\sigma_0(\tilde{r}(t))}{\tilde{r}(t)}+\frac{\varepsilon_0\psi_\sigma(t)}{\tilde{r}(t)}$ does not change sign when $t\in[0,T]$.

Obviously, in the case where $\frac{\partial\sigma_0(r)}{\partial r}+\frac{\sigma_0(r)}{r}\geq 0$,$r\in G$, constant ε_0 is an arbitrary positive number.

PROOF.– (Propositions 12.1–12.3)

Let us consider $\frac{g'(r-\tilde{r}(t))}{\varepsilon}\omega(r,\varphi,t,\varepsilon)$ and $\frac{g'(\varphi-\tilde{\varphi}(t))}{\varepsilon}\omega(r,\varphi,t,\varepsilon)$.

Consider function $f(x,y)=\frac{c_1x^{2\alpha_1}}{\varepsilon}\exp\{-\frac{x^2+y^2}{\varepsilon}\}$ and find the maximum. It is obvious unequality: $f(x,y)=\frac{c_1x^{2\alpha_1}}{\varepsilon}\exp\{-\frac{x^2+y^2}{\varepsilon}\}\leq\frac{c_1x^{2\alpha_1}}{\varepsilon}\exp\{-\frac{x^2}{\varepsilon}\}$.

Differentiating the function $f(x)=\frac{c_1x^{2\alpha_1}}{\varepsilon}\exp\{-\frac{x^2}{\varepsilon}\}$, we obtain:

$$\frac{\partial}{\partial x}f(x)=c_1(\frac{2\alpha_1x^{2\alpha_1-1}}{\varepsilon}-\frac{2x^{2\alpha_1+1}}{\varepsilon^2})\exp\{-\frac{x^2}{\varepsilon}\}=c_1\frac{2x^{2\alpha_1-1}}{\varepsilon}(\alpha_1-\frac{x^2}{\varepsilon})\exp\{-\frac{x^2}{\varepsilon}\}.$$

From this expression, we get: $x^2=\alpha_1\varepsilon$, $x=\pm\sqrt{\alpha_1\varepsilon}$. Obviously, the minimum of this function is reached at point 0. The function is even. Therefore, at the point $x=\sqrt{\alpha_1\varepsilon}$ is maximum. The value of the function at this point is equal $\frac{c_1(\alpha_1\varepsilon)^{\alpha_1}}{\varepsilon}\exp\{-\alpha_1\}=\frac{c_1\alpha_1{}^{\alpha_1}}{\varepsilon^{1-\alpha_1}}\exp\{-\alpha_1\}$.

Then, we have inequality:

$$\frac{c_1 x^{2\alpha_1}}{\varepsilon}\exp\{-\frac{x^2+y^2}{\varepsilon}\}\le\frac{c_1\alpha_1^{\alpha_1}}{\varepsilon^{1-\alpha_1}}\exp\{-\alpha_1\}.$$

Note that in the domain determined by the ratio $\{x : x^2 \ge -\varepsilon\ln\varepsilon\}$, the maximum value of the function is reached at the boundary of the domain at the point $x_{\max}=\sqrt{-\varepsilon\ln\varepsilon}$ and is equal $f(\sqrt{-\varepsilon\ln\varepsilon})=c_1(-\varepsilon\ln\varepsilon)^{\alpha_1}$.

When the inequality $x^2\le\varepsilon^{1+\alpha}$ is satisfied, the maximum value of the function is reached at the point $x_{\max}=\varepsilon^{(1+\alpha)/2}$ and is equal $f(x)=c_1\varepsilon^{(1+\alpha)\alpha_1-1}\exp\{-\varepsilon^{\alpha}\}$.

Thus, there are inequalities in the region G_ε^2:

$$\frac{g'(r-\tilde{r}(t))}{\varepsilon}\omega(r,\varphi,t,\varepsilon)\le\frac{c_1\alpha_1^{\alpha_1}}{\varepsilon^{1-\alpha_1}}e^{-\alpha_1},$$

$$\frac{g'(\varphi-\tilde{\varphi}(t))}{\varepsilon}\omega(r,\varphi,t,\varepsilon)\le\frac{c_1\alpha_1^{\alpha_1}}{\varepsilon^{1-\alpha_1}}e^{-\alpha_1}. \qquad [12.14b]$$

There are inequalities in the region G_ε^1:

$$\frac{g'(r-\tilde{r}(t))}{\varepsilon}\omega(r,\varphi,t,\varepsilon)\le c_1(-\varepsilon\ln\varepsilon)^{\alpha_1},$$

$$\frac{g'(\varphi-\tilde{\varphi}(t))}{\varepsilon}\omega(r,\varphi,t,\varepsilon)\le c_1(-\varepsilon\ln\varepsilon)^{\alpha_1}.$$

Then, we have

$$|f_1|\le\varepsilon_0\varepsilon\psi_f(t),$$

$$\left|\frac{\partial f_1}{\partial r}\right|=\left|-\varepsilon_0\frac{g'(r-\tilde{r}(t))}{\varepsilon}\psi_f(t)\omega(r,\varphi,t,\varepsilon)\right|\le\psi_f(t)\varepsilon_0 c_1(-\varepsilon\ln\varepsilon)^{\alpha_1},$$

$$\left|\frac{\partial f_1}{\partial\varphi}\right|=\left|-\varepsilon_0\frac{2(\varphi-\tilde{\varphi}(t))}{\varepsilon}\psi_f(t)\omega(r,\varphi,t,\varepsilon)\right|\le\psi_f(t)\varepsilon_0 c_1(-\varepsilon\ln\varepsilon)^{\alpha_1},$$

$$\left|\frac{\partial f_1}{\partial t}\right| = \left|\varepsilon_0 \psi_f{}'(t)\omega(r,\varphi,t,\varepsilon) + \right.$$

$$\left. + \frac{1}{\varepsilon}\varepsilon_0 \psi_f(t)\omega(r,\varphi,t,\varepsilon)[\tilde{\varphi}'(t)g'(\varphi-\varphi(t)) + \tilde{r}'(t)g'(r-\tilde{r}(t))]\right| \le$$

$$\le \varepsilon_0 \varepsilon \left|\psi_f{}'(t)\right| + \varepsilon_0 c_1 (-\varepsilon \ln \varepsilon)^{\alpha_1} \left|\psi_f(t)(\tilde{\varphi}'(t) + \tilde{r}'(t))\right|.$$

Considering system [12.6], we see that all inequalities satisfied with precision $O(\varepsilon_0 \varepsilon\ ) + O(\varepsilon_0 (-\varepsilon \ln \varepsilon)^{\alpha_1}) = O(\varepsilon_0 (-\varepsilon \ln \varepsilon)^{\alpha_1})$ for any bounded functions $\sigma_0(r)$, $v_0(r)$, $\dfrac{d\sigma_0(r)}{dr}$, $\dfrac{dv_0(r)}{dr}$ and hence the solution of system [12.6] in the domain G_ε^1 can be found in form [12.7].

Consider the area G_ε^α. In this area, for $\alpha > 0$, the inequalities are as follows:

$$1 \ge \omega(r,\varphi,t,\varepsilon) \ge \exp\{-\frac{2\varepsilon^{2+2\alpha}}{\varepsilon}\} = \exp\{-2\varepsilon^{1+2\alpha}\} = 1 - 2\varepsilon^{1+2\alpha} + o(\varepsilon^{1+2\alpha}).$$

Taking into account the properties of the function $f(x)$, when the ratio $\{x : x^2 \le \varepsilon^{1+\alpha}\}$ reaches the maximum value $x_{\max} = \varepsilon^{(1+\alpha)/2}$ and is equal $f(x) = c_1 \varepsilon^{(1+\alpha)\alpha_1 - 1} \exp\{-\varepsilon^\alpha\}$, we can write the inequalities as follows:

$$\left|\frac{\partial f_1}{\partial r}\right| = \left|-\varepsilon_0 \frac{g'(r-\tilde{r}(t))}{\varepsilon}\psi_f(t)\omega(r,\varphi,t,\varepsilon)\right| \le \psi_f(t)\varepsilon_0 c_1 \varepsilon^{(1+\alpha)\alpha_1 - 1}\exp\{-\varepsilon^\alpha\},$$

$$\left|\frac{\partial f_1}{\partial \varphi}\right| = \left|-\varepsilon_0 \frac{g'(\varphi-\tilde{\varphi}(t))}{\varepsilon}\psi_f(t)\omega(r,\varphi,t,\varepsilon)\right| \le \psi_f(t)\varepsilon_0 c_1 \varepsilon^{(1+\alpha)\alpha_1 - 1}\exp\{-\varepsilon^\alpha\},$$

$$\left|\frac{\partial f_1}{\partial t}\right| = \left|\varepsilon_0 \psi_f{}'(t)\omega(r,\varphi,t,\varepsilon) + \frac{1}{\varepsilon}\varepsilon_0 \psi_f(t)\omega(r,\varphi,t,\varepsilon)[\tilde{\varphi}'(t)g'(\varphi-\varphi(t)) + \right.$$

$$\left. + \tilde{r}'(t)g'(r-\tilde{r}(t))]\right| \le \varepsilon_0 \varepsilon \left|\psi_f{}'(t)\right| +$$

$$+ \varepsilon_0 c_1 \varepsilon^{(1+\alpha)\alpha_1 - 1}\exp\{-\varepsilon^\alpha\}\left|\psi_f(t)(\tilde{\varphi}'(t) + \tilde{r}'(t))\right|.$$

Analyzing the last inequalities, we see that in system [12.6] the quantities $\dfrac{\partial f_1}{\partial \varphi}$, $\dfrac{\partial f_1}{\partial r}$ have the order of smallness $O(\varepsilon_0 \varepsilon^{(1+\alpha)\alpha_1 - 1})$. In addition, the equality holds:

$\frac{\partial f_1}{\partial t} = \varepsilon_0 \psi_f{}'(t)\omega(r,\varphi,t,\varepsilon) + O(\varepsilon_0 \varepsilon^{(1+\alpha)\alpha_1 - 1})$. In this case, the function $\omega(r,\varphi,t,\varepsilon)$ can be considered equal to 1 with accuracy $O(\varepsilon^{1+2\alpha})$. Then, from system [12.6] we can get the following system:

$$\begin{cases} \frac{\partial \sigma_1(r,\phi,t)}{\partial t} = -\sigma_0(r)\frac{u_1(r,\phi,t)}{r} - u_1(r,\phi,t)\frac{\partial \sigma_0(r)}{\partial r} - \frac{1}{r}\sigma_1(r,\phi,t)u_1(r,\phi,t) + \\ + O(\varepsilon_0 \varepsilon^{(1+\alpha)\alpha_1 - 1}), \\ \frac{\partial u_1(r,\phi,t)}{\partial t} = \frac{2v_0(r)v_1(r,\phi,t)}{r} - 3B\sigma_1(r,\phi,t)\frac{\partial \sigma_0(r)}{\partial r} + \frac{3}{r}BD\sigma_0(r)\sigma_1(r,\phi,t) + \\ + \frac{3}{2r}BD\sigma_1^2(r,\phi,t) + \frac{v_1^2(r,\phi,t)}{r} + O(\varepsilon_0 \varepsilon^{(1+\alpha)\alpha_1 - 1}), \\ \frac{\partial v_1(r,\phi,t)}{\partial t} = -u_1(r,\phi,t)\frac{\partial v_0(r)}{\partial r} - \frac{u_1(r,\phi,t)v_0(r)}{r} - \frac{u_1(r,\phi,t)v_1(r,\phi,t)}{r} + \\ + O(\varepsilon_0 \varepsilon^{(1+\alpha)\alpha_1 - 1}). \end{cases} \quad [12.15]$$

Now substitute [12.6] into [12.15]. Decomposing all the functions that depend on r in the Taylor series in the neighborhood $\tilde{r}(t)$ and given that $|r - \tilde{r}(t)| \le \varepsilon^{1+\alpha}$, we obtain a system of relations:

$$\psi_\sigma{}'(t) = -\psi_u(t)\left(\frac{\partial \sigma_0(\tilde{r}(t))}{\partial r} + \frac{\sigma_0(\tilde{r}(t))}{\tilde{r}(t)} + \frac{\varepsilon_0 \psi_\sigma(t)}{\tilde{r}(t)}\right) + O(\varepsilon_0 \varepsilon^{(1+\alpha)\alpha_1 - 1}),$$

$$\begin{cases} \psi_u{}'(t) = 2\frac{v_0(\tilde{r}(t))}{\tilde{r}(t)}\psi_v(t) + \varepsilon_0 \frac{\psi^2{}_v(t)}{\tilde{r}(t)} - \\ -3B\psi_\sigma(t)\frac{\partial \sigma_0(\tilde{r}(t))}{\partial r} + \frac{3BD}{\tilde{r}(t)}\sigma_0(\tilde{r}(t))\psi_\sigma(t) + \\ + \frac{3BD}{2\tilde{r}(t)}\varepsilon_0 \psi^2{}_\sigma(t) + O(\varepsilon_0 \varepsilon^{(1+\alpha)\alpha_1 - 1}), \\ \psi_v{}'(t) = -\psi_u(t)\left(\frac{\partial v_0(\tilde{r}(t))}{\partial r} + \frac{v_0(\tilde{r}(t))}{\tilde{r}(t)} + \frac{\varepsilon_0 \psi_v(t)}{\tilde{r}(t)}\right) + O(\varepsilon_0 \varepsilon^{(1+\alpha)\alpha_1 - 1}). \end{cases} \quad [12.16]$$

Based on physical considerations, we can formulate the following conditions: $\psi_\sigma(t) \ge 0$, $\sigma_0(r) \ge 0$, $r_0 \le \tilde{r}(t) \le R$, where R is some constant. In addition, the functions $\sigma_0(r), v_0(r)$ are continuously differentiable according to the conditions of the theorem. Then, we can show that the Cauchy problem for system [12.16] has a solution for any initial conditions.

Since the function $\psi_\sigma(t)$ determines the magnitude of the density perturbation (together with the constant), ε_0, we will be interested in such solutions of system [12.16] in which the perturbation of the density does not decrease significantly with time. Then, we can talk about the soliton nature of the solutions. It is known that solitons are structurally stable isolated waves propagating in a nonlinear medium.

Let us analyze the solutions of system [12.16] for different functions $\sigma_0(r)$ and $v_0(r)$. As we see from the first equation of system [12.16], the sign $\psi_\sigma'(t)$ depends on the functions $\psi_u(t)$ and functions of the form:

$$\frac{\partial\sigma_0(\tilde{r}(t))}{\partial r}+\frac{\sigma_0(\tilde{r}(t))}{\tilde{r}(t)}+\frac{\varepsilon_0\psi_\sigma(t)}{\tilde{r}(t)}. \quad [12.17]$$

If function [12.17] is non-negative, and $\psi_u(t)<0$ is satisfied: $\psi_\sigma'(t)\geq 0$. Note that for any finite function $\frac{\partial\sigma_0(r)}{\partial r}$ and $\sigma_0(r)$, we can choose such a large enough value that function [12.17] will be positive.

Let in any region $\frac{\partial\sigma_0(r)}{\partial r}+\frac{\sigma_0(r)}{r}\geq 0$. Then, function [12.17] is non-negative for an arbitrary non-negative value.

Let $\psi_u(t)=0$, $t\in[t_1,t_2]$. Then, from the first and second equations of system [12.16], we get: $\psi_\sigma(t)=\psi_\sigma=const$, $\psi_v(t)=\psi_u(t)=0$. From the second equation of system [12.16], we get condition:

$$\frac{\partial\sigma_0(\tilde{r}(t))}{\partial r}-\frac{D}{\tilde{r}(t)}\sigma_0(\tilde{r}(t))-\frac{D}{2\tilde{r}(t)}\varepsilon_0\psi_\sigma(t)=O(\varepsilon^\alpha).$$

The last condition is satisfied if

$$\sigma_0(r)=r^D(C+\int\frac{D}{2r}\varepsilon_0\psi_\sigma r^{-D}dr)=r^D(C-\frac{D\varepsilon_0\psi_\sigma}{2D}r^{-D})=Cr^D-\frac{\varepsilon_0\psi_\sigma}{2},$$

where C is such a constant that $\sigma_0(r)\geq 0$. Thus, we proved Proposition 12.2.

Let in some area $\frac{\partial\sigma_0(r)}{\partial r} = O(\varepsilon_1)$. We choose a monotonically increasing non-negative function that satisfies the conditions: $\tilde{r}(0) = r_0$, $\frac{1}{r_0} < \varepsilon_1$. Then, $\frac{1}{\tilde{r}(t)} < \varepsilon_1$ Therefore, it follows from the first equation of system [12.16] that $\psi_\sigma'(t) = O(\varepsilon_1)$. The sign of the function $\psi_u(t)$ can be any.

Let in some area $\frac{\partial\sigma_0(r)}{\partial r} + \frac{\sigma_0(r)}{r} < 0$. Then, obviously, we can choose such small values ε_0 and functions $\tilde{r}(t)$ such that the inequality holds: $\frac{\partial\sigma_0(\tilde{r}(t))}{\partial r} + \frac{\sigma_0(\tilde{r}(t))}{\tilde{r}(t)} + \frac{\varepsilon_0\psi_\sigma(t)}{\tilde{r}(t)} \le 0$. In this case, if the condition $\psi_u(t) > 0$ is satisfied, the following inequality is satisfied: $\psi_\sigma'(t) \ge 0$. Analyzing the second equation of system [12.16], we see that when the condition $\frac{\partial v_0(\tilde{r}(t))}{\partial r} + \frac{v_0(\tilde{r}(t))}{\tilde{r}(t)} \le 0$ is satisfied, the inequalities $\psi_u'(t) \ge 0$ and $\psi_v'(t) \ge 0$.

Thus, all the conditions formulated in Proposition 12.3 are obtained. The proof is complete.

PROPOSITION 12.4.– Let $\gamma_s = 2$. Then, the partial solution of system [12.6] can be found in form [12.7], and $u_1(r,\varphi,t) = v_1(r,\varphi,t) = 0$, $\psi_\sigma(t) = \psi_\sigma = const$, $\tilde{\varphi}'(t) = \frac{v_0(r)}{r} = \Omega = const$ $\tilde{r}(t) = \tilde{r} = const$,
$\Phi_1(r,\varphi,z,t) = -2B\varepsilon_0\psi_\sigma\omega(r,\varphi,t,\varepsilon) = -2B\sigma_1(r,\varphi,t,\varepsilon)$.

PROOF.–

Let $v_1 \equiv 0$, $\gamma_s = 2$. Then, from [12.10], we get:

$$\varepsilon_0\omega(r,\varphi,t,\varepsilon)(\psi_\sigma'(t) + \sigma_0\frac{\psi_u(t)}{r} + \psi_u(t)\frac{\partial\sigma_0}{\partial r} + \frac{1}{r}\varepsilon_0\psi_\sigma(t)\psi_u(t)\omega(r,\varphi,t,\varepsilon)) +$$
$$+\varepsilon_0\omega(r,\varphi,t,\varepsilon)(\frac{1}{\varepsilon}\psi_\sigma(t)[\tilde{\varphi}'(t)g'(\varphi-\varphi(t)) + \tilde{r}'(t)g'(r-\tilde{r}(t))] -$$
$$-\frac{v_0}{r}\frac{g'(\varphi-\tilde{\varphi}(t))}{\varepsilon}\psi_\sigma(t) + \sigma_0(-\frac{g'(r-\tilde{r}(t))}{\varepsilon}\psi_u(t)) +$$
$$+\frac{1}{r}(-\varepsilon_0\psi_u(t)r\frac{g'(r-\tilde{r}(t))}{\varepsilon}\psi_\sigma(t)\omega(r,\varphi,t,\varepsilon) - \psi_\sigma(t)r\varepsilon_0\frac{g'(r-\tilde{r}(t))}{\varepsilon}\psi_u(t)\omega(r,\varphi,t,\varepsilon) = 0.$$

If $g'(\varphi-\varphi(t))=0, \varphi=\varphi(t)$, $g'(r-\tilde{r}(t))=0, r=\tilde{r}(t)$, then

$$\varepsilon_0\omega(r,\varphi,t,\varepsilon)(\psi_\sigma'(t)+\sigma_0\frac{\psi_u(t)}{r}+\psi_u(t)\frac{\partial\sigma_0}{\partial r}+\frac{1}{r}\varepsilon_0\psi_\sigma(t)\psi_u(t)\omega(r,\varphi,t,\varepsilon))=0.$$

From the last equation, we get:

$$\frac{\varepsilon_0\omega(r,\varphi,t,\varepsilon)}{\varepsilon}g'(\varphi-\varphi(t))\psi_\sigma(t)(\tilde{\varphi}'(t)-\frac{v_0}{r})+$$
$$+\frac{\varepsilon_0\omega(r,\varphi,t,\varepsilon)}{\varepsilon}g'(r-\tilde{r}(t))\psi_\sigma(t)(\tilde{r}'(t)-\sigma_0\psi_u(t)/\psi_\sigma(t)-2\varepsilon_0\psi_u(t)\omega(r,\varphi,t,\varepsilon))=0.$$

We get a system:

$$\begin{cases}\varepsilon_0\omega(\psi_\sigma'(t)+\sigma_0\frac{\psi_u(t)}{r}+\psi_u(t)\frac{\partial\sigma_0}{\partial r}+\frac{1}{r}\varepsilon_0\psi_\sigma(t)\psi_u(t)\omega)=0,\\ \frac{\varepsilon_0\omega}{\varepsilon}g'(\varphi-\varphi(t))\psi_\sigma(t)(\tilde{\varphi}'(t)-\frac{v_0}{r})=0,\\ \frac{\varepsilon_0\omega}{\varepsilon}g'(r-\tilde{r}(t))\psi_\sigma(t)(\tilde{r}'(t)-\sigma_0\psi_u(t)/\psi_\sigma(t)-2\varepsilon_0\psi_u(t)\omega)=0.\end{cases} \quad [12.18]$$

From [12.11] and [12.12], we get:

$$\frac{\partial\Phi_1}{\partial r}=-\varepsilon_0\omega(r,\varphi,t,\varepsilon)\psi_u'(t)-$$
$$-\psi_u(t)\frac{g'(r-\tilde{r}(t))}{\varepsilon}\varepsilon_0\omega(r,\varphi,t,\varepsilon)(\tilde{r}'(t)-2B\psi_\sigma(t)/\psi_u(t)-\varepsilon_0\psi_u(t)\omega(r,\varphi,t,\varepsilon))-$$
$$-\psi_u(t)\frac{g'(\varphi-\tilde{\varphi}(t))}{\varepsilon}\varepsilon_0\omega(r,\varphi,t,\varepsilon)(\tilde{\varphi}'(t)-\frac{v_0}{r}), \quad [12.19]$$

$$\frac{1}{r}\frac{\partial\Phi_1}{\partial\varphi}=-\varepsilon_0\omega(r,\varphi,t,\varepsilon)(\psi_u(t)\frac{\partial v_0}{\partial r}+\frac{\psi_u(t)v_0}{r})+$$
$$+\frac{g'(\varphi-\tilde{\varphi}(t))}{\varepsilon}\varepsilon_0\omega(r,\varphi,t,\varepsilon)\frac{2B}{r}\psi_\sigma(t). \quad [12.20]$$

Let

$$\Phi_1(r,\varphi,z,t)=(\tilde{r}'(t)\psi_u(t)-2B\psi_\sigma(t))\varepsilon_0\omega(r,\varphi,t,\varepsilon)=\tilde{r}'(t)u_1(r,\varphi,t,\varepsilon)-$$
$$-2B\sigma_1(r,\varphi,t,\varepsilon).$$

Then, from [12.19] and [12.20], we get:

$$0=-\varepsilon_0\omega(r,\varphi,t,\varepsilon)\psi_u'(t)+\psi_u(t)\frac{g'(r-\widetilde{r}(t))}{\varepsilon}\varepsilon_0\omega(r,\varphi,t,\varepsilon)\varepsilon_0\psi_u(t)\omega(r,\varphi,t,\varepsilon)- \\ -\psi_u(t)\frac{g'(\varphi-\widetilde{\varphi}(t))}{\varepsilon}\varepsilon_0\omega(r,\varphi,t,\varepsilon)(\widetilde{\varphi}'(t)-\frac{v_0}{r}), \quad [12.21]$$

$$0=-\varepsilon_0\omega(r,\varphi,t,\varepsilon)(\psi_u(t)\frac{\partial v_0}{\partial r}+\frac{\psi_u(t)v_0}{r})+ \\ +\frac{g'(\varphi-\widetilde{\varphi}(t))}{\varepsilon}\varepsilon_0\omega(r,\varphi,t,\varepsilon)\frac{\widetilde{r}'(t)\psi_u(t)}{r}. \quad [12.22]$$

Consider some cases. Let $u_1=0$. Then, from systems [12.18] and [12.21]–[12.22], we obtain the following relation:

$$\psi_\sigma'(t)=0\,,\widetilde{\varphi}'(t)-\frac{v_0}{r}=0\,,\widetilde{r}'(t)=0 \quad [12.23]$$

Equations [9.19]–[9.20] are obviously satisfied. Let there be a relation: $\frac{v_0(r)}{r}=\Omega=const$. This condition means that the gas disk of the galaxy rotates like a solid. Note that the surface adiabatic index $\gamma_s=2$ corresponds to the volume index of adiabatic $\gamma=2$, which can be found using Hunter's formula: $\gamma_s=3-\frac{2}{\gamma}$ (this indicator corresponds to the gas type of ionized plasma). The proof is complete.

Therefore, for the case $\gamma_s=2$, we have found a partial solution of system [12.7] in the form of perturbations, which are an undamped wave moving in a circle with a constant angular velocity: $\widetilde{\varphi}'(t)=\frac{v_0(r)}{r}=const$. The perturbation of the surface density can be arbitrary.

Let $u_1\neq 0$. Analyzing relationships [12.18]–[12.20], we see that in this case, we can only talk about an approximate solution. Note that we will have small parameters: perturbation of the surface density and the velocities. Further analysis will significantly depend on the relationship between these small parameters and the properties of the function g. However, it should be noted that the accuracy with which all equations [12.18]–[12.20] must be satisfied must be less than the magnitude of the surface density perturbation, otherwise we will not be able to talk about the existence of a soliton.

We analyze system [12.18] and equations [12.21]–[12.22]. The equation is satisfied as follows:

$$\tilde{\varphi}'(t) = \frac{v_0(r)}{r} = \Omega = const\ . \qquad [12.24]$$

To simplify further calculations, we assume that $\psi_u(t) = \psi_u = const$ $\psi_\sigma(t) = \psi_\sigma = const$, $\sigma_0(r) = \sigma_0 = const$. Suppose that we have relation:

$$\varepsilon_0 \psi_u = \varepsilon^{1-\alpha_1}\ . \qquad [12.25]$$

From the first equation of system [12.18] we get

$$\varepsilon_0 \omega(r,\varphi,t,\varepsilon)(\sigma_0 \frac{\psi_u}{r} + \frac{1}{r}\varepsilon_0 \psi_\sigma \psi_u \omega(r,\varphi,t,\varepsilon)) =$$

$$= \varepsilon_0 \omega(r,\varphi,t,\varepsilon)\psi_u(t)\frac{\sigma_0}{r} + \frac{1}{r}\varepsilon_0 \psi_\sigma \omega(r,\varphi,t,\varepsilon)\psi_u \varepsilon_0 \omega(r,\varphi,t,\varepsilon) \le$$

$$\le \frac{\sigma_0}{r}\varepsilon^{1-\alpha_1} + \frac{1}{r}\varepsilon_0 \psi_\sigma \varepsilon^{1-\alpha_1} \le \varepsilon^{1-\alpha_1}\ .$$

The last inequality took into account the smallness of the value $\frac{\sigma_0}{r}$ resulting from the observations, as well as smallness $\frac{1}{r}\varepsilon_0 \psi_\sigma$. Therefore, the first equation of system [12.18] is satisfied with an accuracy not exceeding $\varepsilon^{1-\alpha_1}$. The second equation is satisfied under condition [12.24]. From the third equation, we have:

$$\frac{\varepsilon_0 \omega}{\varepsilon} g'(r - \tilde{r}(t))\psi_\sigma(\tilde{r}'(t) - \sigma_0 \psi_u / \psi_\sigma - 2\varepsilon_0 \psi_u \omega) =$$

$$= \frac{\varepsilon_0 \omega}{\varepsilon} g'(r - \tilde{r}(t))\psi_\sigma(\tilde{r}'(t) - \sigma_0 \psi_u / \psi_\sigma - 2\varepsilon_0 \psi_u \omega) =$$

$$= \frac{\varepsilon_0 \omega}{\varepsilon} g'(r - \tilde{r}(t))\psi_\sigma(\tilde{r}'(t) - \sigma_0 \psi_u / \psi_\sigma) - \frac{\varepsilon_0 \omega}{\varepsilon} g'(r - \tilde{r}(t))\psi_\sigma 2\varepsilon_0 \psi_u \omega,$$

$$\left|\frac{\varepsilon_0\omega}{\varepsilon}g'(r-\tilde{r}(t))\psi_\sigma 2\varepsilon_0\psi_u\omega\right|\le\frac{c_1\alpha_1^{\alpha_1}}{\varepsilon^{1-\alpha_1}}e^{-\alpha_1}\psi_\sigma 2\varepsilon_0^2|\psi_u|\le$$

$$\le\frac{c_1\alpha_1^{\alpha_1}}{\varepsilon^{1-\alpha_1}|\psi_u|}e^{-\alpha_1}\psi_\sigma 2\varepsilon^{2(1-\alpha_1)}\le 2c_1\alpha_1^{\alpha_1}e^{-\alpha_1}\frac{\psi_\sigma}{|\psi_u|}\varepsilon^{1-\alpha_1},$$

$$\left|\frac{\varepsilon_0\omega}{\varepsilon}g'(r-\tilde{r}(t))\psi_\sigma(\tilde{r}'(t)-\sigma_0\psi_u/\psi_\sigma)-\frac{\varepsilon_0\omega}{\varepsilon}g'(r-\tilde{r}(t))\psi_\sigma 2\varepsilon_0\psi_u\omega\right|\le$$

$$\le\left|\frac{\varepsilon_0\omega}{\varepsilon}g'(r-\tilde{r}(t))\psi_\sigma(\tilde{r}'(t)-\sigma_0\psi_u/\psi_\sigma)\right|+2c_1\alpha_1^{\alpha_1}e^{-\alpha_1}\frac{\psi_\sigma}{|\psi_u|}\varepsilon^{1-\alpha_1}.$$

Let

$$\tilde{r}'(t)-\sigma_0\psi_u(t)/\psi_\sigma(t)=0\,. \qquad [12.26]$$

Then, the third equation of system [12.18] is satisfied with precision

$$2c_1\alpha_1^{\alpha_1}e^{-\alpha_1}\frac{\psi_\sigma}{|\psi_u|}\varepsilon^{1-\alpha_1}.$$

Consider equation [12.21]. From the inequalities in the region G_ε^2 follows the inequality: $\left|\psi_u(t)\frac{g'(r-\tilde{r}(t))}{\varepsilon}\varepsilon_0\omega(r,\varphi,t,\varepsilon)\varepsilon_0\psi_u(t)\omega(r,\varphi,t,\varepsilon)\right|\le c_1\alpha_1^{\alpha_1}e^{-\alpha_1}\varepsilon^{1-\alpha_1}$.

Therefore, equation [12.21] is satisfied with accuracy $c_1\alpha_1^{\alpha_1}e^{-\alpha_1}\varepsilon^{1-\alpha_1}$.

From relation [12.22], we have

$$\psi_u(t)\varepsilon_0\omega(r,\varphi,t,\varepsilon)(\frac{\partial v_0}{\partial r}+\frac{v_0}{r})\le\varepsilon^{1-\alpha_1}(\frac{\partial v_0}{\partial r}+\frac{v_0}{r})=2\Omega\varepsilon^{1-\alpha_1}\le\varepsilon^{1-\alpha_1}$$

Consider the expression $\frac{g'(\varphi-\tilde{\varphi}(t))}{\varepsilon}\varepsilon_0\omega(r,\varphi,t,\varepsilon)\frac{\tilde{r}'(t)\psi_u}{r}$. From inequality [12.14b], the following inequality follows:

$$\left|\frac{g'(\varphi-\tilde{\varphi}(t))}{\varepsilon r}\varepsilon_0\omega(r,\varphi,t,\varepsilon)\tilde{r}'(t)\psi_u(t)\right|\le\frac{c_1\alpha_1^{\alpha_1}}{\varepsilon^{1-\alpha_1}r}e^{-\alpha_1}|\tilde{r}'(t)|\varepsilon^{1-\alpha_1}=\frac{c_1\alpha_1^{\alpha_1}|\tilde{r}'(t)|}{r}e^{-\alpha_1}.$$

Hence, $\frac{c_1\alpha_1^{\alpha_1}e^{-\alpha_1}}{r}|\tilde{r}'(t)| \le \varepsilon^{1-\alpha_1}$ or $\frac{|\tilde{r}'(t)|}{r} \le \frac{\varepsilon^{1-\alpha_1}e^{-\alpha_1}}{c_1\alpha_1^{\alpha_1}}$ and taking into account [12.26], we get $\frac{\sigma_0|\psi_u|}{r\psi_\sigma} \le \frac{\varepsilon^{1-\alpha_1}e^{-\alpha_1}}{c_1\alpha_1^{\alpha_1}}$. Then, we get an additional restriction on the area in which we consider our system:

$$e^{-\alpha_1}c_1\alpha_1^{\alpha_1}\frac{\sigma_0|\psi_u|}{\varepsilon^{1-\alpha_1}\psi_\sigma} \le r \le R \qquad [12.27]$$

Therefore, when relation [12.27] is true, the trajectories of solitons are determined by equations [12.24] and [12.26], from which we obtain: $\tilde{r}(t) = c_r + \sigma_0\psi_u t/\psi_\sigma + O(\varepsilon^{1-\alpha_1}))$, $\tilde{\varphi}(t) = c_\varphi + \Omega t$,

$$(\tilde{\varphi}(t) - c_\varphi)/\Omega = (\tilde{r}(t) - c_r)\psi_\sigma/(\sigma_0\psi_u), \qquad [12.28]$$

$$\tilde{r}(t) = (\tilde{\varphi}(t) - c_\varphi)\sigma_0\psi_u/(\psi_\sigma\Omega) + c_r.$$

We see that the last equality is the equation of the Archimedes spiral.

PROPOSITION 12.5.– Let $\gamma_s = 2$. Then, the partial solution of system [12.6] can be found in form [12.7] in the region $G = \{(r,\varphi): e^{-\alpha_1}c_1\alpha_1^{\alpha_1}\frac{\sigma_0|\psi_u|}{\varepsilon^{1-\alpha_1}\psi_\sigma} \le r \le R\}$ with an accuracy not exceeding the value $\max\{2c_1\alpha_1^{\alpha_1}e^{-\alpha_1}\frac{\psi_\sigma}{|\psi_u|}, c_1\alpha_1^{\alpha_1}e^{-\alpha_1}, 1\}\varepsilon^{1-\alpha_1}$ and additional conditions are satisfied: $v_1(r,\varphi,t) = 0$, $\psi_u(t) = \psi_u = const$, $\psi_\sigma(t) = \psi_\sigma = const$, $\tilde{\varphi}(t) = c_\varphi + \Omega t$

$$\tilde{r}(t) = c_r + \sigma_0\psi_u t/\psi_\sigma,\ \sigma_0(r) = \sigma_0 = const,\ \frac{v_0(r)}{r} = \Omega = const,$$

$$\Phi_1(r,\varphi,z,t) = \varepsilon_0\omega(r,\varphi,t,\varepsilon)(\sigma_0\psi_u^2 - 2B\psi_\sigma^2)/\psi_\sigma.$$

Consider the case of an arbitrary parameter γ_s. Note that equation [12.10] does not depend on the parameter γ_s and its analysis has already been performed above. Therefore, we consider only equations [12.11] and [12.12]. We write these equations in the form:

$$\frac{1}{r}\frac{\partial \Phi_1}{\partial \varphi} = -\varepsilon_0\omega(\psi_v{}'(t) + \psi_u(t)\frac{\partial v_0}{\partial r} + \frac{\psi_u(t)v_0}{r} + \\
+\frac{\psi_u(t)\varepsilon_0\psi_v(t)\omega}{r}) - \varepsilon_0\omega\frac{g'(r-\tilde{r}(t))}{\varepsilon}\psi_v(t)(\tilde{r}'(t) - \psi_u(t)\varepsilon_0\omega) - \\
-\frac{g'(\varphi-\tilde{\varphi}(t))}{\varepsilon}\varepsilon_0\omega\psi_v(t)(\tilde{\varphi}'(t) - \frac{v_0}{r} - \frac{B\gamma_s}{r}\sigma_0{}^{\gamma_s-2}\psi_\sigma(t)/\psi_v(t) - \\
-\frac{B\gamma_s}{r}(\gamma_s-2)\sigma_0^{\gamma_s-3}\varepsilon_0\psi_\sigma^2(t)/\psi_v(t)\omega - \frac{1}{r}\varepsilon_0\psi_v(t)\omega). \qquad [12.29]$$

Let

$$\Phi_1(r,\varphi,z,t) = (\tilde{r}'(t)\psi_u(t) - B\gamma_s\sigma_0{}^{\gamma_s-2}(r)\psi_\sigma(t))\varepsilon_0\omega(r,\varphi,t,\varepsilon) = \\
= \tilde{r}'(t)u_1(r,\varphi,t,\varepsilon) - B\gamma_s\sigma_0{}^{\gamma_s-2}\sigma_1(r,\varphi,t,\varepsilon), \qquad [12.30]$$

$$\psi_u(t) = \psi_u = const,\ \psi_\sigma(t) = \psi_\sigma = const,\ \sigma_0(r) = \sigma_0 = const,\ \psi_v(t) = 0.$$

Then, equations [12.28] and [12.29] can be written in the form

$$0 = -\varepsilon_0\omega(r,\varphi,t,\varepsilon)(\psi_u{}'(t) - \frac{3}{2r}BD(\gamma_s-1)\sigma_0{}^{\gamma_s-2}\psi_\sigma(t) + B\gamma_s(\gamma_s-2)\psi_\sigma(t)\sigma_0^{\gamma_s-3}\frac{\partial\sigma_0}{\partial r}) + \\
+\psi_u(t)\frac{g'(r-\tilde{r}(t))}{\varepsilon}\varepsilon_0\omega(B\gamma_s(\gamma_s-2)\sigma_0^{\gamma_s-3}\varepsilon_0\psi_\sigma^2(t)/\psi_u(t)\omega + \\
+\varepsilon_0\psi_u(t)\omega) - \psi_u(t)\frac{g'(\varphi-\tilde{\varphi}(t))}{\varepsilon}\varepsilon_0\omega(\tilde{\varphi}'(t) - \frac{v_0}{r}), \qquad [12.31]$$

$$0 = -\varepsilon_0\omega(\psi_u(t)\frac{\partial v_0}{\partial r} + \frac{\psi_u(t)v_0}{r}) - \frac{g'(\varphi-\tilde{\varphi}(t))}{\varepsilon}\varepsilon_0\omega(-\frac{\tilde{r}'(t)\psi_u(t)}{r} - \\
-\frac{B\gamma_s}{r}(\gamma_s-2)\sigma_0^{\gamma_s-3}\varepsilon_0\psi_\sigma^2(t)\omega), \qquad [12.32]$$

$$0 = \varepsilon_0\omega(r,\varphi,t,\varepsilon)\frac{3}{2r}BD(\gamma_s-1)\sigma_0{}^{\gamma_s-2}\psi_\sigma(t) + \\
+\psi_u(t)\frac{g'(r-\tilde{r}(t))}{\varepsilon}\varepsilon_0\omega(B\gamma_s(\gamma_s-2)\sigma_0^{\gamma_s-3}\varepsilon_0\psi_\sigma^2(t)/\psi_u(t)\omega + \varepsilon_0\psi_u(t)\omega), \qquad [12.33]$$

$$0=-\varepsilon_0\omega(\psi_u(t)\frac{\partial v_0}{\partial r}+\frac{\psi_u(t)v_0}{r})-\frac{g'(\varphi-\tilde{\varphi}(t))}{\varepsilon}\varepsilon_0\omega(-\frac{\tilde{r}'(t)\psi_u(t)}{r}-$$
$$-\frac{B\gamma_s}{r}(\gamma_s-2)\sigma_0^{\gamma_s-3}\varepsilon_0\psi_\sigma^2(t)\omega). \quad [12.34]$$

Consider equation [12.33]. Taking into account inequalities [12.14b], we can write the following inequalities:

$$\left|\psi_u(t)\frac{g'(r-\tilde{r}(t))}{\varepsilon}\varepsilon_0\omega(r,\varphi,t,\varepsilon)\varepsilon_0\psi_u(t)\omega(r,\varphi,t,\varepsilon)\right|\le c_1\alpha_1^{\alpha_1}e^{-\alpha_1}\varepsilon^{1-\alpha_1},$$

$$\left|\varepsilon_0\psi_\sigma\omega(r,\varphi,t,\varepsilon)\frac{3}{2r}BD(\gamma_s-1)\sigma_0^{\gamma_s-2}\right|\le\varepsilon^{1-\alpha_1}\frac{\psi_\sigma}{\psi_u}\frac{3}{2r}BD(\gamma_s-1)\sigma_0^{\gamma_s-2},$$

$$\left|\frac{g'(r-\tilde{r}(t))}{\varepsilon}B\gamma_s(\gamma_s-2)\sigma_0^{\gamma_s-3}\varepsilon_0^2\psi_\sigma^2(t)\omega^2\right|\le$$
$$\le\frac{c_1\alpha_1^{\alpha_1}}{\varepsilon^{1-\alpha_1}}e^{-\alpha_1}B\gamma_s(\gamma_s-2)\sigma_0^{\gamma_s-3}\varepsilon_0^2\psi_\sigma^2\omega\le$$

$$\le\frac{c_1\alpha_1^{\alpha_1}}{\varepsilon^{1-\alpha_1}}e^{-\alpha_1}B\gamma_s(\gamma_s-2)\sigma_0^{\gamma_s-3}\varepsilon^{2(1-\alpha_1)}\psi_\sigma^2/\psi_u^2=$$
$$=c_1\alpha_1^{\alpha_1}e^{-\alpha_1}B\gamma_s(\gamma_s-2)\sigma_0^{\gamma_s-3}\varepsilon^{1-\alpha_1}\frac{\psi_\sigma^2}{\psi_u^2}.$$

Consider [12.34]. We have

$$\left|\psi_u(t)\varepsilon_0\omega(r,\varphi,t,\varepsilon)(\frac{\partial v_0}{\partial r}+\frac{v_0}{r})\right|\le\varepsilon^{1-\alpha_1}\left|\frac{\partial v_0}{\partial r}+\frac{v_0}{r}\right|=2\Omega\varepsilon^{1-\alpha_1}\le\varepsilon^{1-\alpha_1}.$$

The expression $\frac{g'(\varphi-\tilde{\varphi}(t))}{\varepsilon}\varepsilon_0\omega(r,\varphi,t,\varepsilon)\frac{\tilde{r}'(t)\psi_u}{r}$ has already been considered above and obtained the ratio [12.27]. It remains to consider the expression $\frac{g'(\varphi-\tilde{\varphi}(t))}{\varepsilon}\frac{B\gamma_s}{r}(\gamma_s-2)\sigma_0^{\gamma_s-3}\varepsilon_0^2\psi_\sigma^2(t)\omega^2$. We have inequalities

$$\left|\frac{g'(\varphi-\tilde{\varphi}(t))}{\varepsilon}\varepsilon_0\omega\frac{B\gamma_s}{r}(\gamma_s-2)\sigma_0^{\gamma_s-3}\varepsilon_0\psi_\sigma^2(t)\omega\right|\leq$$

$$\leq\frac{c_1\alpha_1^{\alpha_1}}{\varepsilon^{1-\alpha_1}}e^{-\alpha_1}\frac{B\gamma_s}{r}(\gamma_s-2)\sigma_0^{\gamma_s-3}\varepsilon_0^2\psi_\sigma^2(t)\omega\leq$$

$$\leq\frac{c_1\alpha_1^{\alpha_1}}{\varepsilon^{1-\alpha_1}}e^{-\alpha_1}\frac{B\gamma_s}{r}(\gamma_s-2)\sigma_0^{\gamma_s-3}\varepsilon^{2(1-\alpha_1)}\psi_\sigma^2/\psi_u^2=$$

$$=c_1\alpha_1^{\alpha_1}e^{-\alpha_1}\frac{B\gamma_s}{r}(\gamma_s-2)\sigma_0^{\gamma_s-3}\frac{\psi_\sigma^2}{\psi_u^2}\varepsilon^{1-\alpha_1}.$$

Therefore, equations [12.33]–[12.34] are satisfied with an accuracy not exceeding $\max\{c_1\alpha_1^{\alpha_1}e^{-\alpha_1}B\gamma_s(\gamma_s-2)\sigma_0^{\gamma_s-3}\frac{\psi_\sigma^2}{\psi_u^2},2c_1\alpha_1^{\alpha_1}e^{-\alpha_1}\frac{\psi_\sigma}{|\psi_u|},1\}\varepsilon^{1-\alpha_1}$.

PROPOSITION 12.6.– Partial solution of system [12.6] can be found in the form [12.7] in the region $G=\{(r,\varphi):e^{-\alpha_1}c_1\alpha_1^{\alpha_1}\frac{\sigma_0|\psi_u|}{\varepsilon^{1-\alpha_1}\psi_\sigma}\leq r\leq R\}$ with an accuracy not exceeding $\max\{c_1\alpha_1^{\alpha_1}e^{-\alpha_1}B\gamma_s(\gamma_s-2)\sigma_0^{\gamma_s-3}\frac{\psi_\sigma^2}{\psi_u^2},2c_1\alpha_1^{\alpha_1}e^{-\alpha_1}\frac{\psi_\sigma}{|\psi_u|},1\}\varepsilon^{1-\alpha_1}$. In this solution

$$v_1(r,\varphi,t)=0\,,\psi_u(t)=\psi_u=const\,,\psi_\sigma(t)=\psi_\sigma=const\,,\tilde{\varphi}(t)=c_\varphi+\Omega t\,,$$

$$\tilde{r}(t)=c_r+\sigma_0\psi_u t/\psi_\sigma\,,\ \sigma_0(r)=\sigma_0=const\,,\frac{v_0(r)}{r}=\Omega=const\,,$$

$$\Phi_1(r,\varphi,z,t)=\varepsilon_0\omega(r,\varphi,t,\varepsilon)(\sigma_0\psi_u^2-2B\psi_\sigma^2)/\psi_\sigma.$$

As is known (Morozov and Khoperskov 2005), the perturbation of the gravitational potential satisfies the Poisson equation:

$$\frac{\partial^2\Phi_1}{\partial r^2}+\frac{1}{r}\frac{\partial\Phi_1}{\partial r}+\frac{\partial^2\Phi_1}{r^2\partial\varphi^2}+\frac{\partial^2\Phi_1}{\partial z^2}=4\pi G\sigma_1\delta(z)\,, \qquad [12.35]$$

where G is the gravitational constant, $G=6{,}67*10^{-11}\frac{\text{Н м}^2}{\text{kg}^2}$, $\delta(z)$ – any function.

Let us consider regions:

$$G_{\varepsilon}^{1}=\{(r,\varphi,t):g(r-\widetilde{r}(t))+g(\varphi-\widetilde{\varphi}(t))\geq -2\varepsilon\ln\varepsilon, t\in[0,T], r\leq R\},$$

$$G_{\varepsilon}^{\alpha}=\{(r,\varphi,t):g(r-\widetilde{r}(t))\leq \varepsilon^{1+\alpha}, g(\varphi-\widetilde{\varphi}(t))<\varepsilon^{1+\alpha}, t\in[0,T], r\leq R\}$$

We will assume that the perturbation of the gravitational potential has the form:

$$\Phi_1(r,\varphi,t)=\widetilde{r}'(t)u_1(r,\varphi,t,\varepsilon)-2B\sigma_1(r,\varphi,t,\varepsilon)=\varepsilon_0(\widetilde{r}'(t)\psi_u(t)-\\ -2B\psi_\sigma(t))w(r,\varphi,t,\varepsilon). \qquad [12.36]$$

Suppose that ε_0 is a function of z $\varepsilon_0(z)$ and at the point $z=0$, this function reaches the maximum value: $\varepsilon_0(z)\leq\varepsilon_0(0)=\varepsilon_0$.

Then, from equation [12.35], we obtain:

$$\varepsilon_0(z)(\widetilde{r}'(t)\psi_u-2B\psi_\sigma)\left(\frac{\partial^2 w}{\partial r^2}+\frac{1}{r}\frac{\partial w}{\partial r}+\frac{\partial^2 w}{r^2\partial\varphi^2}\right)+\\ +(\widetilde{r}'(t)\psi_u-2B\psi_\sigma)w\frac{\partial^2\varepsilon_0(z)}{\partial z^2}=4\pi G\delta(z)\psi_\sigma\varepsilon_0(z)w. \qquad [12.37]$$

Consider region G_{ε}^{1}. Note that there are obvious inequalities in this area: $|w|\leq\varepsilon^2$. We have:

$$\frac{\partial w}{\partial r}=-\frac{g'(r-\widetilde{r}(t))}{\varepsilon}\omega(r,\varphi,t,\varepsilon),$$

$$\frac{\partial^2 w}{\partial r^2}=-\frac{g''(r-\widetilde{r}(t))}{\varepsilon}\omega(r,\varphi,t,\varepsilon)+\frac{g'^2(r-\widetilde{r}(t))}{\varepsilon^2}\omega(r,\varphi,t,\varepsilon),$$

$$\frac{\partial^2 w}{\partial\varphi^2}=-\frac{g''(\varphi-\widetilde{\varphi}(t))}{\varepsilon}\omega(r,\varphi,t,\varepsilon)+\frac{g'^2(\varphi-\widetilde{\varphi}(t))}{\varepsilon^2}\omega(r,\varphi,t,\varepsilon).$$

Let us analyze in the region G_ε^1 functions $\frac{\partial f_1}{\partial r}, \frac{\partial^2 f_1}{\partial r^2}, \frac{\partial^2 f_1}{\partial \varphi^2}$. Note that in the region $\{x : x^2 \geq -2\varepsilon \ln \varepsilon\}$, the maximal value of the function $f(x,y) = \frac{c_1 x^{2\alpha_1}}{\varepsilon} \exp\{-\frac{x^2+y^2}{\varepsilon}\}$ is

$$f(-2\varepsilon \ln \varepsilon) = \frac{c_1(-2\varepsilon \ln \varepsilon)^{\alpha_1}}{\varepsilon} \exp\{-\frac{-2\varepsilon \ln \varepsilon}{\varepsilon}\} = c_1(-2\varepsilon \ln \varepsilon)^{\alpha_1} \varepsilon .$$

Thus, $\left| \frac{g'(r-\tilde{r}(t))}{\varepsilon} \omega(r,\varphi,t,\varepsilon) \right| \leq c_1(-2\varepsilon \ln \varepsilon)^{\alpha_1} \varepsilon$,

$$\left| \frac{g''(r-\tilde{r}(t))}{\varepsilon} \omega(r,\varphi,t,\varepsilon) \right| \leq c_2(-2\varepsilon \ln \varepsilon)^{\alpha_2} \varepsilon ,$$

$$\frac{g'^2(r-\tilde{r}(t))}{\varepsilon^2} \omega(r,\varphi,t,\varepsilon) \leq \frac{c_1 g^{2\alpha_1}(r-\tilde{r}(t))}{\varepsilon^2} \omega(r,\varphi,t,\varepsilon) .$$

Consider the function $f(x,y) = \frac{c_1 x^{4\alpha_1}}{\varepsilon^2} \exp\{-\frac{x^2+y^2}{\varepsilon}\}$. Extremum of this function is ï $f(-2\varepsilon \ln \varepsilon) = \frac{c_1(-2\varepsilon \ln \varepsilon)^{2\alpha_1}}{\varepsilon^2} \exp\{-\frac{-2\varepsilon \ln \varepsilon}{\varepsilon}\} = c_1(-2\varepsilon \ln \varepsilon)^{2\alpha_1}$.

Thus, $\frac{g'^2(r-\tilde{r}(t))}{\varepsilon^2} \omega(r,\varphi,t,\varepsilon) \leq c_1 2^{2\alpha_1}(-\varepsilon \ln \varepsilon)^{2\alpha_1}$ and we have inequalities:

$$\left| \frac{\partial w}{\partial r} \right| = \left| -\frac{g'(r-\tilde{r}(t))}{\varepsilon} \omega(r,\varphi,t,\varepsilon) \right| \leq c_1(-2\varepsilon \ln \varepsilon)^{\alpha_1} \varepsilon ,$$

$$\left| \frac{\partial^2 w}{\partial r^2} \right| = \left| -\frac{g''(r-\tilde{r}(t))}{\varepsilon} \omega(r,\varphi,t,\varepsilon) + \frac{g'^2(r-\tilde{r}(t))}{\varepsilon^2} \omega(r,\varphi,t,\varepsilon) \right| \leq$$
$$\leq c_2(-2\varepsilon \ln \varepsilon)^{\alpha_2} \varepsilon + c_1 2^{2\alpha_1}(-\varepsilon \ln \varepsilon)^{2\alpha_1} ,$$

$$\left| \frac{\partial^2 w}{\partial \varphi^2} \right| \leq c_2(-2\varepsilon \ln \varepsilon)^{\alpha_2} \varepsilon + c_1 2^{2\alpha_1}(-\varepsilon \ln \varepsilon)^{2\alpha_1} .$$

Then, we have:

$$\left|\frac{\partial^2 w}{\partial r^2}+\frac{1}{r}\frac{\partial w}{\partial r}+\frac{\partial^2 w}{r^2\partial\varphi^2}\right| \le c_1 2^{2\alpha_1}(-\varepsilon\ln\varepsilon)^{2\alpha_1}+c_2 2^{\alpha_2}(-\varepsilon\ln\varepsilon)^{\alpha_2}\varepsilon+$$
$$+c_1 2^{\alpha_1}(-\varepsilon\ln\varepsilon)^{\alpha_1}\varepsilon/r+(c_1 2^{2\alpha_1}(-\varepsilon\ln\varepsilon)^{2\alpha_1}+c_2 2^{\alpha_2}(-\varepsilon\ln\varepsilon)^{\alpha_2}\varepsilon)/r^2=$$
$$=(c_1 2^{2\alpha_1}(-\varepsilon\ln\varepsilon)^{2\alpha_1}+c_2 2^{\alpha_2}(-\varepsilon\ln\varepsilon)^{\alpha_2}\varepsilon)(1+1/r^2)+c_1 2^{\alpha_1}(-\varepsilon\ln\varepsilon)^{\alpha_1}\varepsilon/r.$$

Let the equality hold:

$$(\tilde{r}'(t)\psi_u-2B\psi_\sigma)\frac{\partial^2\varepsilon_0(z)}{\partial z^2}=4\pi G\delta(z)\psi_\sigma\varepsilon_0(z)\,. \qquad [12.38]$$

Obviously, we can choose the function $\varepsilon_0(z)$ so that equation [12.38] holds. The logical conditions are

$$\varepsilon_0'(0)=0\,,\varepsilon_0(0)=\varepsilon_0\,. \qquad [12.39]$$

Then, equation [12.35] holds with an accuracy not exceeding $\varepsilon_0(z)(\tilde{r}'(t)\psi_u-2B\psi_\sigma)(c_1 2^{2\alpha_1}(-\varepsilon\ln\varepsilon)^{2\alpha_1}+c_2 2^{\alpha_2}(-\varepsilon\ln\varepsilon)^{\alpha_2}\varepsilon)(1+1/r^2)+$ $+c_1 2^{\alpha_1}(-\varepsilon\ln\varepsilon)^{\alpha_1}\varepsilon/r)$.

Consider region G_ε^α. In this region

$$\exp\{-2\varepsilon^{\alpha-1}\}\le\omega(r,\varphi,t,\varepsilon)=\exp\{-\frac{g(r-\tilde{r}(t))+g(\varphi-\tilde{\varphi}(t))}{\varepsilon}\}\le 1\,.$$

The following inequalities follow from the properties of the function g:

$$\left|\frac{g''(r-\tilde{r}(t))}{\varepsilon}\right|\le\frac{\left|c_2 g^{\alpha_2}((r-\tilde{r}(t))\right|}{\varepsilon}\le\frac{c_2\varepsilon^{\alpha\alpha_2}}{\varepsilon}=c_2\varepsilon^{\alpha\alpha_2-1}\,,$$

$$\frac{\left|g'^2(r-\tilde{r}(t))\right|}{\varepsilon^2}\le\frac{c_1^2 g^{2\alpha_1}((r-\tilde{r}(t)))}{\varepsilon^2}\le\frac{c_1^2\varepsilon^{2\alpha_1\alpha}}{\varepsilon^2}=c_1^2\varepsilon^{2(\alpha_1\alpha-1)}\,.$$

Similar inequalities will be for $g''(\varphi-\tilde{\varphi}(t))$ and $g'^2(\varphi-\tilde{\varphi}(t))$ Thus,

$$\left|\varepsilon_0(z)(\tilde{r}'(t)\psi_u - 2B\psi_\sigma)(\frac{\partial^2 w}{\partial r^2}+\frac{1}{r}\frac{\partial w}{\partial r}+\frac{\partial^2 w}{r^2\partial\varphi^2})\right| \le$$

$$\le \varepsilon_0(\tilde{r}'(t)\psi_u - 2B\psi_\sigma)((c_2\varepsilon^{\alpha\alpha_2-1}+c_1{}^2\varepsilon^{2(\alpha_1\alpha-1)})(1+1/r^2)+c_1\varepsilon^{\alpha_1\alpha-1}).$$

If equation [12.38] is satisfied, then equation [12.37] is satisfied with an accuracy that does not exceed

$$\varepsilon_0(\tilde{r}'(t)\psi_u - 2B\psi_\sigma)((c_2\varepsilon^{\alpha\alpha_2-1}+c_1{}^2\varepsilon^{2(\alpha_1\alpha-1)})(1+1/r^2)+c_1\varepsilon^{\alpha_1\alpha-1}).$$

PROPOSITION 12.7.– Let the perturbation of the gravitational potential be determined by relation [12.36], let $\varepsilon_0=\varepsilon_0(z)$ be the function, which is the solution of the Cauchy problem [12.38]–[12.39]. Then, there is the Poisson equation [12.35] with an accuracy not exceeding the value

$$\tilde{\varepsilon}=\begin{cases}\varepsilon_0(z)(\tilde{r}'(t)\psi_u-2B\psi_\sigma)(c_1 2^{2\alpha_1}(-\varepsilon\ln\varepsilon)^{2\alpha_1}+c_2 2^{\alpha_2}(-\varepsilon\ln\varepsilon)^{\alpha_2}\varepsilon)(1+1/r^2)+\\ +c_1 2^{\alpha_1}(-\varepsilon\ln\varepsilon)^{\alpha_1}\varepsilon/r),\ (r,\varphi)\in G_\varepsilon^1\\ \varepsilon_0(\tilde{r}'(t)\psi_u-2B\psi_\sigma)((c_2\varepsilon^{\alpha\alpha_2-1}+c_1{}^2\varepsilon^{2(\alpha_1\alpha-1)})(1+1/r^2)+c_1\varepsilon^{\alpha_1\alpha-1}),\\ (r,\varphi)\in G_\varepsilon^\alpha\end{cases}$$

where α is a number, such that $\alpha\alpha_1>1$ and $\alpha\alpha_2>1$.

Note that when the condition $\tilde{r}'(t)\psi_u-2B\psi_\sigma=0$ is satisfied, the Poisson equation holds exactly. The perturbation of the gravitational potential is equal to 0. Comparing the last equality with relation [12.26], we have: $\tilde{r}'(t)=\sigma_0\psi_u/\psi_\sigma=2B\psi_\sigma/\psi_u$. From the last equation we obtain the condition for surface density:

$$\sigma_0=2B\psi_\sigma^2/\psi_u^2 \qquad [12.40]$$

Therefore, we considered the possibility of finding solutions of the system of equations of gas dynamics of galaxies in the form of separate structurally stable waves, which are perturbations of surface density and equilibrium velocity components in form [12.7]. Note that the perturbation of the gravitational potential was considered to be zero.

Note that in practice (Friedman 2007), the surface density behaves differently. For a number of galaxies, individual areas (rings) of increase in surface density are observed at the same time, as well as cases when it decreases with distance from the center of the galaxy almost exponentially or is close to a constant.

In the above statements, only certain specific situations are considered. It is of considerable interest to trace what happens to a single wave of the soliton type when it passes through the regions of maximum or minimum surface density, i.e. the values of variables at which the function $\dfrac{\partial \sigma_0(r)}{\partial r}$ changes sign.

Analyzing equation [12.20], we can see that in such cases there are pulsations of amplitude – when passing through the density extremum, a single wave first decreases its amplitude, but after some time under certain conditions it can begin to increase again (if, for example, the amplitude of the wave decreased to a critically possible small value at which it can no longer physically exist).

Note that the single waves mentioned in our work can play a significant role in such processes when the intensity of the waves is high and collective mechanisms come into play. After all, the interaction of numerous even small perturbations can lead to significant effects on the system.

12.4. Analysis of the localized wave trajectories

Consider equation [12.10]. If $g'(\varphi-\varphi(t))=0, \varphi=\varphi(t)$, $g'(r-\tilde{r}(t))=0, r=\tilde{r}(t)$, then:

$$\varepsilon_0\omega(r,\varphi,t,\varepsilon)(\psi_\sigma{}'(t)+\sigma_0\frac{\psi_u(t)}{r}+\psi_u(t)\frac{\partial\sigma_0}{\partial r}+\frac{\psi_v(t)}{r}\frac{\partial\sigma_0}{\partial\varphi}+$$

$$+\frac{1}{r}\varepsilon_0\psi_\sigma(t)\psi_u(t)\omega(r,\varphi,t,\varepsilon))=0.$$

Thus

$$\frac{\varepsilon_0\omega(r,\varphi,t,\varepsilon)}{\varepsilon}g'(\varphi-\varphi(t))\psi_\sigma(t)(\tilde{\varphi}'(t)-\frac{v_0}{r}-\psi_v(t)\frac{\sigma_0}{r\psi_\sigma(t)}-\frac{2\varepsilon_0}{r}\psi_v(t)\omega(r,\varphi,t,\varepsilon))+$$

$$+\frac{\varepsilon_0\omega(r,\varphi,t,\varepsilon)}{\varepsilon}g'(r-\tilde{r}(t))\psi_\sigma(t)(\tilde{r}'(t)-\sigma_0\psi_u(t)/\psi_\sigma(t)-2\varepsilon_0\psi_u(t)\omega(r,\varphi,t,\varepsilon))=0.$$

From the last equation, we get

$$\frac{\varepsilon_0\omega(r,\varphi,t,\varepsilon)}{\varepsilon}g'(\varphi-\varphi(t))\psi_\sigma(t)(\tilde{\varphi}'(t)-\frac{v_0}{r}-\psi_v(t)\frac{\sigma_0}{r\psi_\sigma(t)}-$$
$$-\frac{2\varepsilon_0}{r}\psi_v(t)\omega(r,\varphi,t,\varepsilon))=0,$$

$$\frac{\varepsilon_0\omega(r,\varphi,t,\varepsilon)}{\varepsilon}g'(r-\tilde{r}(t))\psi_\sigma(t)(\tilde{r}'(t)-\sigma_0\psi_u(t)/\psi_\sigma(t)-2\varepsilon_0\psi_u(t)\omega(r,\varphi,t,\varepsilon))=0.$$

Thus, we get the following system of equations:

$$\begin{cases}\psi_\sigma'(t)+\sigma_0\dfrac{\psi_u(t)}{r}+\psi_u(t)\dfrac{\partial\sigma_0}{\partial r}+\dfrac{\psi_v(t)}{r}\dfrac{\partial\sigma_0}{\partial\varphi}+\\ +\dfrac{1}{r}\varepsilon_0\psi_\sigma(t)\psi_u(t)\omega(r,\varphi,t,\varepsilon)=0,\\ \tilde{\varphi}'(t)-\dfrac{v_0}{r}-\psi_v(t)\dfrac{\sigma_0}{r\psi_\sigma(t)}-\dfrac{2\varepsilon_0}{r}\psi_v(t)\omega(r,\varphi,t,\varepsilon)=0,\\ \tilde{r}'(t)-\sigma_0\psi_u(t)/\psi_\sigma(t)-2\varepsilon_0\psi_u(t)\omega(r,\varphi,t,\varepsilon)=0.\end{cases} \quad [12.41]$$

Similarly, analyzing equations [12.11]–[12.12], we obtain, respectively

$$\frac{\partial\Phi_1}{\partial r}=-\varepsilon_0\omega(r,\varphi,t,\varepsilon)(\psi_u'(t)-\frac{2v_0\psi_v(t)}{r}-\frac{3}{2r}BD(\gamma_s-1)\sigma_0^{\gamma_s-2}\psi_\sigma(t)+$$
$$+\frac{\varepsilon_0\psi^2{}_v(t)\omega(r,\varphi,t,\varepsilon)}{r}+B\gamma_s(\gamma_s-2)\psi_\sigma(t)\sigma_0^{\gamma_s-3}\frac{\partial\sigma_0}{\partial r})-$$

$$-\psi_u(t)\frac{g'(r-\tilde{r}(t))}{\varepsilon}\varepsilon_0\omega(r,\varphi,t,\varepsilon)(\tilde{r}'(t)-B\gamma_s\sigma_0^{\gamma_s-2}\psi_\sigma(t)/\psi_u(t)-$$
$$-B\gamma_s(\gamma_s-2)\sigma_0^{\gamma_s-3}\varepsilon_0\psi_\sigma^2(t)/\psi_u(t)\omega(r,\varphi,t,\varepsilon)- \quad [12.42]$$
$$-\varepsilon_0\psi_u(t)\omega(r,\varphi,t,\varepsilon))-\psi_u(t)\frac{g'(\varphi-\tilde{\varphi}(t))}{\varepsilon}\varepsilon_0\omega(r,\varphi,t,\varepsilon)(\tilde{\varphi}'(t)-\frac{v_0}{r}-\frac{\psi_v(t)}{r}\varepsilon_0\omega(r,\varphi,t,\varepsilon)),$$

$$\frac{1}{r}\frac{\partial \Phi_1}{\partial \varphi}=-\varepsilon_0\omega(r,\varphi,t,\varepsilon)(\psi_v{}'(t)+\psi_u(t)\frac{\partial v_0}{\partial r}+\frac{\psi_u(t)v_0}{r}+\frac{\psi_v(t)}{r}\frac{\partial v_0}{\partial \varphi}+$$
$$+\frac{\psi_u(t)\varepsilon_0\psi_v(t)\omega(r,\varphi,t,\varepsilon)}{r}+$$
$$+\frac{B\gamma_s}{r}(\gamma_s-2)\psi_\sigma(t)\sigma_0^{\gamma_s-3}\frac{\partial \sigma_0}{\partial \varphi})-\varepsilon_0\omega(r,\varphi,t,\varepsilon)\frac{g'(r-\tilde{r}(t))}{\varepsilon}\psi_v(t)(\tilde{r}'(t)- \qquad [12.43]$$
$$-\psi_u(t)\varepsilon_0\omega(r,\varphi,t,\varepsilon))-$$
$$-\frac{g'(\varphi-\tilde{\varphi}(t))}{\varepsilon}\varepsilon_0\omega(r,\varphi,t,\varepsilon)\psi_v(t)(\tilde{\varphi}'(t)-\frac{v_0}{r}-\frac{B\gamma_s}{r}\sigma_0{}^{\gamma_s-2}\psi_\sigma(t)/\psi_v(t)-$$
$$-\frac{B\gamma_s}{r}(\gamma_s-2)\sigma_0^{\gamma_s-3}\varepsilon_0\psi_\sigma^2(t)/\psi_v(t)\omega(r,\varphi,t,\varepsilon)-\frac{1}{r}\varepsilon_0\psi_v(t)\omega(r,\varphi,t,\varepsilon)).$$

Let $\Phi_1=0$.

Then, from equations [12.41]–[12.43], we get

$$\begin{cases}\psi_\sigma{}'(t)+\sigma_0\dfrac{\psi_u(t)}{r}+\psi_u(t)\dfrac{\partial \sigma_0}{\partial r}+\dfrac{\psi_v(t)}{r}\dfrac{\partial \sigma_0}{\partial \varphi}+\\ +\dfrac{1}{r}\varepsilon_0\psi_\sigma(t)\psi_u(t)\omega(r,\varphi,t,\varepsilon)=0,\\ \tilde{\varphi}'(t)-\dfrac{v_0}{r}-\psi_v(t)\dfrac{\sigma_0}{r\psi_\sigma(t)}-\dfrac{2\varepsilon_0}{r}\psi_v(t)\omega(r,\varphi,t,\varepsilon)=0,\\ \tilde{r}'(t)-\sigma_0\psi_u(t)/\psi_\sigma(t)-2\varepsilon_0\psi_u(t)\omega(r,\varphi,t,\varepsilon)=0.\end{cases} \qquad [12.44]$$

$$\psi_u{}'(t)-\frac{2v_0\psi_v(t)}{r}-\frac{3}{2r}BD(\gamma_s-1)\sigma_0{}^{\gamma_s-2}\psi_\sigma(t)+$$
$$+\frac{\varepsilon_0\psi^2{}_v(t)\omega(r,\varphi,t,\varepsilon)}{r}+B\gamma_s(\gamma_s-2)\psi_\sigma(t)\sigma_0^{\gamma_s-3}\frac{\partial \sigma_0}{\partial r}=0,$$

$$\tilde{r}'(t)-B\gamma_s\sigma_0{}^{\gamma_s-2}\psi_\sigma(t)/\psi_u(t)-B\gamma_s(\gamma_s-2)\sigma_0^{\gamma_s-3}\varepsilon_0\psi_\sigma^2(t)/\psi_u(t)\omega(r,\varphi,t,\varepsilon)-$$
$$-\varepsilon_0\psi_u(t)\omega(r,\varphi,t,\varepsilon)=0, \qquad [12.45]$$

$$\tilde{\varphi}'(t)-\frac{v_0}{r}-\frac{\psi_v(t)}{r}\varepsilon_0\omega(r,\varphi,t,\varepsilon)=0.$$

From equation [12.43], we get

$$\psi_v'(t)+\psi_u(t)\frac{\partial v_0}{\partial r}+\frac{\psi_u(t)v_0}{r}+\frac{\psi_v(t)}{r}\frac{\partial v_0}{\partial \varphi}+\frac{\psi_u(t)\varepsilon_0\psi_v(t)\omega(r,\varphi,t,\varepsilon)}{r}+$$
$$+\frac{B\gamma_s}{r}(\gamma_s-2)\psi_\sigma(t)\sigma_0^{\gamma_s-3}\frac{\partial\sigma_0}{\partial\varphi}=0,$$

$$\tilde{r}'(t)-\psi_u(t)\varepsilon_0\omega(r,\varphi,t,\varepsilon)=0\,,$$

$$\tilde{\varphi}'(t)-\frac{v_0}{r}-\frac{B\gamma_s}{r}\sigma_0{}^{\gamma_s-2}\psi_\sigma(t)/\psi_v(t)-$$
$$-\frac{B\gamma_s}{r}(\gamma_s-2)\sigma_0^{\gamma_s-3}\varepsilon_0\psi_\sigma^2(t)/\psi_v(t)\omega(r,\varphi,t,\varepsilon)-\frac{1}{r}\varepsilon_0\psi_v(t)\omega(r,\varphi,t,\varepsilon)=0. \quad [12.46]$$

Let $\psi_v(t)\neq 0$, $\psi_u(t)\neq 0$.

Consider systems [12.44]–[12.46] and analyze equations that contain $\tilde{r}'(t)$.

Let us analyze the equations containing $\tilde{r}'(t)$. Then, we receive conditions:

$$\psi_u(t)\varepsilon_0\omega(r,\varphi,t,\varepsilon)=B\gamma_s\sigma_0{}^{\gamma_s-2}\psi_\sigma(t)/\psi_u(t)+$$
$$+B\gamma_s(\gamma_s-2)\sigma_0^{\gamma_s-3}\varepsilon_0\psi_\sigma^2(t)/\psi_u(t)\omega(r,\varphi,t,\varepsilon)+$$
$$+\varepsilon_0\psi_u(t)\omega(r,\varphi,t,\varepsilon)=\sigma_0\psi_u(t)/\psi_\sigma(t)+2\varepsilon_0\psi_u(t)\omega(r,\varphi,t,\varepsilon)$$

From the last equation, we get

$$B\gamma_s\sigma_0{}^{\gamma_s-2}\psi_\sigma(t)/\psi_u(t)+B\gamma_s(\gamma_s-2)\sigma_0^{\gamma_s-3}\varepsilon_0\psi_\sigma^2(t)/\psi_u(t)\omega(r,\varphi,t,\varepsilon)=0\,, \quad [12.47]$$

$$\sigma_0\ +(\gamma_s-2)\varepsilon_0\psi_\sigma(t)\omega(r,\varphi,t,\varepsilon)=0\,.$$

$$\sigma_0\psi_u(t)/\psi_\sigma(t)+\varepsilon_0\psi_u(t)\omega(r,\varphi,t,\varepsilon)=0\,,$$

$$\sigma_0+\varepsilon_0\psi_\sigma(t)\omega(r,\varphi,t,\varepsilon)=0\,. \quad [12.48]$$

Comparing [12.47] and [12.48], we see that $\gamma_s=3$. This result is very important because it corresponds to the case when the gas is in a state of ionized plasma.

In addition, ratio [12.48] shows that the surface density perturbation is negative.

Let us analyze the equations containing $\tilde{\varphi}'(t)$ in systems [12.44]–[12.46]. We get

$$\psi_v(t)\frac{\sigma_0}{r\psi_\sigma(t)}+\frac{2\varepsilon_0}{r}\psi_v(t)\omega(r,\varphi,t,\varepsilon)=\frac{\psi_v(t)}{r}\varepsilon_0\omega(r,\varphi,t,\varepsilon)$$

From the last equation, we have

$$\psi_v(t)\frac{\sigma_0}{r\psi_\sigma(t)}+\frac{\varepsilon_0}{r}\psi_v(t)\omega(r,\varphi,t,\varepsilon)=0.$$

We get equation [12.48]. Thus, we have the following system:

$$\psi_\sigma'(t)+\sigma_0\frac{\psi_u(t)}{r}+\psi_u(t)\frac{\partial\sigma_0}{\partial r}+\frac{\psi_v(t)}{r}\frac{\partial\sigma_0}{\partial\varphi}+\frac{1}{r}\varepsilon_0\psi_\sigma(t)\psi_u(t)\omega(r,\varphi,t,\varepsilon)=0,$$

$$\begin{aligned}
&\psi_u'(t)-\frac{2v_0\psi_v(t)}{r}-\frac{3}{r}BD\sigma_0\,\psi_\sigma(t)-\frac{\varepsilon_0\psi^2{}_v(t)\omega(r,\varphi,t,\varepsilon)}{r}+\\
&+3B\psi_\sigma(t)\frac{\partial\sigma_0}{\partial r}=0,\\
&\psi_v'(t)+\psi_u(t)\frac{\partial v_0}{\partial r}+\frac{\psi_v(t)}{r}\frac{\partial v_0}{\partial\varphi}+\frac{\psi_u(t)v_0}{r}+\\
&+\frac{\psi_u(t)\varepsilon_0\psi_v(t)\omega(r,\varphi,t,\varepsilon)}{r}=0,\\
&\tilde{\varphi}'(t)-\frac{v_0}{r}-\frac{\psi_v(t)}{r}\varepsilon_0\omega(r,\varphi,t,\varepsilon)=0,\\
&\tilde{r}'(t)-\psi_u(t)\varepsilon_0\omega(r,\varphi,t,\varepsilon)=0,\\
&\sigma_0+\varepsilon_0\psi_\sigma(t)\omega(r,\varphi,t,\varepsilon)=0.
\end{aligned}\qquad [12.49]$$

We will consider the system in the area

$$G_\varepsilon^\alpha=\{(r,\varphi,t):g(r-\tilde{r}(t))+g(\varphi-\tilde{\varphi}(t))<\varepsilon^{1/\alpha_1}\}.$$

There are obvious inequalities in this area:

$$1\geq\omega(r,\varphi,t,\varepsilon)=1-\varepsilon^{1/\alpha_1-1}+O(\varepsilon^{2/\alpha_1-2}),\ g(r-\tilde{r}(t))<\varepsilon^{1/\alpha_1}.$$

From here, we get

$$|r-\tilde{r}(t)|<g^{-1}(\varepsilon^{1/\alpha_1}) \text{ or } r=\tilde{r}(t)+\alpha g^{-1}(\varepsilon^{1/\alpha_1}),\ \alpha_1\in(-1,1).$$

Then

$$\frac{\partial\sigma_0(r)}{\partial r}=\frac{\partial\sigma_0(\tilde{r}(t)+\alpha g^{-1}(\varepsilon^{1/\alpha_1}))}{\partial r}=\frac{\partial\sigma_0(\tilde{r}(t))}{\partial r}+\alpha g^{-1}(\varepsilon^{1/\alpha_1})\frac{\partial^2\sigma_0(\tilde{r}(t))}{\partial r^2}+$$
$$+o(\alpha g^{-1}(\varepsilon^{1/\alpha_1})).$$

Similarly, we can write other schedules. Then, we get the following system, which will be satisfied with accuracy $O(g^{-1}(\varepsilon^{1/\alpha_1}))$:

$$\psi_\sigma'(t)+\sigma_0\frac{\psi_u(t)}{\tilde{r}(t)}+\psi_u(t)\frac{\partial\sigma_0}{\partial r}+\frac{\psi_v(t)}{\tilde{r}(t)}\frac{\partial\sigma_0}{\partial\varphi}+\frac{1}{\tilde{r}(t)}\varepsilon_0\psi_\sigma(t)\psi_u(t)=0,$$

$$\psi_u'(t)-\frac{2v_0\psi_v(t)}{\tilde{r}(t)}-\frac{3}{\tilde{r}(t)}BD\sigma_0\ \psi_\sigma(t)-\frac{\varepsilon_0\psi^2{}_v(t)}{\tilde{r}(t)}+3B\psi_\sigma(t)\frac{\partial\sigma_0}{\partial r}=0,$$

$$\psi_v'(t)+\psi_u(t)\frac{\partial v_0}{\partial r}+\frac{\psi_v(t)}{\tilde{r}(t)}\frac{\partial v_0}{\partial\varphi}+\frac{\psi_u(t)v_0}{\tilde{r}(t)}+\frac{\psi_u(t)\varepsilon_0\psi_v(t)}{\tilde{r}(t)}=0,$$

$$\tilde{\varphi}'(t)-\frac{v_0}{\tilde{r}(t)}-\frac{\psi_v(t)}{\tilde{r}(t)}\varepsilon_0=0,\ \tilde{r}'(t)-\psi_u(t)\varepsilon_0=0,$$
$$\sigma_0(\tilde{r}(t),\tilde{\varphi}(t),t)+\varepsilon_0\psi_\sigma(t)=0.$$

Taking into account the last equations and D=0, we get

$$\begin{cases}\psi_u'(t)-\dfrac{2v_0\psi_v(t)}{\tilde{r}(t)}-\dfrac{\varepsilon_0\psi^2{}_v(t)}{\tilde{r}(t)}-3B\dfrac{\partial\sigma_0}{\partial r}\sigma_0(\tilde{r}(t),\tilde{\varphi}(t))/\varepsilon_0=0,\\ \psi_v'(t)+\psi_u(t)\dfrac{\partial v_0}{\partial r}+\dfrac{\psi_v(t)}{\tilde{r}(t)}\dfrac{\partial v_0}{\partial\varphi}+\dfrac{\psi_u(t)v_0}{\tilde{r}(t)}+\dfrac{\psi_u(t)\varepsilon_0\psi_v(t)}{\tilde{r}(t)}=0\\ \tilde{\varphi}'(t)-\dfrac{v_0}{\tilde{r}(t)}-\dfrac{\psi_v(t)}{\tilde{r}(t)}\varepsilon_0=0,\\ \tilde{r}'(t)-\psi_u(t)\varepsilon_0=0,\end{cases} \qquad [12.50]$$

$$\begin{cases} \dfrac{\partial \sigma_0}{\partial t} + \dfrac{v_0 \partial \sigma_0}{r \partial \varphi} = 0, \\ -\dfrac{{v_0}^2}{r} = -\dfrac{\partial \Phi_0}{\partial r} - B\gamma_s \sigma_0 \dfrac{\partial \sigma_0}{\partial r} + \dfrac{3}{2r} DB{\sigma_0}^2, \\ \dfrac{\partial v_0}{\partial t} + \dfrac{v_0}{r} \dfrac{\partial v_0}{\partial \varphi} = -\sigma_0 \dfrac{3B}{r} \dfrac{\partial \sigma_0}{\partial \varphi}. \end{cases} \quad [12.51]$$

Thus, we obtained the system of equations, solving which we can find the trajectories of localized waves.

12.5. Numerical results

We will solve system [12.49] in Mathcad 14.0, where we use the identifiers of the functions y0(u), y1(u), y2(u) and y3 (u), instead of the functions $\psi_u(t)$, $\psi_v(t)$ $\tilde{\varphi}(t)$, $\tilde{r}(t)$, respectively. In this case, we will simulate two solitons that have different velocities at the initial time. The initial conditions here are as follows. The first two values specify the radial and azimuthal components of the soliton velocity; the other two values are the initial coordinates of the soliton in the polar system, coordinate, radius and angle. We will consider the surface density as a function that depends only on the radius and choose it in the form:

$$g_0(r,\varphi) = C + Be^{-\alpha r} + \sum_{i=0}^{k} a_i e^{-\frac{(r-R_i)^2}{l_i}}.$$

The simulation result is shown in Figure 12.2. The two solitons (blue and red trajectories in the color version of the image) start from the points (0,380), (0,390); the azimuthal components are the same, but the radial ones are different and equal to 5 and 54, respectively. Interestingly, the amplitude of oscillations does not change. This can be seen in the graph of Figure 12.1.

The initial condition for the waves is: first soliton (the blue trajectory in Figure 12.2): y0(0)=5, y1(0)=180, y2(0)=0, y3(0)=380, and second soliton (the red trajectory in Figure 12.2): y0(0)=58, y1(0)=180, y2(0)=0, y3(0)=390, T1=16.28, Eps=0.2, R=185, R1=285, R2=385, R3=485.

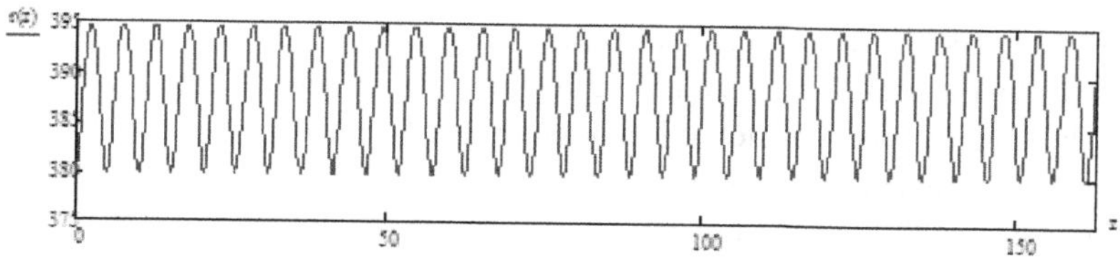

Figure 12.1. *Function $\tilde{r}(t)$ for the first soliton (blue trajectories in Figure 12.2). For a color version of this figure, see www.iste.co.uk/koroliouk/computational.zip*

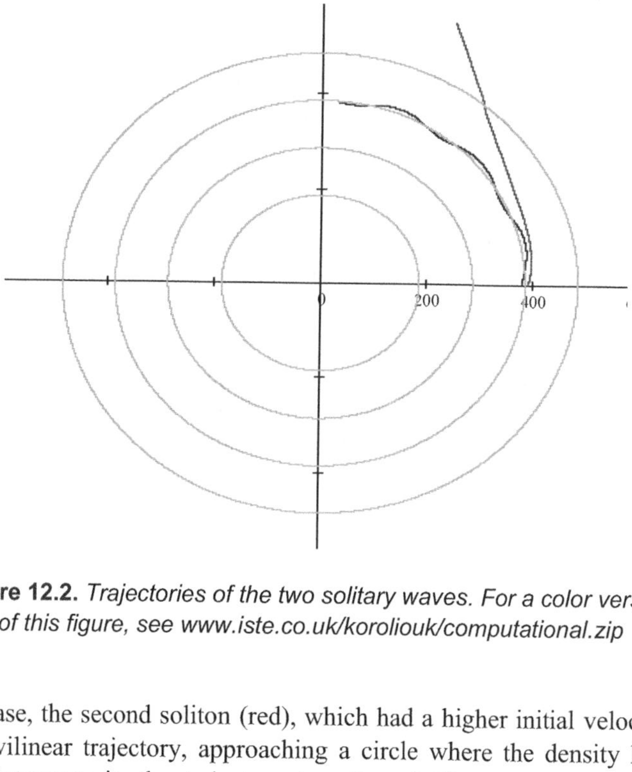

Figure 12.2. *Trajectories of the two solitary waves. For a color version of this figure, see www.iste.co.uk/koroliouk/computational.zip*

In this case, the second soliton (red), which had a higher initial velocity, moves along a curvilinear trajectory, approaching a circle where the density has a local maximum, intersects it, the trajectory is refracted like a ray of light and then becomes virtually rectilinear. We see that for a large T the soliton moves away from the center, asymptotically approaching the motion along some straight line, moving away from the center.

The corresponding behavior of a slower soliton can be interpreted in two ways: approximation to the region of maximum density and repulsion from the region of minimum. Note that we actually have an antisoliton here, a negative perturbation of the density. Since the density changes continuously, the soliton is smoothly knocked out. If we had a clear boundary of the regions of the density change, we would obviously have the phenomenon of embossing like Snelius's law. Thus, the soliton in a continuous medium will move in the direction of maximum density in areas where it is a smooth function and will be reflected in the presence of a sharp change in density.

Let us consider another example. Consider the case when the density has circular areas with small values: $g_0(r,\varphi)=11-g\sum_{i=1}^{4}a_i e^{-\frac{(r-R_i)^2}{15}}$

The graph of this function can be seen in Figure 12.3 The initial conditions for the numerical experiment: first soliton (blue trajectory in Figure 12.4): y0(0)=5, y1(0)=180,y2(0)=0, y3(0)=380, and second soliton (red trajectory in Figure 12.4): y0(0)=35, y1(0)=50,y2(0)=0, y3(0)=430.

Other parameters are the same as in the previous example. In Figure 12.4, we see that antisolitons at the corresponding initial velocities cannot penetrate regions with low density and are reflected in the same way as light quanta. The red antisoliton is reflected from both circular regions. This behavior of negative perturbations of the medium (localized regions of lower density) is in good agreement with the general properties of solitons, which are known to repel as elastic solids. Having a computer program, we can easily investigate the trajectories of solitary waves under any initial conditions.

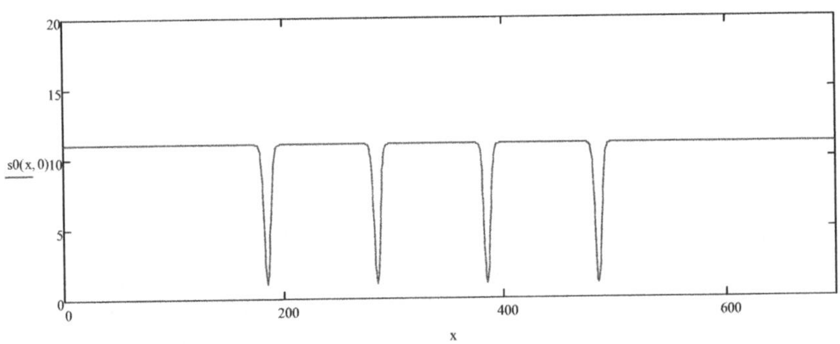

Figure 12.3. *Graph of the surface density. For a color version of this figure, see www.iste.co.uk/koroliouk/computational.zip*

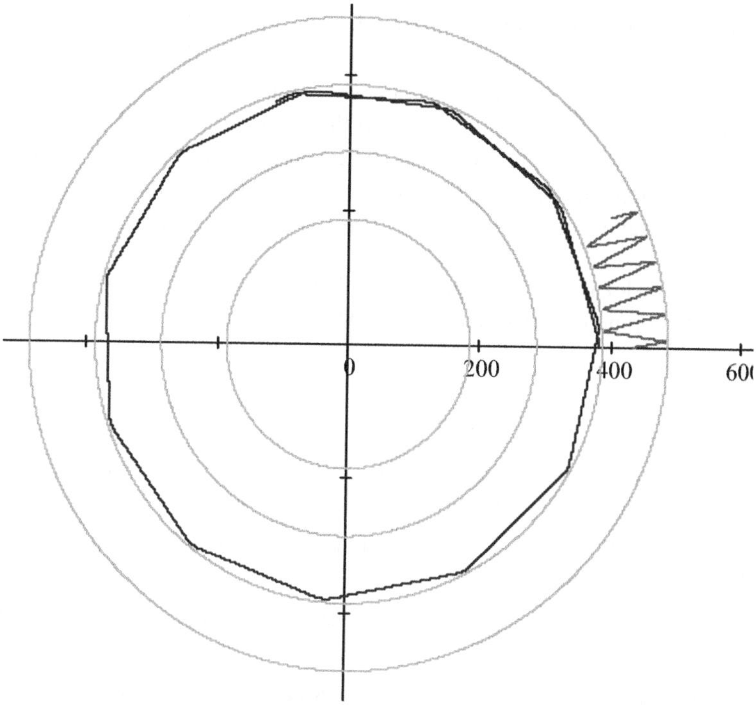

Figure 12.4. *Example of the solitary wave trajectory. For a color version of this figure, see www.iste.co.uk/koroliouk/computational.zip*

12.6. Conclusion

Thus, we consider modeling of the localized soliton-like perturbations in media where the characteristics are described by the shallow water equations. The specificity of the approach, which is used here to study for the trajectories of the corresponding perturbations, is the assumption of the exact solutions in the corresponding systems of the equations and using the approximate approach.

As a result of using the method of T-representations, the system of differential equations is obtained. From this system, the functions that describe the trajectories of localized waves are obtained. The system is solved using the Mathcad environment. It is shown that the corresponding software package is effective for cases when function compositions are used.

Solutions for different cases of surface density are found. It is shown that there are only antisolitons. The antisolitons are moved in the nonlinear environments like light quanta, although the laws of reflection are essentially nonlinear. The theoretical analysis has shown that solitons occur when the gas is in a state of ionized plasma. The obtained results correlate well with experimental studies, which confirms the adequacy of the applied approach to modeling and the possibility of using it for the development of wave monitoring software (see, for example, Lyashko et al. 2021).

In the future, localized perturbations will be investigated using point sources such as the Delta functions (Lyashko et al. 2021).

12.7. References

Antonova, R.A., Zhvania, B.P., Lominadze, J.G., Nanobashvili J.I., Petviashvili, V.I. (1983). On drift solitons in a shallow rotating fluid. *JETP Let.*, 37(11), 545–548 [in Russian].

Asselmeyer-Maluga, T. and Kró, J. (2020). Dark matter as gravitational solitons in the weak field limit. *Universe*, 6(12), 234.

Basak, S., Dutta, R., Kar, P.S. (2021). A new model with solitary waves: Solution, stability and quasinormal modes. *Eur. Phys. J. Plus.*, 136, 618.

Bazeia, D., Belendryasova, E., Gani, V.A. (2018). Scattering of kinks of the sinh-deformed φ4 model. *Eur. Phys. J. C*, 78, 340.

Bila, N., Mansfield, E., Clarkson, P. (2006). Symmetry group analysis of the shallow water and semi-geostrophic equations. *Q. J. Mech. Appl. Math.*, 59, 95–123.

Bomba, A. and Turbal, Y. (2015). Data analysis method and problems of identification of trajectories of solitary waves. *J. Autom. Inf. Sci.*, 47, 13–23.

Bomba, A., Turbal, Y., Radovenyuk, O.V., Turbal, M. (2015). Mathematical and computer modeling of seismic processes based on the soliton approach. *Technol. Audit Prod. Reserves*, 2/5(22), 26–30.

Chabchoub, A., Hoffmann, N., Onorato, M., Akhmediev, N. (2012) Super rogue waves: Observation of a higher-order breather in water waves. *Phys. Rev. X*, 2, 011015.

Chesnokov, A.A. (2008). Simmetrii uravnenij teorii melkoj vody na vra- shchayushchejsya ploskosti. *SibZHIM.*, 11(3), 135–146 [in Russian].

Friedman, A.M. (2007). Prediction and discovery of new structures in spiral galaxies. *Adv. in Phys. Sci.*, 117(2), 121–148 [in Russian].

Klyushin, D.A., Lyashko, S.I., Lyashko, N.I., Bondar, O.S., Tymoshenko, A.A. (2020). Generalized optimization of processes of drug transport in tumors. *Cybern. Sys. Anal.*, 56(5), 758–765.

Lyashko, S.I., Samoilenko, V.H., Samoilenko, Yu.I., Gapyak, I.V., Orlova, M.S. (2021). Asymptotic stepwise solutions of the Korteweg–de Vries equation with a singular perturbation and their accuracy. *Math. Model. Comput.*, 8(3), 410–421.

Mielke, E.W. (2019). Soliton model of dark matter and natural inflation. *J. Phys. Conf. Ser.*, 1208, 012012.

Morozov, A.G. and Khoperskov, A.V. (2005). *Fizika diskov*. Volgograd University Press [in Russian].

Petviashvili, V.I. and Pokhotelov, O.A. (2016). *Solitary Waves in Plasmas and in the Atmosphere*. Routledge, London.

Turbal, Y., Turbal, M., Bomba, A., Radoveniuk, O. (2014). Method of earthquake prediction based on the soliton mechanisms of some shocks. *J. Environ. Sci. Eng. B.*, 3, 151–155.

Turbal, Y., Turbal, M., Bomba, A., Sokh, A. (2015). T-transformation method for studing the multi-solitone solutions of the Korteveg–de Vries type equations. *J. Math. Sys. Sci.*, 4, 164–169.

Ureña-López, L.A. (2019). Brief review on scalar field dark matter models. *Front. Astron. Space Sci.*, 6, 47.

13

Instrument Element and Grid Middleware in Metrology Problems

Pavlo NEYEZHMAKOV[1], Stanislav ZUB[1], Sergiy LYASHKO[2], Irina YALOVEGA[3] and Nataliia LYASHKO[4]

[1] *Ukrainian National Scientific Centre, Institute of Metrology, Kharkiv, Ukraine*

[2] *Taras Shevchenko National University of Kyiv, Ukraine*

[3] *Simon Kuznets Kharkiv National University of Economics, Ukraine*

[4] *V.M. Glushkov Institute of Cybernetics, National Academy of Sciences of Ukraine, Kyiv, Ukraine*

13.1. Introduction

Let us consider the new opportunities provided by cybernetics and its new grid tool for solving the problems in metrology. In a globalized world, metrological activities are designed to ensure the traceability of measurements in geographically dispersed regions, as well as a high level of confidence in these data in global networks (Neyezhmakov 2011; Agarwal and Srivastava 2017).

Coordination and management of measurement processes, often separated by long distances, require support from modern information technology. It is believed that such support can be provided through Internet technology. However, this system does not provide the possibility to control measurement and computational

Computational Methods and Mathematical Modeling in Cyberphysics and Engineering Applications 1, coordinated by Dmitri KOROLIOUK, Sergiy LYASHKO and Nikolaos LIMNIOS.

processes, and the transmitted data are constantly at risk of falsification and even irreversible loss.

Problems of remote interaction with instruments, as well as protection of measurement data in global networks, are being solved (Matsumoto and Yoshida 2007; Sand et al. 2007; Neyezhmakov et al. 2009; Fotohi and Effatparvar 2013; Manoj et al. 2015); however, these solutions are always partial, they are adapted to solve a particular problem and require significant special efforts. Is there a way out? The answer is yes. One of the newest tools, which embodies the latest advances in cybernetics, is the grid.

It provides a spatially distributed operating environment with a flexible, secure and coordinated allocation of resources to perform tasks in dynamically forming virtual organizations (VOs) (Simon et al. 2001; Foster et al. 2002). It was created to manage and exchange scientific data, as well as to ensure their collective use (Glushkov 1990; Zub et al. 2006).

Physically, the same hardware (i.e. the same telecommunications elements used by the Internet), a qualitatively new system of globally distributed computing, can function, providing the highest possible level of security to date, which even eliminates painful phenomena such as viruses and hacker attacks.

There are many implementations of this concept in the United States and the European Union. However, in recent years, the leading place among the implementations of the grid concept is occupied by gLite middleware.

The fact is that gLite middleware is used to service the most complex physical (while measuring) installation created by mankind – the Large Hadron Collider (LHC) at CERN (European Council for Nuclear Research) (Bos et al. 2005; Zub et al. 2006; Zub 2014a, 2014b). There, for the first time, they faced the problem of limited computing resources to process the vast amount of information generated in experiments with the LHC.

At the press conference at CERN announcing the discovery of the Higgs boson, CERN General Director Rolf-Dieter Heuer said: "Without the worldwide distributed computing network we would not have been able to obtain the results we need. The computing power and networking involved by CERN have become a very important part of the research."

The structure of the classical grid based on gLite middleware is quite well known (Zub et al. 2006; Demichev et al. 2007). Now, gLite middleware is becoming widespread not only for high-energy physics, but also for complex problems in rigid body mechanics (Hryhorieva et al. 2007; Lyashko et al. 2022), biology (Gerasin et al. 2012), medicine and other fields of science and technology.

13.2. Security in the grid

Of particular importance for metrology is the exclusion of unintentional or intentional distortions of data, measurement results, transmitted over networks and stored on remote devices.

It should be emphasized that safety is a component of the very foundation of the grid. Its components include special secure data transfer protocols such as GSIFTP, Secure RFIO, Gsidcap and the authorization system, which in turn is multi-level and provides a deep differentiation of access rights to grid resources.

Thus, the grid security system has all the currently known means of cryptography and is provided with a coherent system of authorization and allocation of rights, based on electronic certificates.

13.3. Grid element for measuring instruments

As already noted, the LHC is not only the most complex physical, but also the most complex measurement setup created by mankind. It would seem that this example has solved all the problems of linking measuring instruments (MIs) and the grid. However, this is not entirely so. The LHC Computing Grid Project (LCG) and its successor gLite middleware were created in parallel with the configuration of LHC detectors. Therefore, the processing of detector data is not a software solution, but rather a hardware–software solution, which was possible only for LCG/gLite developers. But today, “ordinary” grid users need to integrate MIs with the grid more and more often in order to build information and measurement systems with increased protection of data transmitted over long distances, as well as for remote access to MIs, in principle, from any site in the grid.

To solve these problems, the Grid-Enabled Remote Instrumentation with Distributed Control and Computation (GRIDCC) project was initiated. The aim of the project was to use the capabilities of the grid for secure remote collaboration of a team of researchers to ensure the monitoring and management of MI data generated and stored on distributed scientific equipment using traditional grid resources.

Within this project, a special tool has been developed, the so-called IE (or instrument element), which is successfully used by various scientific collaborations for remote interaction between MIs and the grid environment.

As can be seen from Figure 13.1 (Lelli et al. 2007), GridCC significantly expands the capabilities of a traditional grid, the components of which are located in the lower right corner of the figure.

A relatively new element is a virtual control room (VCR) that provides a multiuser interface, designed for joint work of a group of researchers with a complex information-measuring system with the distribution of functions.

The key element of GRIDCC is, of course, the IE itself. The storage element (SE) icon inside the IE shows that the IE was designed on the basis of the SE and inherited its properties. Thus, the IE, as a software module, is a server, and its interaction with the SE may be traditional for a classic grid.

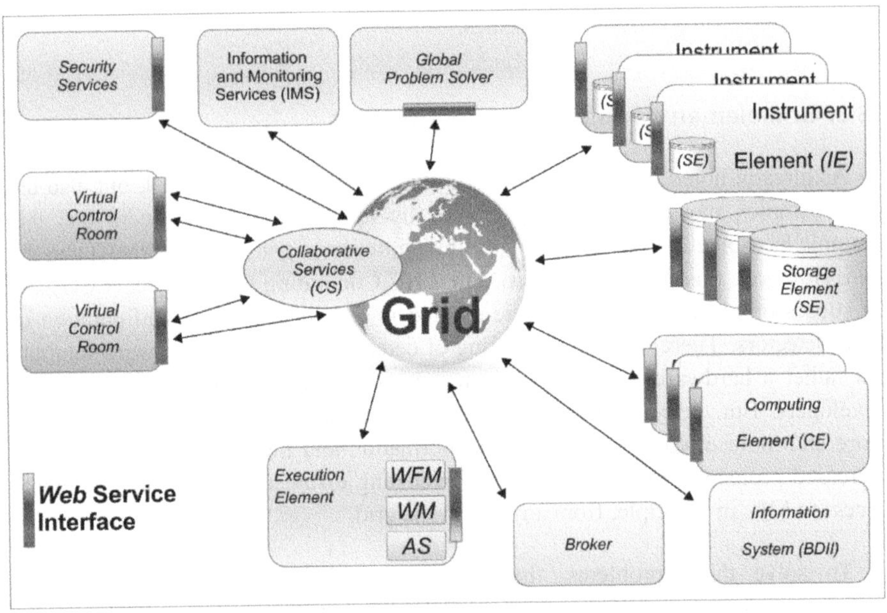

Figure 13.1. *GRIDCC project architecture. For a color version of this figure, see www.iste.co.uk/koroliouk/computational.zip*

The IE consists of a linked set of services, which provide all functionality for configuration, management and monitoring of measuring instruments located behind the IE interface, which provides their interaction with the grid.

The IE provides the following functional requirements: a unified model of a measurement instrument, standard grid access to the instruments and the possibility of interaction between different instruments belonging to different institutes in VOs.

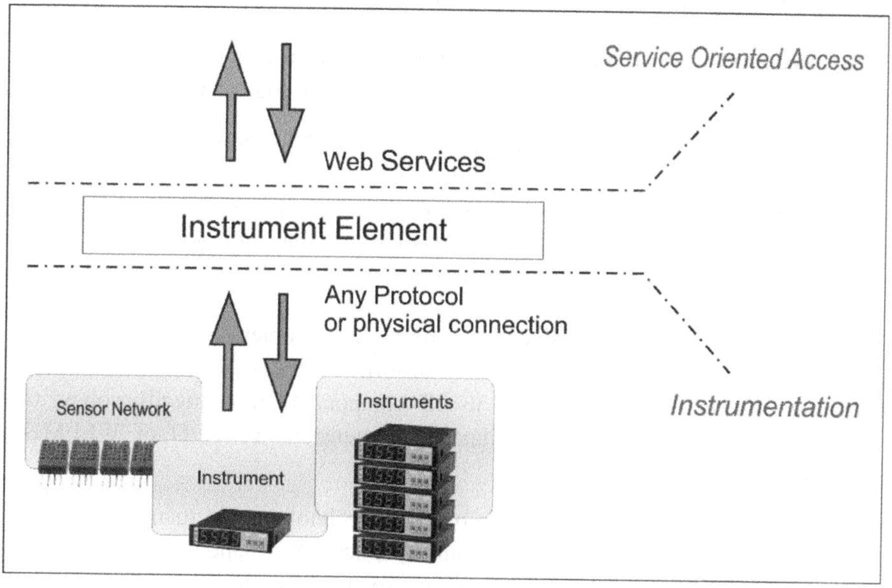

Figure 13.2. *Scheme of integration of MIs in grid. For a color version of this figure, see www.iste.co.uk/korolliouk/computational.zip*

An IE user can have one of the following roles: observer, operator or administrator.

When developing an abstract MI model, the following classification of MIs was introduced: dummy instrument, smart instrument and smart instrument in an ad hoc network.

The unified MI model should be set by describing the instructions based on XML (Lelli et al. 2007). Thus, the MI in the grid is represented by its model, and is a grid service that "provides services" on the measurement of the corresponding quantities.

It can be seen from Figure 13.2 that it is the IE between grid and the MI. On the one hand, we see sets of sensors, an MI or a group of MIs that use the control tools they contain and are physically connected to a computer with the installed IE. On the other hand, the user via grid has access to the IE through a secured web interface.

13.4. Grid and some problems of metrology

Let us consider the problematic issues of metrology, in which the use of the grid can lead to a significant increase in the efficiency of metrological activity:

1) Long-term storage and transfer of measurement data, as well as downloading new versions of software for MIs are the subject of attention of legal metrology, which establishes certain requirements for these processes. In the WELMEC Software Guide 7.2 of the European Cooperation in Legal Metrology (WELMEC), these requirements are formulated in Extensions L, T, D.

The grid was specially designed to solve such problems, as the aim when developing grid tools was to prevent any distortion or loss of experimental data. Therefore, the use of the grid and its IE component automatically solves the problems of legal metrology, formulated in Extensions L, T, D of WELMEC Software Guide 7.2.

2) Remote comparisons. Modern information technologies in many cases allow remote comparisons of MIs (Matsumoto and Yoshida 2007; Sand et al. 2007; Manoj et al. 2015). There are international programs for time synchronization, navigation and in other areas that use remote comparisons. The grid can take these approaches to a whole new level, both qualitatively and quantitatively, providing greater processing performance while protecting measurement data.

3) Key comparisons of measurement standards. Such comparisons are often time-consuming, labor-intensive and costly procedures. The grid can provide more efficient tools than the conventional Internet for performing remote comparisons of national measurement standards, unified computational procedures and secure exchange of information.

Much of the work, and in some cases all of it, can be done remotely. These opportunities are especially promising in light of the inevitable transition to the measurement standards based on quantum effects and in connection with the redefinition of the base units of the International System of Units (SI).

4) Instruments with built-in OS. The complexity of MI software is constantly growing. MIs with the built-in Linux OS find more and more use. Some of them fall

under the requirements of legal metrology. The complexity of testing such systems and the problems of protection of measurement data is an urgent task. The above mentioned requirements L, T, D of WELMEC Software Guide 7.2 should be met, which is a new and difficult task for such systems (Thiel et al. 2011).

The available grid tools in combination with the instrument element IE provide all the possibilities for the efficient and safe use of MIs with the built-in Linux OS.

As another approach to solving measurement problems by applying complex software, we propose to use the "thin client" concept. In other words, the instruments should have minimum software that ensures the integration of the instrument into the grid system, and complex processing of "raw" data should be performed in the grid environment. The choice of approach is left to the developers of the measuring instrument.

5) Reference software. One of the main methods of researching MI software is its comparison with the "reference" software. At the same time, the concept of "reference" software is not sufficiently defined. In addition, the comparison can occur both by the results of code execution and by a literal match of the code.

In the long run, the grid can give the concept of "reference" software a transparent and more acceptable meaning and perform comparisons at a new higher level. In our opinion, the leading national metrology institutes can set and fulfill the task of creating a repository of reference metrological software as one of the resources of the grid by analogy with gLite middleware repositories (Zub et al. 2006). This is the software that should be used in key comparisons of measurement standards.

6) Reference test data. Together with the previous task, the current problem is the creation of a grid resource of "reference" test data or, more broadly, test tasks (Cook et al. 1999). Similar tasks have been used for several years to test clusters that are part of the computer support system for LHC experiments based on the grid.

7) Smart electrical grid. Recently, the so-called "smart meters" have attracted great attention, and allow us to save energy resources. As it emphasizes in Hossain (2010), one of the most important goals of the European Metrology Research Programme (EMRP) "is to improve metrological infrastructure in Europe in order to ensure the successful implementation of the project of intelligent power electric networks". The term "grid" in the expression "smart electrical grid" means a power electric network that, in our opinion, can also be managed and served through the grid.

It should be noted that the existing implementation of the concept of the grid – gLite middleware – along with the IE provides a complete set of tools to create programs that solve the listed tasks in this area (Lelli et al. 2007).

Let us suggest that a magnetic dipole is a small rigid body with axial symmetry, and its magnetic moment is also directed along the axis of symmetry, i.e. symmetries of the magnetic and mass distributions coincide.

13.5. Discussion and conclusion

A number of up-to-date tasks of metrology require adequate support based on the most modern information technologies. Such a technology that meets the global challenges of the time and represents a new stage in the development of cybernetic systems, according to our belief, is the grid. In metrology, solving the problem of integrating measurement standards, individual measurement instruments, sensors (Neyezhmakov et al. 2018) and sensor networks into a global network is an extremely important task. The issues of reliability of interaction with remote instruments, as well as safety and confidentiality of data exchange, are one of the stated problems of the metrological digital infrastructure. Similar problems were formulated for the IE developers. Based on the properties of the IE, which has very high reliability and all the necessary functionality for solving metrology problems related to the organization of secure remote joint work of metrological digital infrastructure propose to use it as a main element of National Metrology Institutes information architecture.

The creation of a grid IE component makes this system an optimal environment for effective monitoring, management and service of measuring resources. Such a system provides the highest level of security for the transmission, storage and processing of measurement data, and a high level of reliability of measurement results, and also opens up new possibilities for metrological traceability. Moreover, for digital transformation in metrology, integrating the digital infrastructure of NMIs and Metrology Cloud with EGI and EOSC can be a good solution. However, in this case, it is necessary to eliminate the contradiction between the "openness" and "confidentiality" of measurement data.

13.6. References

Agarwal, M. and Srivastava, G.M.S. (2017). Cloud computing: A paradigm shift in the way of computing. *Inter. J. Modern Educ. Comput. Sci.*, 9(12), 38–48.

Bos, K., Brook, N., Duellmann, D., Eck, C., Fisk, I., Foster, D., Gibbard, B., Grandi, C. Grey, F., Harvey, J. et al. (2005). LHC computing grid – Technical design report. CERN-LHCC-2005-024; LCG-TDR-001. CERN, Geneva [Online]. Available at: https://ncsli.org/store/viewproduct.aspx?id =16817607.

Cook, H.R., Cox, M.G., Dainton, M.P., Harris, P.H. (1999). Methodology for testing spreadsheets and other packages used in metrology. Report to the National Measurement System Policy Unit, Department of Trade & Industry. NPL Report CMSC 07/00.

Demichev, A.P., Ilin, V.A., Kryukov, A.P. (2007). *Introduction to Grid Technologies*. SINP MSU, Moscow.

Foster, I., Kesselman, C., Nick, J., Tuecke, S. (2002). The physiology of the grid: An open grid services architecture for distributed systems integration. *Globus* [Online]. Available at: http://www.globus.org/research/papers/ogsa.pdf.

Fotohi, R. and Effatparvar, M. (2013). A cluster based job scheduling algorithm for grid computing. *IJITCS*, 5(12), 70–77.

Gerasin, S.N., Yakovlev, S.V., Yalovega, I.G. (2012). Qualitative analysis of one class of population models with commensal interaction. *Cybern. Syst. Anal.*, 48, 192–199.

Glushkov, V.M. (1990). *Cybernetics, Computer Technology, Informatics. The Electronic Computing Machine – The Technical Base of Cybernetics*. Naukova Dumka, Kyiv.

GRIDCC Project (2004). GRIDCC [Online]. Available at: http://www.gridcc.org.

Hossain, K. (2010). Global challenges for metrology. *NCSL International Workshop and Symposium* [Online]. Available at: https://ncsli.org/store/viewproduct.aspx?id =16817607.

Hryhorieva, L.V., Kozoriz, V.V., Liashko, S.I. (2007). Maple-modeling of the dynamics of a body with a fixed point in the field of magnetic and electric forces. *Re. Nat. Acad. Sci. Ukr.*, 8, 45–48.

Kozorez, V.V. (1981). *Dynamic Systems of Free Magnetically Interacting Bodies*. Naukova Dumka, Kyiv.

Lelli, F., Frizziero. E., Gulmini, M., Maron, G., Orlando, S., Petrucci, A., Squizzato, S. (2007). The many faces of the integration of instruments and the grid. *Int. J. Web Grid Services*, 3(3), 239–266.

Lyashko, S.I., Zub, S.S., Yalovega, I.G., Lyashko, V.S. (2022). Mathematical model of permanent magnets and superconducting coils. *Cybern. Syst. Anal.*, 58, 490.

Manoj, V., Thomas, A., Chandrasekaran, K. (2015). Single sign-on in cloud federation using CloudSim. *IJCNIS*, 7(6), 50–58.

Matsumoto, H. and Yoshida, H. (2007). Recent status of remote calibration (e-trace) system. *PTB-BIPM Workshop "Impact IT in Metrology"*, Berlin.

Neyezhmakov, P.I. (2011). 20 years of COOMET: We measure together for a better tomorrow. *OIML Bull.*, LII(3), 32–37.

Neyezhmakov, P.I., Zub, S.I., Zisky, N., Neumann, J. (2009). Application of technologies of safe data transfer in metrology. *Mathematics, Statistics and Computation to Support Measurement Quality International Seminar*, St. Petersburg.

Neyezhmakov, P.I., Prokopov, A., Skliarov, V. (2018). On the accounting for the influence of the Earth's atmosphere on the results of distance measurement realized by lasers. *J. Phys.: Conf. Ser. 1065 142008*.

Sand, A., Parkin, G., Stevens, M. (2007). A generic approach to distributed instrumentation. *PTB-BIPM Workshop "Impact IT in Metrology"*, Berlin.

Simon, M.D., Heflinger, L.O., Geim, A.K. (2001). Diamagnetically stabilized magnet levitation. *Am. J. Phys.*, 6(69), 702–713.

Thiel, F., Grottker, U., Richter, D. (2011). The challenge for legal metrology of operating systems embedded in measuring instruments. *OIML Bull.*, LII(1), 7–12.

Zub, S.S. (2014a). Magnetic levitation in Orbitron system. *Pro. Atom. Sci. Technol.*, 5(93), 31–34.

Zub, S.S. (2014b). Stable orbital motion of magnetic dipole in the field of permanent magnets. *Phys. D*, (275), 67–73.

Zub, S.S., Levchuk, L., Sorokin, P., Soroka, D. (2006). Grid middleware configuration at the KIPT CMS Linux cluster. *Nuclear Instr. Methods Phys. Res. A*, 559(1), 35–37.

14

Differential Evolution for Best Uniform Spline Approximation

Larysa VAKAL[1] and Evgen VAKAL[2]

[1] *V.M. Glushkov Institute of Cybernetics, National Academy of Sciences of Ukraine, Kyiv, Ukraine*

[2] *Taras Shevchenko National University of Kyiv, Ukraine*

Differential evolution is a population-based stochastic search technique. It has exhibited remarkable results on many optimization problems due to its simplicity and powerful search. Mutation, crossover and selection operators are used at each generation to move its population towards the global optimum. We propose the differential evolution algorithm for solving the problem of best uniform spline approximation in the fixed-knots case.

As the differential evolution algorithm performance crucially depends on choosing a trial vector generation strategy (i.e. mutation and crossover operators) and their associated control parameters, different strategies and combinations of control parameter values are compared. The comparison results are given. For the considered problem, the best trial vector generation strategies are determined.

The effectiveness of the proposed algorithm is demonstrated on numerical examples which involve approximating functions by splines of different degrees and defects.

Computational Methods and Mathematical Modeling in Cyberphysics and Engineering Applications 1,
coordinated by Dmitri KOROLIOUK, Sergiy LYASHKO and Nikolaos LIMNIOS.

14.1. Introduction

Evolutionary algorithms are powerful population-based stochastic search techniques. In these algorithms, each individual of a population represents a search point in the space of potential solutions to a given problem. The population is randomly initialized and evolves towards better and better regions of the search space by means of mutation, crossover and selection operators (Bäck and Schwefel 1993). The evolutionary algorithms have been successfully applied to solve numerous problems in diverse fields (e.g. Bhargava 2013; Vakal 2015; Vakal 2016a).

Differential evolution (DE), proposed by Storn and Price (1997), is one of the best floating-point encoded evolutionary algorithms for solving global optimization problems over continuous spaces. Due to its simplicity and powerful search, DE has exhibited remarkable performance in many problems (e.g. Price et al. 2005; Das and Suganthan 2011; Vakal 2016b; Vakal et al. 2017; Vakal and Vakal 2019). The success of DE implementation crucially depends on choosing the mutation and crossover operators and their associated control parameters.

In this study, the DE algorithm is developed to solve the problem of best uniform spline approximation with fixed knots. This problem can be considered as a problem of minimizing a function of many variables. We need to study different mutation and crossover operators, choose the best of them and fine-tune their associated control parameters.

14.2. Problem statement

Denote by $\widetilde{\Delta}_r$ a set of knots on an interval $[\alpha, \beta]$:

$$\widetilde{\Delta}_r = \{\alpha = \widetilde{x}_0 < \widetilde{x}_1 < \ldots < \widetilde{x}_r < \widetilde{x}_{r+1} = \beta\}. \qquad [14.1]$$

DEFINITION 14.1 (Berdyshev and Petrak 1999).– *A function* $s(x)$ *is called a polynomial spline of degree n and defect k* ($0 \le k \le n$) *with knots* $\widetilde{x}_1, \ldots, \widetilde{x}_r$, *if and only if it possesses the following two properties*:

– $s(x)$ *is a polynomial of degree n* ($n \ge 0$) *on each of the intervals* $[\widetilde{x}_{i-1}, \widetilde{x}_i]$, $i = 1, 2, \ldots, r+1$;

– $s(x)$ *has continuous derivatives up to order* $n-k$ *on* $[\alpha, \beta]$.

Denote by $S_{n,k}(\widetilde{\Delta}_r)$ the set of polynomial splines of power n and defect k with fixed knots [14.1]. It is a linear space of dimension $D=n+rk+1$. Every polynomial spline $s(x)\in S_{n,k}(\widetilde{\Delta}_r)$ can be represented as the sum of a polynomial and a linear combination of elementary spline function $(x-\widetilde{x}_i)_+^j$ (Stechkin and Subbotin 1976; Berdyshev and Petrak 1999):

$$s(x)=\sum_{i=0}^{n}a_i(x-\widetilde{x}_0)^i+\sum_{i=1}^{r}\sum_{j=n-k+1}^{n}c_{ij}(x-\widetilde{x}_i)_+^j, \qquad [14.2]$$

where $a_0,a_1,\ldots,a_n,c_{1,n-k+1},\ldots,c_{rn}$ are real numbers and

$$(x-\widetilde{x}_i)_+^j=\begin{cases}(x-\widetilde{x}_i)^j, & x>\widetilde{x}_i\\ 0, & x\le\widetilde{x}_i\end{cases}.$$

Let $f(x)$ be a given real-valued function defined on a set of points Δ_m:

$$\Delta_m=\{\alpha=x_1<x_2<\ldots<x_{m-1}<x_m=\beta\},\ m>n+rk+1.$$

For $f(x)$, we are interested in computing best spline approximation $s^*(x)\in S_{n,k}(\widetilde{\Delta}_r)$ with respect to the uniform (Chebyshev) norm, i.e.

$$\max_{x_l\in\Delta_m}\left|f(x_l)-s^*(x_l)\right|=\min_{s\in S_{n,k}(\widetilde{\Delta}_r)}\max_{x_l\in\Delta_m}\left|f(x_l)-s(x_l)\right|. \qquad [14.3]$$

Denote by ρ the error of best uniform spline approximation

$$\rho=\max_{x_l\in\Delta_m}\left|f(x_l)-s^*(x_l)\right|.$$

There may be more than one best spline approximation s^* for f. A full characterization theory for best spline approximation is given in Rice (1967).

14.3. Review of methods for spline approximation

To compute best uniform spline approximations, some methods and algorithms were developed. In Schumaker (1968), a method for solving problem [14.3] based on an exchange process of the Remez variety (e.g. Remez 1969; Kalenchuk-Porkhanova 2009) was given. Such an approach, which assumes a "normal" error curve with $D+1$ alternating extrema, suffers a disadvantage due to the non-Chebyshev nature of the approximating function. For some $f(x)$, an error curve has fewer than $D+1$ alternating extrema (Esch and Eastman 1969).

Some researchers (Barrodale and Young 1966; Esch and Eastman 1969; Vakal 2017) have used a linear programming approach for computing s^*. This approach is not based explicitly on any characterization process, and thus avoids the above difficulties. As a rule, a dual maximum problem of linear programming is chosen as a main problem because it has significantly fewer constraints than a primal problem.

For $f(x) \in C[\alpha, \beta]$, the existence and uniqueness of best spline approximation $s^*(x) \in S_{n,1}(\widetilde{\Delta}_r)$, for which the error $f - s^*$ has certain alternation properties, are proven if and only if $r \leq n+1$ (Nürnberger and Sommer 1983b). Such best approximations play an important role in developing a Remez-type algorithms for computing best spline approximations. In Nürnberger and Sommer (1983a), such an algorithm is proposed. The sequence constructed in it converges to a best approximation, if $r \leq n+1$, and at least to a nearly best approximation, if $r > n+1$.

The spline s^* can also be obtained as a Chebyshev point for an incompatible inconsistent system of linear equations using special methods for solving such systems (Stechkin and Subbotin 1976).

As a rule, the above algorithms find splines of defect 1, although splines with a larger defect are sometimes better (Stechkin and Subbotin 1976).

Problem [14.3] can also be considered as a problem of minimizing a multivariable function λ:

$$\lambda(a_0, a_1, \ldots, a_n, c_{1,n-k+1}, \ldots, c_{rn}) \to \min, \qquad [14.4]$$

where

$$\lambda = \max_{x_l \in \Delta_m} \left| f(x_l) - s(x_l) \right|.$$

To solve problem [14.4], the evolutionary algorithms dealing with a real-parameter optimization problem can be used, for example DE. In the next section, we describe the DE algorithm to solve problem [14.4] for splines of an arbitrary defect $k \leq n$.

14.4. Algorithm

DE is a novel efficient method developed to solve an optimization problem in a continuous space. It uses a population of D-dimensional vectors which encode the candidate solutions. In the First International IEEE Competition on Evolutionary Optimization, it was shown that DE is one of the fastest evolutionary algorithms (Storn and Price 1997).

DE starts with the randomly generated vector population. Then, DE iteratively uses the trial vector generation strategy (i.e. mutation and crossover operators) and the selection operator to evolve the population until a stopping criterion is met (Lu et al. 2014). For each vector, so-called target vectors, the mutation operator produces a mutant vector using some other vectors from the current population generation. The crossover operator mixes the mutant vector coordinates with the target vector coordinates to generate a so-called trial vector. The mutation and crossover operators aim to diversify the search, providing a wider overview of the search space and a higher probability of localizing the global extremum of the objective function, which is an optimization criteria. The selection operator compares the objective function values of the trial vector and the target vector to decide which of these vectors should become a member of the next generation.

To solve optimization problem [14.4], we can describe the DE algorithm scheme as follows.

1) Initialization. A population of vectors $V_i = (v_{i1}, \ldots, v_{iD})$, $i = 1, \ldots, NP$, is created, where NP is the population size, $D = n + rk + 1$ is the number of unknown spline coefficients and $v_{i1}, \ldots, v_{iD}$ are random real numbers from a user-specified interval. By default, the interval $[-1, 1]$ is considered.

2) Estimation. For vectors V_i, $i = 1, \ldots, NP$, the values of the objective function F are calculated by the following formula:

$$F(V_i) = \max_{1 \leq l \leq m} \left| f(x_l) - \sum_{p=1}^{n+1} v_{ip} (x_l - \tilde{x}_0)^{p-1} - \sum_{p=1}^{r} \sum_{j=1}^{k} v_{i,n+1+k(p-1)+j} (x_l - \tilde{x}_p)_+^{n-k+j} \right|.$$

3) Mutation. For each target vector V_i, a mutant vector $\widetilde{V}_i$ is generated. There are several schemes for creating $\widetilde{V}_i$, $i=\overline{1, NP}$ (e.g. Price et al. 2005; Wang et al. 2011; Lu et al. 2014). The five widely used mutant schemes are

– "rand/1": $\widetilde{V}_i = V_{r1} + FM \cdot (V_{r2} - V_{r3})$;

– "best/1": $\widetilde{V}_i = V_{best} + FM \cdot (V_{r1} - V_{r2})$;

– "target-to-best/1: $\widetilde{V}_i = V_i + FM \cdot (V_{best} - V_i) + FM \cdot (V_{r1} - V_{r2})$;

– "rand/2": $\widetilde{V}_i = V_{r1} + FM \cdot (V_{r2} - V_{r3}) + FM \cdot (V_{r4} - V_{r5})$;

– "best/2": $\widetilde{V}_i = V_{best} + FM \cdot (V_{r1} - V_{r2}) + FM \cdot (V_{r3} - V_{r4})$.

In the above formulas, $V_{r1}, V_{r2}, V_{r3}, V_{r4}, V_{r5}$ $(r1 \neq r2 \neq r3 \neq r4 \neq r5 \neq i)$ are randomly selected vectors from the current generation, V_{best} is the best vector at the current generation and $FM \in (0, 2]$ is a real constant called the scaling factor or the mutation force.

4) Crossover. There exist two schemes for producing U_i, exponential and binomial, and the latter one is more frequently used. According to the binomial scheme (Storn and Price 1997), the coordinates u_{ij} of the trial vector are computed by the formula

$$u_{ij} = \begin{cases} \widetilde{v}_{ij}, & \text{if } randb(j) \leq CR \text{ or } j = j_{rand} \\ v_{ij}, & \text{if } randb(j) > CR \text{ and } j \neq j_{rand} \end{cases},$$

where $i=\overline{1, NP}$, $j=\overline{1, D}$, $randb(j)$ is a random number drawn uniformly between 0 and 1, $j_{rand} \in [1, D]$ is a randomly selected integer, and $CR \in [0, 1]$ is a user-specified constant called the crossover constant or the crossover rate.

In conjunction with the aforementioned mutation schemes, the binomial crossover forms five trial vector strategies which are denoted "rand/1/bin", "best/1/bin", "target-to-best/1/bin", "rand/2/bin" and "best/2/bin".

5) Selection. If $F(U_i)<F(V_i)$, the trial vector U_i replaces the target vector V_i in the following generation.

6) Stopping criterion. The DE algorithm stops if one of the following conditions is satisfied:

– the maximum number of generations G is exceeded,

– the value $F(V_{best})$ is less than a given constant ε,

– the evolutionary process stagnates, i.e.

$$\max_{1\le i\le NP} F(V_i) - \min_{1\le i\le NP} F(V_i) < \delta \cdot \min_{1\le i\le NP} F(V_i),$$

where G, ε and δ are user-specified constants. If none of the above conditions are fulfilled, the passage to step 3 is carried out.

DE performance mainly depends on appropriately choosing the trial vector generation strategies and control parameters, i.e. the population size *NP*, the mutation force *FM* and the crossover rate *CR*. The parameter *NP* does not need to be fine-tuned. It is suggested that a reasonable value for *NP* should be between *5D* and *10D*. The parameter *CR* controls which and how many components are mutated in each vector of the current generation. Large *CR* speeds up convergence (Storn and Price 1997). For a given problem, good values of *CR* generally fall into a small range, with which the DE algorithm can perform consistently well. The mutation force *FM* is closely related to the convergence speed. *FM* values typically range from 0.4 to 1 (Das and Suganthan 2011). If $FM > 1$, the probability of escaping from a local optimum increases.

The best values for the control parameters *FM* and *CR* can be different for different optimization problems. Therefore, to successfully solve a given optimization problem, it is necessary to perform a search for the most appropriate trial vector generation strategies and their associated parameter values. Inappropriate choice of the strategies and control parameter settings may lead to premature convergence or stagnation of the DE algorithm (e.g. Price et al. 2005).

We have carried out computational experiments to select the potentially best trial vector generation strategies and their parameter values of *FM* and *CR*. For the considered optimization problem [14.4], we have compared the five above-mentioned strategies with several combinations of *FM* and *CR* values. All variants of DE were run 20 times.

Table 14.1 shows the average number of objective function evaluations (*NOFE*) by applying 10 variants of DE (two best variants for each strategy) to approximation of test functions $f_1(x)=e^x$, $f_2(x)=(1+x)^{-1}$, and $f_3(x)=\sqrt{x}$ by splines with $D=5$ and $D=10$ coefficients. All 10 variants result in an objective function value $\varepsilon \le 1.01\rho$ with the number of population generations less than G (G = 350 if $D=5$ and G = 600 if $D=10$). The average numbers of objective function evaluations allow us to compare the convergence speed of the considered variants of the DE algorithm.

The results of the computational experiments show that the best trial generation strategies are "target-to-best/1/bin" with *FM*=0.7, *CR*=1, "best/1/bin" with *FM*=0.7, *CR*=1, and "best/2/bin" with *FM*=0.5, *CR*=1.

Trial vector generation strategy	D = 5		D = 10	
	Control parameters	***NOFE***	**Control parameters**	***NOFE***
"rand/1/bin"	*FM*=0.5, *CR*=1	6874	*FM*=0.5, *CR*=1	24441
"rand/1/bin"	*FM*=0.6, *CR*=1	9113	*FM*=0.1, *CR*=1	45045
"best/1/bin"	*FM*=0.7, *CR*=1	3962	*FM*=0.7, *CR*=1	17515
"best/1/bin"	*FM*=0.8, *CR*=1	6201	*FM*=0.8, *CR*=1	22253
"target-to-best/1/bin"	*FM*=0.7, *CR*=1	4286	*FM*=0.7, *CR*=1	11715
"target-to-best/1/bin"	*FM*=0.8, *CR*=1	6257	*FM*=0.8, *CR*=1	23330
"rand/2/bin"	*FM*=0.4, *CR*=1	8475	*FM*=0.4, *CR*=1	37753
"rand/2/bin"	*FM*=0.5, *CR*=1	9489	*FM*=0.3, *CR*=1	46425
"best/2/bin"	*FM*=0.5, *CR*=1	4280	*FM*=0.5, *CR*=0.1	11957
"best/2/bin"	*FM*=0.6, *CR*=1	6155	*FM*=0.5, *CR*=0.9	35450

Table 14.1. *The average number of objective function evaluations*

14.5. Experimental results and discussion

To check the effectiveness of the described DE algorithm, computational experiments were performed using MATLAB. The algorithm has been tested on a number of functions. Here, we give only two of these numerical examples. The first involves approximating $f_1(x)=e^x$ on the set $\Delta_{1001}=\{x_i=0.001(i-1), i=\overline{1,1001}\}$

by splines of degree $n=3$ and defects $k=1,2,3$. Table 14.2 shows approximation errors of best spline approximations on the set of equidistant knots $\widetilde{\Delta}_r \subset [0, 1]$.

As stated above, if $r \leq n+1$, there is the spline s^* of degree n and defect $k = 1$, which has the error curve of "normal" form with $D+1$ alternating extrema. In the case of the best uniform approximation of function $f_1(x) = e^x$ by the 3rd degree spline of defect $k=1$ with two knots $\widetilde{x}_1 = 1/3$ and $\widetilde{x}_2 = 2/3$, the alternating extrema must be seven. For the spline

$$s^*(x) = 1.0000265 + 0.9996198\,x + 0.4970944\,x^2 + 0.1936726\,x^3 + \\ + 0.0811476\,(x - 1/3)_+^3 + 0.1037494\,(x - 2/3)_+^3$$

computed by the DE algorithm, there are seven points, namely 0, 0.152, 0.331, 0.499, 0.664, 0.819, 0.972, in which the approximation errors equal $\rho = 2.651 \cdot 10^{-5}$ with an alternate change of sign. This spline has continuous derivatives up to order 2 on $[0,1]$. The spline of degree $n=3$ and defect $k = 2$

$$s^*(x) = 0.9999821 + 1.0012368\,x + 0.4867336\,x^2 + 0.2110104\,x^3 + \\ + 0.0678139\,(x-1/3)_+^2 - 0.0097317\,(x-1/3)_+^3 + 0.0939254\,(x-2/3)_+^2 + \\ + 0.0004760\,(x-2/3)_+^3$$

approximates $f_1(x) = e^x$ with error $\rho = 1.790 \cdot 10^{-5}$, but it has continuous derivatives up to order 1 on $[0,1]$. The spline of defect $k=3$

$$s^*(x) = 0.9999942 + 1.0004925\,x + 0.4931657\,x^2 + 0.1976169\,x^3 + \\ + 0.0010432\,(x-1/3)_+ - 0.0017526\,(x-1/3)_+^2 + 0.0767980\,(x-1/3)_+^3 + \\ + 0.0016047\,(x-2/3)_+ - 0.0024760\,(x-2/3)_+^2 + 0.1099434\,(x-2/3)_+^3$$

has a smaller approximation error of $9.227 \cdot 10^{-6}$ than the above splines of defect $k=1$ and $k = 2$, but its derivatives may be discontinuous.

The second example involves approximating $f_2(x) = (1+x)^{-1}$ on $\Delta_{1001} \subset [0, 1]$ by splines of degree $n=3,4$ and defect $k=1$ with fixed equidistant knots [14.1]. Table 14.3 shows approximation errors which are computed using a stochastic DE algorithm and two deterministic algorithms, namely a linear programming algorithm (LP algorithm), described in Vakal (2017), and a Remez-type algorithm (RM algorithm), proposed in Nürnberger and Sommer (1983a). A comparison of the data given in Table 14.3 shows that the approximation errors obtained by the simple DE algorithm are the same as errors calculated by more sophisticated deterministic algorithms.

r	$k=1$	$k=2$	$k=3$
1	$8.225\cdot10^{-5}$	$6.731\cdot10^{-5}$	$4.323\cdot10^{-5}$
2	$2.651\cdot10^{-5}$	$1.790\cdot10^{-5}$	$9.227\cdot10^{-6}$
3	$9.524\cdot10^{-6}$	$6.674\cdot10^{-6}$	$3.053\cdot10^{-6}$
4	$4.224\cdot10^{-6}$	$3.024\cdot10^{-6}$	$1.282\cdot10^{-6}$
5	$2.139\cdot10^{-6}$	$1.563\cdot10^{-6}$	$6.236\cdot10^{-7}$
6	$1.196\cdot10^{-6}$	$8.875\cdot10^{-7}$	$3.419\cdot10^{-7}$
7	$7.194\cdot10^{-7}$	$5.414\cdot10^{-7}$	$2.029\cdot10^{-7}$

Table 14.2. *Approximation errors of* $f_1(x)=e^x$ *by splines of degree 3 and defect k*

r	Spline of degree $n=3$			Spline of degree $n=4$		
	DE algorithm	LP algorithm	RT algorithm	DE algorithm	LP algorithm	RT algorithm
1	$3.25\cdot10^{-4}$	$3.25\cdot10^{-4}$	$3.25\cdot10^{-4}$	$5.1\cdot10^{-5}$	$5.1\cdot10^{-5}$	$5.1\cdot10^{-5}$
2	$8.9\cdot10^{-5}$	$8.6\cdot10^{-5}$	$8.9\cdot10^{-5}$	$1.0\cdot10^{-5}$	$9.9\cdot10^{-6}$	$1.0\cdot10^{-5}$
3	$3.3\cdot10^{-5}$	$3.3\cdot10^{-5}$	$3.3\cdot10^{-5}$	$3.6\cdot10^{-6}$	$3.6\cdot10^{-6}$	$3.6\cdot10^{-6}$
4	$1.5\cdot10^{-5}$	$1.5\cdot10^{-5}$	$1.5\cdot10^{-5}$	$1.7\cdot10^{-6}$	$1.7\cdot10^{-6}$	$1.7\cdot10^{-6}$

Table 14.3. *Spline approximation errors for* $f_2(x)=(1+x)^{-1}$

14.6. Conclusion

In this section, the best uniform approximation of functions by polynomial splines of degree n and defect $k \le n$ with r fixed knots is studied. The approximation problem is considered as a problem of minimizing a function of $n+rk+1$ variables. To solve this minimization problem, the DE algorithm is proposed, which has shown remarkable results on many optimization problems due to its simplicity and power search.

As DE performance depends on appropriately choosing the trial vector generation strategies and their control parameters. Computational experiments were carried out to compare the five most popular strategies and values of the mutation force *FM* and the crossover rate *CR*. The experiments showed that for the considered problem the best strategies are "target-to-best/1/bin" (*FM*=0.7, *CR*=1), "best/1/bin" (*FM*=0.7, *CR*=1) and "best/2/bin" (*FM*=0.5, *CR*=1).

The effectiveness of the DE algorithm is demonstrated on numerical examples involving approximating functions used in the literature on approximation by splines of different degrees and defects.

14.7. References

Bäck, T. and Schwefel, H.-P. (1993). An overview of evolutionary algorithms for parameter optimization. *Evolutionary Computation*, 1(1), 1–23.

Barrodale, J. and Young, A. (1966). A note on numerical procedures for approximation by spline functions. *The Computer Journal*, 9(3), 318–320.

Berdyshev, V.I. and Petrak, L.V. (1999). *Functions Approximation, Compression of Numerical Information, Applications*. Uro RAN, Yekaterinburg [In Russian].

Berdyshev, V.I. and Subbotin, Yu.N. (1979). *Numerical Methods for Functions Approximation*. Sredne-Ural'skoye knizhnoye izdatel'stvo, Sverdlovsk [In Russian].

Bhargava, S. (2013). A note on evolutionary algorithms and its applications. *Adults Learning Mathematics: An International Journal*, 8(1), 31–45.

Das, S. and Suganthan, P. (2011). Differential evolution: A survey of the state-of-the-art. *IEEE Transactions on Evolutionary Computation*, 15(1), 4–31.

Esch, R.E. and Eastman, W.L. (1969). Computational methods for best spline function approximation. *Journal of Approximation Theory*, 2(1), 85–96.

Kalenchuk-Porkhanova, A.A. (2009). Best Chebyshev approximation of functions of one and many variables. *Cybernetics and Systems Analysis*, 45(6), 988–996.

Lu, X., Tang, K., Sendhoff, B., Yao, X. (2014). A new self-adaptation scheme for differential evolution. *Neurocomputing*, 146, 2–16.

Nürnberger, G. and Sommer, M. (1983a). A Remez type algorithm for spline functions. *Numerische Mathematik*, 41(1), 117–146.

Nürnberger, G. and Sommer, M. (1983b). Alternation for best spline approximation. *Numerische Mathematik*, 41(2), 207–221.

Price, K., Storn, R., Lampinen, J. (2005). *Differential Evolution: A Practical Approach to Global Optimization*. Springer-Verlag, Berlin, Heidelberg.

Qin, A.K., Huang, V.L., Suganthan, P.N. (2009). Differential evolution algorithm with strategy adaptation for global numerical optimization. *IEEE Transactions on Evolutionary Computation*, 13(2), 398–417.

Remez, Ye.Ya. (1969). *Fundamentals of Numerical Methods of Chebyshev Approximation.* Naukova Dumka, Kyiv [In Russian].

Rice, J. (1967). Characterization of Chebyshev approximation by splines. *SIAM Journal on Numerical Analysis*, 4(4), 557–567.

Schumaker, L.L. (1968). On computing best spline approximations. Mathematics Research Center Report No. 833, Madison, Wisconsin.

Stechkin, S.B. and Subbotin, Yu.N. (1976). *Splines in Numerical Mathematics.* Nauka, Moscow [In Russian].

Storn, R. and Price, K. (1997). Differential evolution – A simple and efficient heuristic for global optimization over continuous spaces. *Journal of Global Optimization*, 11, 341–359.

Vakal, L.P. (2015). Using genetic algorithm for solving boundary value problems. *Journal of Automation and Information Sciences*, 47(8), 52–62.

Vakal, L.P. (2016a). Solving uniform nonlinear approximation problem using continuous genetic algorithm. *Journal of Automation and Information Sciences*, 48(6), 49–59.

Vakal, L.P. (2016b). Seeking optimal knots for segment approximation. *Journal of Automation and Information Sciences*, 48(11), 68–75.

Vakal, L.P. (2017). Constructing best Chebyshev spline approximations. *Artificial Intelligence*, 2(76), 94–100 [In Ukrainian].

Vakal, L.P. and Vakal, E.S. (2019). Algorithm for the best uniform spline approximation with free knots. *Cybernetics and Systems Analysis*, 55(5), 449–455.

Vakal, L.P., Kalenchuk-Porkhanova, A.A., Vakal, E.S. (2017). Increasing the efficiency of Chebyshev segment fractional rational approximation. *Cybernetics and Systems Analysis*, 53(5), 759–765.

Wang, Y., Cai, Z., Zhang, Q. (2011). Differential evolution with composite trial vector generation strategies and control parameters. *IEEE Transactions on Evolutionary Computation*, 15(1), 55–66.

15

Finding a Nearest Pair of Points Between Two Smooth Curves in Euclidean Space

Vladimir V. SEMENOV[1], **Nataliia LYASHKO**[2], **Stanislav ZUB**[3] **and Yevhen HAVRYLKO**[2]

[1] *Taras Shevchenko National University of Kyiv, Ukraine*

[2] *V.M. Glushkov Institute of Cybernetics, National Academy of Sciences of Ukraine, Kyiv, Ukraine*

[3] *Ukrainian National Scientific Centre, Institute of Metrology, Kharkiv, Ukraine*

15.1. Introduction

Let A and B be subsets in Euclidean space R^n. One of the popular optimization problems is to find the closest pairs of elements of sets A and B (or calculate the distance between A and B):

$$\text{find } a_0 \in A, b_0 \in B : \|a_0 - b_0\| = \min_{a\in A, b\in B} \|a - b\|. \qquad [15.1]$$

An effective solution to the problem [15.1] is a key element in many applied issues (e.g. when creating route schedulers in robotics). Therefore, it is natural that many researchers paid attention to the problem [15.1] for different classes of sets (Gilbert et al. 1988; Gilbert and Foo 1990; Llanas et al. 2000; Lin and Han 2002;

Computational Methods and Mathematical Modeling in Cyberphysics and Engineering Applications 1, coordinated by Dmitri KOROLIOUK, Sergiy LYASHKO and Nikolaos LIMNIOS.

Bauschke et al. 2004; Uteshev and Yashina 2008; Malitsky 2014; Abbasov 2016, 2017a, 2017b).

In this chapter, we will consider the problem [15.1] for two smooth curves in three-dimensional Euclidean space and propose a method for solving it, which uses a model of physical interaction of material points with a given potential energy. A similar idea of using Coulomb's physical force and Newton's equation of mechanics with connections, but to find the closest pairs on two smooth curves in Euclidean space, is implemented in the method of charged balls (Abbasov 2016, 2017a, 2017b). Note that perhaps the most well-known example of using a physical analogy in optimization algorithmics is the well-known heavy ball method (Polyak 2014).

The use of Lagrange and Hamilton equations, which reached a high level of development in classical mechanics, and later became a generally recognized starting point for obtaining equations in modern physics, allows us to maximally generalize the methods of physical analogy to the optimization problems mentioned above. In particular, this made it possible for the third author of this work to realize the idea of using potentials of arbitrary type in an optimization problems set.

15.2. Define the problem and notations

In a three-dimensional Euclidean space $\mathbb{R}^3$, consider two smooth parametrized curves:

$$\vec{x}(\alpha)=\left(x_1(\alpha),x_2(\alpha),x_3(\alpha)\right),\quad \vec{y}(\beta)=\left(y_1(\beta),y_2(\beta),y_3(\beta)\right),$$

where $\vec{x},\vec{y}:\mathbb{R}\to\mathbb{R}^3$ are continuously differential vector-functions of scalar parameters α and β, accordingly. We will look for the nearest pairs of points lying on the curves. That is, consider the problem of minimization

$$\left\|\vec{x}(\alpha)-\vec{y}(\beta)\right\|\to\min_{\alpha,\beta}.$$

This chapter proposes a method for solving this problem, which uses a model of physical interaction of material points with the potential energy of a general form.

To describe the movement of a point along a curve in space, we must know the radius vector of a point on a curve $\vec{x}(t)$ and its velocity $\vec{v}(t)=\dot{\vec{x}}(t)$ at any given time $t>0$. And since the curves are parametrized, the movement along them can be described by the dependence of the parameters α and β on time t.

Let the movement occur under the action of the force of attraction between points that can only move along spatial curves set parametrically. We assume that the potential energy of a system of points depends only on the distance between them:

$$U = U\left(\left\|\vec{y}(\beta) - \vec{x}(\alpha)\right\|\right).$$

Potential energy can be an arbitrary function $r > 0$, but it makes sense to limit ourselves to such a segment of the Laurent series by r:

$$U(r) = U(\|\vec{y} - \vec{x}\|) = \frac{q^2}{r} + U_0 + g \cdot r + k_2 \cdot r^2 + k_3 \cdot r^3.$$

Next, we obtain the Lagrange and Hamilton equations of motion of this system of two points. The approximate integration of Cauchy problems for the derived differential equations gives algorithms for approximation of the nearest pair.

15.3. Lagrange function with energy dissipation

We introduce the following notation: $\mathcal{L}_0$ – Lagrange function without energy dissipation; $\mathcal{L}$ – Lagrange function with energy dissipation; $\vec{r}(\alpha,\beta) = \vec{y}(\beta) - \vec{x}(\alpha)$ – radius vector between two points on given curves. For the derivatives, we will use the notation

$$\vec{x}_\alpha' = \frac{d\vec{x}}{d\alpha}, \vec{y}_\beta' = \frac{d\vec{y}}{d\beta} \text{ and } \vec{x}_\alpha'' = \frac{d^2\vec{x}}{d\alpha^2}, \vec{y}_\beta'' = \frac{d^2\vec{y}}{d\beta^2}.$$

Then, the Lagrange function of the generalized coordinates α and β, and velocities $\dot{\alpha}$ and $\dot{\beta}$, will be as follows:

$$\mathcal{L}_0(\alpha,\dot{\alpha};\beta,\dot{\beta}) = \frac{m_\alpha}{2}\left\|\vec{x}_\alpha'\right\|^2 \dot{\alpha}^2 + \frac{m_\beta}{2}\left\|\vec{y}_\beta'\right\|^2 \dot{\beta}^2 - U(r), \qquad [15.2]$$

where $r = \|\vec{r}\|$, $m_\alpha > 0$, $m_\beta > 0$.

Energy dissipation is introduced through the exponential coefficient (Kazakov 2008)

$$\mathcal{L}(\alpha,\dot{\alpha};\beta,\dot{\beta};t) = e^{\lambda t}\mathcal{L}_0(\alpha,\dot{\alpha};\beta,\dot{\beta}), \qquad [15.3]$$

where λ is the positive constant.

NOTE.– The Lagrange function [15.3] describes the motion of interacting particles with masses m_α and m_β, which, in addition, are subject to friction forces proportional to the velocities of particles.

15.4. Lagrange equation

We obtain the equation of Lagrange motion (Kazakov 2008) for our system of two particles:

$$\frac{d}{dt}\left(\frac{\partial \mathcal{L}}{\partial \dot{q}^{\nu}}\right)-\frac{\partial \mathcal{L}}{\partial q^{\nu}}=0. \qquad [15.4]$$

Let us find derivatives $\frac{\partial \mathcal{L}}{\partial q^{\nu}}$, $\frac{\partial \mathcal{L}}{\partial \dot{q}^{\nu}}$ and $\frac{d}{dt}\left(\frac{\partial \mathcal{L}}{\partial \dot{q}^{\nu}}\right)$. We obtain

$$\begin{cases} \dfrac{\partial \mathcal{L}}{\partial \alpha}=e^{\lambda t}\dfrac{\partial \mathcal{L}_0}{\partial \alpha}, \dfrac{\partial \mathcal{L}}{\partial \beta}=e^{\lambda t}\dfrac{\partial \mathcal{L}_0}{\partial \beta}, \\ \dfrac{\partial \mathcal{L}}{\partial \dot{\alpha}}=e^{\lambda t}\dfrac{\partial \mathcal{L}_0}{\partial \dot{\alpha}}, \dfrac{\partial \mathcal{L}}{\partial \dot{\beta}}=e^{\lambda t}\dfrac{\partial \mathcal{L}_0}{\partial \dot{\beta}}, \\ \dfrac{d}{dt}\left(\dfrac{\partial \mathcal{L}}{\partial \dot{\alpha}}\right)=\dfrac{d}{dt}\left(e^{\lambda t}\dfrac{\partial \mathcal{L}_0}{\partial \dot{\alpha}}\right), \dfrac{d}{dt}\left(\dfrac{\partial \mathcal{L}}{\partial \dot{\beta}}\right)=\dfrac{d}{dt}\left(e^{\lambda t}\dfrac{\partial \mathcal{L}_0}{\partial \dot{\beta}}\right). \end{cases} \qquad [15.5]$$

In the Lagrange function without dissipation $\mathcal{L}_0$ [15.2], the first and third terms depend on the parameter α. Therefore, we have

$$\frac{\partial \mathcal{L}_0}{\partial \alpha}=\frac{\partial}{\partial \alpha}\left(\frac{m_\alpha}{2}\left\|\vec{x}_\alpha^{'}\right\|^2\dot{\alpha}^2-U(r)\right)=m_\alpha\langle\vec{x}_\alpha^{''},\vec{x}_\alpha^{'}\rangle\dot{\alpha}^2+\frac{1}{r}\frac{\partial U}{\partial r}\langle\vec{y}-\vec{x},\vec{x}_\alpha^{'}\rangle. \qquad [15.6]$$

Similarly

$$\frac{\partial \mathcal{L}_0}{\partial \beta}=m_\beta\langle\vec{y}_\beta^{''},\vec{y}_\beta^{'}\rangle\dot{\beta}^2-\frac{1}{r}\frac{\partial U}{\partial r}\langle\vec{y}-\vec{x},\vec{y}_\beta^{'}\rangle. \qquad [15.7]$$

In the Lagrange function $\mathcal{L}_0$, only the first term depends on $\dot{\alpha}$. We have

$$\frac{\partial \mathcal{L}_0}{\partial \dot{\alpha}}=m_\alpha\left\|\vec{x}_\alpha^{'}\right\|^2\dot{\alpha}. \qquad [15.8]$$

Similarly

$$\frac{\partial \mathcal{L}_0}{\partial \dot{\beta}} = m_\beta \left\| \vec{y}_\beta' \right\|^2 \dot{\beta}. \tag{15.9}$$

Calculate the time derivative of $\frac{\partial \mathcal{L}_0}{\partial \dot{\alpha}}$:

$$\begin{aligned} \frac{d}{dt}\left(\frac{\partial \mathcal{L}_0}{\partial \dot{\alpha}}\right) &= m_\alpha \frac{d}{dt}\left(\left\| \vec{x}_\alpha' \right\|^2 \dot{\alpha}\right) = \\ &= m_\alpha \left(\frac{d}{dt}\left(\left\| \vec{x}_\alpha' \right\|^2\right)\dot{\alpha} + \left\| \vec{x}_\alpha' \right\|^2 \ddot{\alpha}\right) = \\ &= m_\alpha \left(2\langle \vec{x}_\alpha'', \vec{x}_\alpha' \rangle \dot{\alpha}^2 + \left\| \vec{x}_\alpha' \right\|^2 \ddot{\alpha}\right). \end{aligned} \tag{15.10}$$

Similarly, let us calculate the time derivative of $\frac{\partial \mathcal{L}_0}{\partial \dot{\beta}}$:

$$\frac{d}{dt}\left(\frac{\partial \mathcal{L}_0}{\partial \dot{\beta}}\right) = m_\beta \left(2\langle \vec{y}_\beta'', \vec{y}_\beta' \rangle \dot{\beta}^2 + \left\| \vec{y}_\beta' \right\|^2 \ddot{\beta}\right). \tag{15.11}$$

Substituting [15.6]–[15.11] in [15.5], we obtain

$$\begin{cases} \frac{\partial \mathcal{L}}{\partial \alpha} = e^{\lambda t}\left(\frac{\partial \mathcal{L}_0}{\partial \alpha}\right) = e^{\lambda t}\left(m_\alpha \langle \vec{x}_\alpha'', \vec{x}_\alpha' \rangle \dot{\alpha}^2 + \frac{1}{r}\frac{\partial U}{\partial r}\langle \vec{y} - \vec{x}, \vec{x}_\alpha' \rangle\right), \\ \frac{\partial \mathcal{L}}{\partial \beta} = e^{\lambda t}\left(\frac{\partial \mathcal{L}_0}{\partial \beta}\right) = e^{\lambda t}\left(m_\beta \langle \vec{y}_\beta'', \vec{y}_\beta' \rangle \dot{\beta}^2 - \frac{1}{r}\frac{\partial U}{\partial r}\langle \vec{y} - \vec{x}, \vec{y}_\beta' \rangle\right), \\ \frac{\partial \mathcal{L}}{\partial \dot{\alpha}} = e^{\lambda t}\frac{\partial \mathcal{L}_0}{\partial \dot{\alpha}} = e^{\lambda t} m_\alpha \left\| \vec{x}_\alpha' \right\|^2 \dot{\alpha}, \\ \frac{\partial \mathcal{L}}{\partial \dot{\beta}} = e^{\lambda t}\frac{\partial \mathcal{L}_0}{\partial \dot{\beta}} = e^{\lambda t} m_\beta \left\| \vec{y}_\beta' \right\|^2 \dot{\beta}, \\ \frac{d}{dt}\left(e^{\lambda t}\frac{\partial \mathcal{L}_0}{\partial \dot{\alpha}}\right) = e^{\lambda t}\left(\lambda m_\alpha \left\| \vec{x}_\alpha' \right\|^2 \dot{\alpha} + m_\alpha\left(2\langle \vec{x}_\alpha'', \vec{x}_\alpha' \rangle \dot{\alpha}^2 + \left\| \vec{x}_\alpha' \right\|^2 \ddot{\alpha}\right)\right), \\ \frac{d}{dt}\left(e^{\lambda t}\frac{\partial \mathcal{L}_0}{\partial \dot{\beta}}\right) = e^{\lambda t}\left(\lambda m_\beta \left\| \vec{y}_\beta' \right\|^2 \dot{\beta} + m_\beta\left(2\langle \vec{y}_\beta'', \vec{y}_\beta' \rangle \dot{\beta}^2 + \left\| \vec{y}_\beta' \right\|^2 \ddot{\beta}\right)\right). \end{cases} \tag{15.12}$$

So, by substituting the first, third and fifth equations with [15.12] in [15.4], we obtain

$$e^{\lambda t}\left(\lambda m_\alpha \left\|\vec{x}_\alpha^{'}\right\|^2 \dot{\alpha} + m_\alpha \left(2\langle \vec{x}_\alpha^{''}, \vec{x}_\alpha^{'}\rangle \dot{\alpha}^2 + \left\|\vec{x}_\alpha^{'}\right\|^2 \ddot{\alpha}\right)\right) -$$

$$-e^{\lambda t}\left(m_\alpha \langle \vec{x}_\alpha^{''}, \vec{x}_\alpha^{'}\rangle \dot{\alpha}^2 + \frac{1}{r}\frac{\partial U}{\partial r}\langle \vec{y} - \vec{x}, \vec{x}_\alpha^{'}\rangle\right) = 0.$$

By shorting on $e^{\lambda t}$, we come to the equation

$$\lambda m_\alpha \left\|\vec{x}_\alpha^{'}\right\|^2 \dot{\alpha} + m_\alpha \langle \vec{x}_\alpha^{''}, \vec{x}_\alpha^{'}\rangle \dot{\alpha}^2 + m_\alpha \left\|\vec{x}_\alpha^{'}\right\|^2 \ddot{\alpha} - \frac{1}{r}\frac{\partial U}{\partial r}\langle \vec{y} - \vec{x}, \vec{x}_\alpha^{'}\rangle = 0.$$

Similarly, by substituting the second, fourth, sixth equations with [15.12] in [15.4] and shortening on $e^{\lambda t}$, we have

$$\lambda m_\beta \left\|\vec{y}_\beta^{'}\right\|^2 \dot{\beta} + m_\beta \langle \vec{y}_\beta^{''}, \vec{y}_\beta^{'}\rangle \dot{\beta}^2 + m_\beta \left\|\vec{y}_\beta^{'}\right\|^2 \ddot{\beta} + \frac{1}{r}\frac{\partial U}{\partial r}\langle \vec{y} - \vec{x}, \vec{y}_\beta^{'}\rangle = 0.$$

We obtain the Lagrange equation of particle motion on curves:

$$\begin{cases} m_\alpha \left\|\vec{x}_\alpha^{'}\right\|^2 \ddot{\alpha} + \lambda m_\alpha \left\|\vec{x}_\alpha^{'}\right\|^2 \dot{\alpha} + m_\alpha \langle \vec{x}_\alpha^{''}, \vec{x}_\alpha^{'}\rangle \dot{\alpha}^2 - \\ \qquad - \dfrac{1}{r}\dfrac{\partial U}{\partial r}\langle \vec{y} - \vec{x}, \vec{x}_\alpha^{'}\rangle = 0, \\ m_\beta \left\|\vec{y}_\beta^{'}\right\|^2 \ddot{\beta} + \lambda m_\beta \left\|\vec{y}_\beta^{'}\right\|^2 \dot{\beta} + m_\beta \langle \vec{y}_\beta^{''}, \vec{y}_\beta^{'}\rangle \dot{\beta}^2 + \\ \qquad + \dfrac{1}{r}\dfrac{\partial U}{\partial r}\langle \vec{y} - \vec{x}, \vec{y}_\beta^{'}\rangle = 0 \end{cases}. \qquad [15.13]$$

NOTE.– The partial case of the system [15.13] is obtained in Abbasov (2017a). In the cited work, a system of two point charges (different signs) was considered, which moved along the curves under the action of the Coulomb force and the friction force, which is proportional to the speed.

15.5. Hamiltonian equations

Let us construct the Hamilton equation of motion of the particle system. Hamilton's function is obtained from the Lejander transformation according to the classical scheme (Kazakov 2008):

$$\mathcal{H}_0 = \left(\sum_{i=1}^{n} p_i \dot{q}_i - \mathcal{L}_0(q, \dot{q})\right)\Bigg|_{\dot{q}=\dot{q}(p,q)}. \qquad [15.14]$$

When constructing a Hamilton function, variables q and p are considered as independent. We put

$$\begin{cases} p_\alpha = \dfrac{\partial \mathcal{L}_0}{\partial \dot{\alpha}} = m_\alpha \left\| \vec{x}_\alpha' \right\|^2 \dot{\alpha}, \\ p_\beta = \dfrac{\partial \mathcal{L}_0}{\partial \dot{\beta}} = m_\beta \left\| \vec{y}_\beta' \right\|^2 \dot{\beta}. \end{cases} \quad [15.15]$$

Let us rewrite [15.15] differently, solving the relation relative to $\dot{\alpha}$ and $\dot{\beta}$:

$$\left\{ \dot{\alpha} = \left\| \vec{x}_\alpha' \right\|^{-2} \frac{p_\alpha}{m_\alpha}, \dot{\beta} = \left\| \vec{y}_\beta' \right\|^{-2} \frac{p_\beta}{m_\beta}. \right. \quad [15.16]$$

Substituting [15.16] in [15.14], we have

$$\mathcal{H}_0(\alpha, p_\alpha; \beta, p_\beta) = \frac{p_\alpha^2}{2m_\alpha} \left\| \vec{x}_\alpha' \right\|^{-2} + \frac{p_\beta^2}{2m_\beta} \left\| \vec{y}_\beta' \right\|^{-2} + U(r). \quad [15.17]$$

In [15.17], the first and third terms depend on the α parameter. We have

$$\frac{\partial}{\partial \alpha}\left(\left\| \vec{x}_\alpha' \right\|^{-2} \right) = \frac{-\dfrac{\partial}{\partial \alpha} \langle \vec{x}_\alpha', \vec{x}_\alpha' \rangle}{\left\| \vec{x}_\alpha' \right\|^4} = -2 \frac{\langle \vec{x}_\alpha', \vec{x}_\alpha'' \rangle}{\left\| \vec{x}_\alpha' \right\|^4},$$

$$\frac{\partial U(r)}{\partial \alpha} = -\frac{1}{r} \frac{\partial U}{\partial r} \langle \vec{y} - \vec{x}, \vec{x}_\alpha' \rangle.$$

The equation of the Hamilton system for P_α has the form

$$\dot{p}_\alpha = -\frac{\partial \mathcal{H}_0}{\partial \alpha} = -\frac{\partial}{\partial \alpha}\left(\frac{p_\alpha^2}{2m_\alpha} \left\| \vec{x}_\alpha' \right\|^{-2} + \frac{p_\beta^2}{2m_\beta} \left\| \vec{y}_\beta' \right\|^{-2} + U(r) \right) = \\ = \frac{p_\alpha^2}{m_\alpha} \frac{\langle \vec{x}_\alpha', \vec{x}_\alpha'' \rangle}{\left\| \vec{x}_\alpha' \right\|^4} + \frac{1}{r} \frac{\partial U}{\partial r} \langle \vec{y} - \vec{x}, \vec{x}_\alpha' \rangle. \quad [15.18]$$

Similarly, for p_β, we have

$$\dot{p}_\beta = -\frac{\partial \mathcal{H}_0}{\partial \beta} = \frac{p_\beta^2}{m_\beta} \frac{\langle \vec{y}_\beta^{\,\prime}, \vec{y}_\beta^{\,\prime\prime} \rangle}{\| \vec{y}_\beta^{\,\prime} \|^4} - \frac{1}{r}\frac{\partial U}{\partial r} \langle \vec{y} - \vec{x}, \vec{y}_\beta^{\,\prime} \rangle. \quad [15.19]$$

Let us derive the Hamilton function with dissipation from [15.14], where

$$\begin{cases} P_\alpha = \dfrac{\partial \mathcal{L}}{\partial \dot{\alpha}} = e^{\lambda t} \dfrac{\partial \mathcal{L}_0}{\partial \dot{\alpha}} = e^{\lambda t} p_\alpha, \\ \dot{Q}_\alpha = \dot{q}_\alpha, \\ \mathcal{L} = e^{\lambda t} \mathcal{L}_0. \end{cases} \quad [15.20]$$

Therefore

$$\mathcal{H}(\alpha, p_\alpha; \beta, p_\beta; t) = e^{\lambda t} \mathcal{H}_0(\alpha, p_\alpha; \beta, p_\beta), \quad [15.21]$$

$$\dot{P}_\alpha = \frac{d}{dt}\left(e^{\lambda t} p_\alpha\right) = e^{\lambda t}\left(\dot{p}_\alpha + \lambda p_\alpha\right). \quad [15.22]$$

Hamilton's equation for P_α according to

$$\dot{P}_\alpha = -\frac{\partial \mathcal{H}}{\partial \alpha} = -e^{\lambda t} \frac{\partial \mathcal{H}_0}{\partial \alpha}$$

has the form

$$e^{\lambda t}\left(\dot{p}_\alpha + \lambda p_\alpha\right) = e^{\lambda t}\left(\frac{p_\alpha^2}{m_\alpha} \frac{\langle \vec{x}_\alpha^{\,\prime}, \vec{x}_\alpha^{\,\prime\prime} \rangle}{\| \vec{x}_\alpha^{\,\prime} \|^4} + \frac{1}{r}\frac{\partial U}{\partial r} \langle \vec{y} - \vec{x}, \vec{x}_\alpha^{\,\prime} \rangle \right). \quad [15.23]$$

Similarly, we obtain

$$e^{\lambda t}\left(\dot{p}_\beta + \lambda p_\beta\right) = e^{\lambda t}\left(\frac{p_\beta^2}{m_\beta} \frac{\langle \vec{y}_\beta^{\,\prime}, \vec{y}_\beta^{\,\prime\prime} \rangle}{\| \vec{y}_\beta^{\,\prime} \|^4} - \frac{1}{r}\frac{\partial U}{\partial r} \langle \vec{y} - \vec{x}, \vec{y}_\beta^{\,\prime} \rangle \right). \quad [15.24]$$

Shortening [15.23] and [15.24] by $e^{\lambda t}$, we obtain

$$\begin{cases} \dot{p}_\alpha = \dfrac{p_\alpha^2}{m_\alpha} \dfrac{\langle \vec{x}_\alpha^{\,'}, \vec{x}_\alpha^{\,''} \rangle}{\| \vec{x}_\alpha^{\,'} \|^4} + \dfrac{1}{r} \dfrac{\partial U}{\partial r} \langle \vec{y} - \vec{x}, \vec{x}_\alpha^{\,'} \rangle - \lambda p_\alpha, \\ \dot{p}_\beta = \dfrac{p_\beta^2}{m_\beta} \dfrac{\langle \vec{y}_\beta^{\,'}, \vec{y}_\beta^{\,''} \rangle}{\| \vec{y}_\beta^{\,'} \|^4} - \dfrac{1}{r} \dfrac{\partial U}{\partial r} \langle \vec{y} - \vec{x}, \vec{y}_\beta^{\,'} \rangle - \lambda p_\beta. \end{cases} \quad [15.25]$$

Combining [15.16] and [15.25], we come to the following Hamilton system

$$\begin{cases} \dot{\alpha} = \dfrac{1}{\left\| \vec{x}_\alpha^{\,'} \right\|^2} \dfrac{p_\alpha}{m_\alpha}, \\ \dot{\beta} = \dfrac{1}{\left\| \vec{y}_\beta^{\,'} \right\|^2} \dfrac{p_\beta}{m_\beta}, \\ \dot{p}_\alpha = \dfrac{p_\alpha^2}{m_\alpha} \dfrac{\langle \vec{x}_\alpha^{\,'}, \vec{x}_\alpha^{\,''} \rangle}{\left\| \vec{x}_\alpha^{\,'} \right\|^4} + \dfrac{1}{r} \dfrac{\partial U}{\partial r} \langle \vec{r}, \vec{x}_\alpha^{\,'} \rangle - \lambda p_\alpha, \\ \dot{p}_\beta = \dfrac{p_\beta^2}{m_\beta} \dfrac{\langle \vec{y}_\beta^{\,'}, \vec{y}_\beta^{\,''} \rangle}{\left\| \vec{y}_\beta^{\,'} \right\|^4} - \dfrac{1}{r} \dfrac{\partial U}{\partial r} \langle \vec{r}, \vec{y}_\beta^{\,'} \rangle - \lambda p_\beta. \end{cases} \quad [15.26]$$

NOTE.– In the case of the Newtonian potential $U(r) = q^2 / r$, the system of ordinary first-order differential equations [15.26] takes the form

$$\begin{cases} \dot{\alpha} = \dfrac{1}{\left\| \vec{x}_\alpha^{\,'} \right\|^2} \dfrac{p_\alpha}{m_\alpha}, \\ \dot{\beta} = \dfrac{1}{\left\| \vec{y}_\beta^{\,'} \right\|^2} \dfrac{p_\beta}{m_\beta}, \\ \dot{p}_\alpha = \dfrac{p_\alpha^2}{m_\alpha} \dfrac{\langle \vec{x}_\alpha^{\,'}, \vec{x}_\alpha^{\,''} \rangle}{\left\| \vec{x}_\alpha^{\,'} \right\|^4} - \dfrac{q^2}{r^3} \langle \vec{r}, \vec{x}_\alpha^{\,'} \rangle - \lambda p_\alpha, \\ \dot{p}_\beta = \dfrac{p_\beta^2}{m_\beta} \dfrac{\langle \vec{y}_\beta^{\,'}, \vec{y}_\beta^{\,''} \rangle}{\left\| \vec{y}_\beta^{\,'} \right\|^4} + \dfrac{q^2}{r^3} \langle \vec{r}, \vec{y}_\beta^{\,'} \rangle - \lambda p_\beta. \end{cases}$$

15.6. Numerical experiments

Let one of the curves be a parabola that is symmetric with respect to the y axis and passing through the origin, and let the second one be the line in the xy plane that does not intersect the parabola:

$$\begin{cases} y = ax^2, a > 0; \\ y = bx + c. \end{cases} \quad [15.27]$$

Let us rewrite the [15.27] system in a form of radius vector, so we can use it in [15.26]:

$$\begin{cases} \vec{x} = (\alpha, b\alpha + c); \\ \vec{y} = (\beta, a\beta^2). \end{cases} \quad [15.28]$$

We also need to define masses and coefficients of friction for material points. To make the example simple, we will go with the following limitations:

$$\begin{cases} m_\alpha = m_\beta = m; \\ \lambda = \gamma / m. \end{cases} \quad [15.29]$$

The potential energy also needs to be defined. Derivative and force can then be easily defined:

$$\begin{cases} U = \kappa \dfrac{r^2}{2}; \\ \dfrac{\partial U}{\partial r} = \kappa r; \\ F_r = -\dfrac{\partial U(r)}{\partial r} = -\kappa r. \end{cases} \quad [15.30]$$

Substituting [15.28]–[15.30] into [15.27], the system can be represented as

$$\begin{cases} \dot{\alpha} = \dfrac{p_\alpha}{m}; \\ \dot{\beta} = \dfrac{1}{(1+4a^2\beta^2)} \dfrac{p_\beta}{m}; \\ \dot{p}_\alpha = 2\kappa(\beta-\alpha) - \gamma\dfrac{p_\alpha}{m}; \\ \dot{p}_\beta = \dfrac{4a^2\beta}{(1+4a^2\beta^2)^2} \dfrac{p_\beta^2}{m} - \\ \quad -2\kappa(\beta-\alpha+2a\beta(a\beta^2-b\alpha-c)) - \gamma\dfrac{p_\beta}{m}. \end{cases} \quad [15.31]$$

Let us consider specific curves for the numeric example:

$$\begin{cases} a = 2; \\ b = 2.5; \\ c = -1. \end{cases} \quad [15.32]$$

The initial conditions can be

$$\begin{cases} \alpha(0) = 0.9; \\ \beta(0) = 0.5; \\ p_\alpha(0) = 0; \\ p_\beta(0) = 0. \end{cases} \quad [15.33]$$

The analytical solution of the system is

$$\begin{cases} \beta_0 = \dfrac{b}{2a} = 0.625; \\ \alpha_0 = \dfrac{\beta_0 + ba\beta_0^2 - bc}{1+b^2} = 0.700431; \\ \vec{x}^a = (0.700431, 0.7510776, 0); \\ \vec{y}^a = (0.625, 0.78125, 0); \\ r^a = 8.124171 \cdot 10^{-2}. \end{cases} \quad [15.34]$$

On the contrary, we have to define physical system arguments. For this example, let us consider the following values:

$$\begin{cases} m = 1; \\ \kappa = 5; \\ \gamma / \kappa \in (0.04 \ldots 0.1). \end{cases} \quad [15.35]$$

As a criterion for stopping, it is logical to use the speed at the minimum point. But there are two material points and two corresponding velocities vectors:

$$|\dot{\vec{x}}(\alpha)|+|\dot{\vec{y}}(\beta)|=|\vec{v}(\alpha)|+|\vec{v}(\beta)|=\|\vec{x}_{\alpha'}\dot{\alpha}\|+\|\vec{y}_{\beta'}\dot{\beta}\|.$$

On the contrary, the distance between points $r(t)$ and the speed of its change are scalar functions. Therefore, as a criterion for stopping, we choose

$$\left|\frac{\Delta r}{\Delta t}\right|<\varepsilon. \qquad [15.36]$$

NOTE.– Despite the fact that the step of integration according to the curve parameter is constant, the step value for the $\Delta\vec{x}_i=\vec{v}(t_i)\Delta t$ coordinates is automatically adjusted by changing the speed.

N	γ/κ	n	$n_{T/2}$	r	$\|\Delta r\|$	$\|\Delta\vec{x}\|$	$\|\Delta\vec{y}\|$
1	0.04	184	88	$8.124180\cdot10^{-2}$	$8.671\cdot10^{-8}$	$8.392\cdot10^{-5}$	$3.468\cdot10^{-5}$
2	0.05	123	59	$8.124401\cdot10^{-2}$	$2.297\cdot10^{-6}$	$2.262\cdot10^{-4}$	$3.827\cdot10^{-4}$
3	0.06	113	54	$8.124191\cdot10^{-2}$	$1.980\cdot10^{-7}$	$1.242\cdot10^{-4}$	$5.499\cdot10^{-5}$
4	0.07	99	46	$8.124173\cdot10^{-2}$	$2.084\cdot10^{-8}$	$5.832\cdot10^{-5}$	$1.299\cdot10^{-7}$
5	0.08	87	40	$8.124171\cdot10^{-2}$	$8.551\cdot10^{-10}$	$5.862\cdot10^{-7}$	$1.111\cdot10^{-5}$
6	0.09	75	33	$8.124186\cdot10^{-2}$	$1.460\cdot10^{-7}$	$9.357\cdot10^{-4}$	$1.016\cdot10^{-3}$
7	0.1.	58	23	$8.124820\cdot10^{-2}$	$6.487\cdot10^{-6}$	$5.810\cdot10^{-3}$	$6.419\cdot10^{-3}$

Table 15.1. *Numerical solutions of the system [15.31] with* $\varepsilon=1\cdot10^{-6}$ *and* $\Delta t_0=0.5$ *for [15.32], [15.33], [15.34] and [15.35], where the stop condition is [15.36]*

15.7. Concluding remarks

This chapter considers the method of searching for the nearest pairs on two smooth curves in a three-dimensional Euclidean space, which uses a model of physical interaction of a system of material points with a given potential energy. The Lagrange and Hamilton equations of points motion are obtained, the approximate integration of which gives various algorithms for approximation of the nearest pair. For parametrically defined curves

$$\vec{x}(\alpha) = \left(x_1(\alpha), x_2(\alpha), x_3(\alpha)\right) \text{ and } \vec{y}(\beta) = \left(y_1(\beta), y_2(\beta), y_3(\beta)\right)$$

the corresponding equations are

$$\begin{cases} m_\alpha \left\|\vec{x}'_\alpha\right\|^2 \ddot{\alpha} + \lambda m_\alpha \left\|\vec{x}'_\alpha\right\|^2 \dot{\alpha} + m_\alpha \langle \vec{x}''_\alpha, \vec{x}'_\alpha \rangle \dot{\alpha}^2 - \dfrac{1}{r}\dfrac{\partial U}{\partial r}\langle \vec{y} - \vec{x}, \vec{x}'_\alpha \rangle = 0, \\ m_\beta \left\|\vec{y}'_\beta\right\|^2 \ddot{\beta} + \lambda m_\beta \left\|\vec{y}'_\beta\right\|^2 \dot{\beta} + m_\beta \langle \vec{y}''_\beta, \vec{y}'_\beta \rangle \dot{\beta}^2 + \dfrac{1}{r}\dfrac{\partial U}{\partial r}\langle \vec{y} - \vec{x}, \vec{y}'_\beta \rangle = 0 \end{cases}$$

and

$$\begin{cases} \dot{\alpha} = \dfrac{1}{\left\|\vec{x}'_\alpha\right\|^2} \dfrac{p_\alpha}{m_\alpha}, \\ \dot{\beta} = \dfrac{1}{\left\|\vec{y}'_\beta\right\|^2} \dfrac{p_\beta}{m_\beta}, \\ \dot{p}_\alpha = \dfrac{p_\alpha^2}{m_\alpha} \dfrac{\langle \vec{x}'_\alpha, \vec{x}''_\alpha \rangle}{\left\|\vec{x}'_\alpha\right\|^4} + \dfrac{1}{r}\dfrac{\partial U}{\partial r}\langle \vec{r}, \vec{x}'_\alpha \rangle - \lambda p_\alpha, \\ \dot{p}_\beta = \dfrac{p_\beta^2}{m_\beta} \dfrac{\langle \vec{y}'_\beta, \vec{y}''_\beta \rangle}{\left\|\vec{y}'_\beta\right\|^4} - \dfrac{1}{r}\dfrac{\partial U}{\partial r}\langle \vec{r}, \vec{y}'_\beta \rangle - \lambda p_\beta, \end{cases}$$

where m_α, m_β, λ are the positive parameters, $r = \left\|\vec{y}(\beta) - \vec{x}(\alpha)\right\|$, and $U = U(r)$ is the potential energy of interaction.

It is possible to generalize the proposed equations. It is interesting to consider similar dynamics to calculate the distance between the curve and the surface, or between two surfaces. We can go even further and generalize these equations to Euclidean spaces of arbitrary dimension.

15.8. References

Abbasov, M.E. (2016). Charged balls method for finding the minimum distance between two plane convex smooth curves in three-dimensional space. *VIII Moscow International Conference on Operations Research (ORM2016)*, Moscow, 17–22 October, Proceedings: Vol. I. – M., MAKS Press, 12–13.

Abbasov, M.E. (2017a). Finding the distance between two smooth curves in three-dimensional space. In *Selected Lectures on Extremal Problems, Part 1*, Malozyomova, V.N. (ed.). Publication House VVM, St. Petersburg (In Russian).

Abbasov, M.E. (2017b). Method of charged balls. In *Selected Lectures on Extremal Problems, Part 1*, Malozyomova, V.N. (ed.). Publishing House VVM, St. Petersburg (In Russian).

Bauschke, H.H., Combettes, P.L., Luke, D.R. (2004). Finding best approximation pairs relative to two closed convex sets in Hilbert spaces. *Journal of Approximation Theory*, 127, 178–192.

Gilbert, E.G. and Foo, C.P. (1990). Computing the distance between general convex objects in three-dimensional space. *IEEE Transactions on Robotics and Automation*, 6(1), 53–61.

Gilbert, E.G., Johnson, D.W., Keerthi, S.S. (1988). A fast procedure for computing the distance between complex objects in three-dimensional space. *IEEE Journal of Robotics and Automation*, 4, 193–203.

Kazakov, K.A. (2008). *Introduction to Theoretical and Quantum Mechanics.* MSU, Moscow (In Russian).

Landau, L.D. and Lifschitz, E.M. (1969). *Mechanics. Electrodynamics.* Science, Moscow (In Russian).

Lin, A. and Han, S.P. (2002). On the distance between two ellipsoids. *SIAM Journal on Optimization*, 13, 298–308.

Llanas, B., Fernandez de Sevilla, M., Feliu, V. (2000). An iterative algorithm for finding a nearest pair of points in two convex subsets of R^n. *Computers and Mathematics with Applications*, 40, 971–983.

Malitsky, Yu. (2014). Douglas-Rachford algorithm for best approximation pair. *VII International Conference "Calculation and Applied Mathematics" dedicated to Academician I. I. Lyashko. Conference Materials*, Kyiv, 123.

Polyak, B.T. (2014). *Introduction to Optimization.* LANAND, Moscow (In Russian).

Uteshev, A.Yu. and Yashina, M.V. (2008). Finding the distance from an ellipsoid in a plane and a quadric. *Reports of the Russian Academy of Sciences*, 419(4), 471–474 (In Russian).

16

Constrained Markov Decision Process for the Industry

Michel BOUSSEMART[1] and Nikolaos LIMNIOS[2]

[1] *IoT Consultants, Adelaide, Australia*

[2] *Sorbonne University Alliance, Université de Technologie de Compiègne, Laboratoire de Mathématiques Appliquées de Compiègne, France*

16.1. Introduction

In engineering science, Markov or semi-Markov processes are a powerful tool for describing phenomena, including transient or asymptotic behavior. This is the case, for example, with dependability activities to compute the reliability or availability of complex systems (see, for example, Barbu et al. (2004) or Limnios and Oprişan (2000)). However, it becomes insufficient when we wish to bring a control that is also optimal. For example, in maintenance activities, is there a way to find optimal policies without having to search by hand?

To overcome this limitation, Markov decision processes were introduced at the end of the 1950s by Bellman (1957) and Howard (1960). The key idea was to introduce a control and economic criteria.

As engineers in the industry try to find compromises every day, constraints are often to be considered in optimization problems, such as resource limitations, or, for maintenance activities, constraints on the failure rate or availability. To address

Computational Methods and Mathematical Modeling in Cyberphysics and Engineering Applications 1,
coordinated by Dmitri KOROLIOUK, Sergiy LYASHKO and Nikolaos LIMNIOS.

this need, a more general class was introduced, namely constrained Markov decision processes, which will be presented in the following sections.

In this chapter, the general framework of Markov decision processes, including the definition of generic constraints, is introduced in section 16.2. As we will see, additional effort is needed to apply these constraints in the industry. For this reason, in section 16.3, we present the case of constraints on availability, with an application to a maintenance case. Finally, in section 16.4, we present one of our significant contributions to the constrained Markov decision process theory by introducing a constraint on the asymptotic failure rate (Bickard et al. (2001); Boussemart and Limnios (2002, 2004)).

16.2. Introduction to constrained Markov decision processes

16.2.1. *Introduction*

Before introducing a formal definition of constrained Markov decision processes (cMDP), let us start with a simple illustration. In Markov processes (see Figure 16.1 (left)), transitions are static and known, whereas in Markov decision processes (see Figure 16.1 (right)), transitions depend on the user's decisions for each observed state. Here, for state 1, depending on the decision made (1 or 2), there are two possible transitions: $p_{12}(1)$ or $p_{12}(2)$.

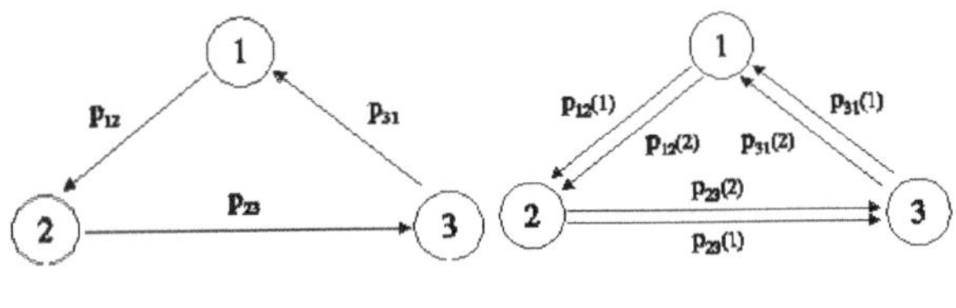

Figure 16.1. *Markov process versus Markov decision process*

To make the best decision at each time, also called the decision epoch, economic criteria must be considered. The principle is first to introduce an immediate reward associated with every possible state and every possible action from this state. As we are interested in making the best decisions for a long term, a global economic function will sum all of the immediate rewards and the problem becomes an optimization one, where the decisions are unknown, i.e. the way to choose the action at every stage.

The general optimization problem over a long term (finite or infinite time) generally implies that we know, at each decision epoch, the whole history of the system, i.e. past states and past decisions, leading to a difficult problem. Fortunately, this is a well-known result showing that, under certain conditions, it is sufficient to consider only stationary policies instead of history-dependent policies. We will emphasis this point in the following sections. Indeed, it is not obvious that, for the optimization problem, we stay in this research space when new elements, such as constraints, are introduced. As it is important to be able to use practical algorithms, we will see how to stay in this stationary decision space.

Now, let us go in depth into the Markov decision process formalism. First, the model including the underlying stochastic processes and notations are introduced. Then, for two major economic criteria, classical algorithms for finding optimal policies and the way to consider constraints are presented.

16.2.2. *Model*

16.2.2.1. *States, decisions and history*

Let us assume that at a point of time $n = 0, 1, \ldots$ (decision epochs), the system is observed. Let $X = (X_n)_{n \in \mathrm{N}}$ denote the sequence of observed states and $E = \{1, \ldots, s\}$ the state space.

Let $Y = (Y_n)_{n \in \mathrm{N}}$ denote the sequence of actions and K_i the set of possible actions when the system is in state $i \in E$.

For any decision epoch n, as a result of observing state X_n and making decision Y_n, there is a reward $w(X_n, Y_n)$. The history of the system up to time n, $H = (H_n)_{n \in \mathrm{N}}$, is defined by

$$\begin{cases} H_0 = X_0 & \text{if} \quad n = 0 \\ H_n = (X_0, Y_0, \ldots, X_{n-1}, Y_{n-1}, X_n) & \text{if} \quad n > 0. \end{cases}$$

REMARK 16.1.– *X, Y and H are stochastic processes.*

16.2.2.2. *Decision rule and policy*

At each decision epoch n, when the history $\{H_n = h_n\}$ is known, the decision rule d_{n+1} specifies a probability distribution $d_{n+1}(\cdot \mid h_n)$ on the set of actions. It specifies how to choose an action. A policy π is a sequence of decision rules $\pi = (d_1, d_2, \ldots)$.

In the general case, a policy depends on the whole history. But such a policy is hard to obtain. The key concept of Markov decision processes is to prove that a small class of policy is sufficient in many cases. The different classes of policies are defined as follows:

– Π^H denotes the set of all policies. $\pi \in \Pi^H$ implies that all decision rules of π are history dependent: this is the greatest class of policies.

– $\Pi^M \subset \Pi^H$ denotes the set of Markovian policies. A policy $\pi = (d_1, d_2, \ldots) \in \Pi^M$ implies that all decision rules d_n do not depend on the past, i.e., for every

$h_n = (i_0, a_0, \ldots, i_{n-1}, a_{n-1}, i)$, we have $d_{n+1}(a \mid h_n) = d_{n+1}(a \mid i)$, for $a \in K_i$. In other words, d_n are Markovian. We let D_M be the set of Markovian decision rules.

– $\Pi^S \subset \Pi^M$ denotes the set of stationary policies. $\pi \in \Pi^S$ implies that at each decision epoch the same decision rule d is applied. Such policy is denoted by $\pi = (d)^\infty$.

A policy $\pi \in \Pi^H$ and an initial distribution μ induce a probability $\mathbf{P}^\pi$ that satisfies for all $n \geq 0$, $h_n = (i_0, a_0, \ldots, i_{n-1}, a_{n-1}, i)$ and $a \in K_i$:

$$\mathbf{P}^\pi(X_0 = i) = \mu(i);$$

$$\mathbf{P}^\pi(Y_n = a \mid H_n = h_n) = d_{n+1}(a \mid h_n);$$

$$\mathbf{P}^\pi(X_{n+1} = j \mid H_n = h_n, Y_n = a) = p_{ij}(a).$$

The last property corresponds to the Markov property. The $\{p_{ij}(a)\}$ are called the transition probabilities of the MDP (in this chapter, we assume that they are invariant in time or stationary).

16.2.2.3. *Notations*

Let us define D_M as the set of Markovian decision rules and $D_M^* \subset D_M$ as the set of deterministic decision rules. A policy in Π^S can be written as $\pi = (d)^\infty$, where $d \in D_M$, and a stationary deterministic policy ($\in \Pi^{SD} \subset \Pi^S$) can be written as $\pi = (d)^\infty$, where $d \in D_M^*$.

Let $d \in D_M$ be a decision rule, and let us define the matrix P_d as

$$P_d(i,j) = \sum_{a \in K_i} p_{ij}(a) d(a \mid i), \quad i, j \in E$$

and the column vector W_d as

$$W_d(i) = \sum_{a \in K_i} w(i,a) d(a \mid i), \quad i \in E.$$

For a general Markovian policy, the associated Markov process is non-homogeneous. When the policy is stationary, the associated Markov process is homogeneous.

For the particular case of an initial distribution $\mu(i) = 1$ for any $i \in E$, we define the corresponding probability or expectation, respectively $\mathbf{P}_i^\pi$ and $\mathbf{E}_i^\pi$.

16.2.3. *Economic criteria*

In the Markov decision process framework, there are two separate families: finite and infinite horizon criteria. For the first one, average rewards on a short period of time are added. In this case, it can be shown that an optimal Markovian policy can be obtained by a backward or Bellman type algorithm (see, for example, White (1993) and Puterman (1994)). This type of criteria is relevant, for example, in robotics when trying to reach a goal. There are also applications in maintenance (see, for example, Huang and Guo (2011)).

For this introductory chapter to cMDP, we will consider infinite horizon criteria that can be used to analyze the long-term system's behavior, such as during maintenance activities.

Before introducing the two main criteria, let us define the immediate expected reward when the state X_n is observed and when the decision Y_n is made:

$$\mathbf{E}_i^{\pi}(w(X_n, Y_n)) = \sum_{j \in E} \sum_{a \in K_j} \mathbf{P}_i^{\pi}(X_n = j, Y_n = a) w(j, a), \quad i \in E.$$

Recall that rewards $w(j, a)$ are known and belong to the input data.

16.2.3.1. *Infinite horizon expected discounted reward*

This criterion is the sum of all immediate expected rewards with a discount factor γ (which can be seen as a currency inflation factor).

DEFINITION 16.1.– *For $\gamma \in]0, 1[$, the infinite horizon expected discounted reward is defined as*

$$v_{\gamma}^{\pi}(i) = \mathbf{E}_i^{\pi}\left[\sum_{k=0}^{\infty} \gamma^k w(X_k, Y_k)\right], \quad i \in E.$$

We introduce the notation v_{γ}^{π} as the corresponding column vector.

16.2.3.2. *Infinite horizon expected average reward*

This criterion can be used when there is no discounted factor γ.

DEFINITION 16.2.– *The infinite horizon expected average reward is defined as:*

$$g^{\pi}(i) = \liminf_{n \to \infty} \frac{1}{n} \mathbf{E}_i^{\pi}\left[\sum_{k=0}^{n-1} w(X_k, Y_k)\right], \quad i \in E.$$

As above, g^{π} is the corresponding column vector.

16.2.3.3. *Other criteria*

Both the aforementioned criteria emphasize the asymptotic behavior of the system. The main drawback is that it is not possible to consider any transient phenomenon, in particular when there are absorbing classes. There is an abundant literature around other criteria, such as Blackwell's criterion (see, for example, Blackwell (1962)). Mahadevan (1996a) reviews the main criteria and links between them. Puterman (1994) gives conditions for the existence of optimal policies. When the system rarely reaches absorbing states, we propose a new criterion based on the quasi-stationary behavior (see Bickard et al. (2001) and Boussemart and Limnios (2002, 2004)).

16.2.4. *Infinite horizon expected discounted reward*

The objective is to find an optimal policy π that maximizes each component of v_γ^π, i.e. to solve the following problem:

$$v_\gamma^* = \max_{\pi \in \Pi^H} v_\gamma^\pi.$$

To prove that we can restrict ourselves to search stationary policies, the idea is to show that if v_γ^* exists, it satisfies the Bellman optimality equation (Definition 16.3 below). Then, for this value, a stationary policy can be obtained (Theorem 16.1).

16.2.4.1. *Optimality equation*

Let us define V as the set of all possible values of vector v_γ^π for all policies $\pi \in \Pi^H$, and the norm $||\cdot||$ is defined as:

$$||v|| = \max_{i \in E} |v(i)|, \quad v \in V.$$

Let us suppose that we apply a stationary deterministic policy $\pi = (d)^\infty \in \Pi^S$. Then, it can be easily shown (see, for example, White (1993)) that:

$$v_\gamma^\pi = \sum_{k=0}^{\infty} \gamma^k (P_d)^k W_d.$$

Thus

$$v_\gamma^\pi = W_d + \gamma P_d v_\gamma^\pi. \qquad [16.1]$$

Equation [16.1] gives us insight to find an optimal solution: as we are trying to find v_γ^π maximal, we will try to maximize the right-hand side of this equation, hence the following definition of the optimality equation:

DEFINITION 16.3.– *Vector $v \in V$ satisfies Bellman's equation if*

$$v(i) = \max_{a \in K_i} \{w(i,a) + \gamma \sum_{j \in E} p_{ij}(a) v(j)\}, \quad i \in E,$$

or, in the matrix form

$$v = \max_{d \in D_M^*} \{W_d + \gamma P_d v\}, \qquad [16.2]$$

or

$$v = Lv, \qquad [16.3]$$

where L is a nonlinear operator defined as

$$Lv = \max_{d \in D_M^*} \{W_d + \gamma P_d v\}, \quad v \in V.$$

REMARK 16.2.– *In the literature, Bellman's equation [16.2] can be found with D_M instead of D_M^*.*

To present algorithms in a simple framework, the following assumptions are made:

ASSUMPTION 16.1.– *There exists a $M \in \mathbf{R}$ such that for all $i \in E$ and $a \in K_i$, $|w(i,a)| < M$;*

ASSUMPTION 16.2.– *We suppose E and K_i are finite for all $i \in E$ and we let $s = |E|$ (number of elements).*

Both assumptions are sufficient conditions for the infinite horizon expected discounted reward to be finite. In the literature, there are some works to extend these assumptions (see, for example, Lippman (1975) and Puterman (1994)).

THEOREM 16.1 (DERMAN 1970).– *Under Assumptions 16.1 and 16.2, v_γ^* exists and satisfies Bellman's equation [16.3]. Moreover, there exists a stationary deterministic policy $\pi^* = (d^*)^\infty \in \Pi^{SD}$ defined by*

$$d^* = \text{argmax}_{d \in D_M^*} \{W_d + \gamma P_d v_\gamma^*\}, \qquad [16.4]$$

optimal for the problem.

In the case when state and action spaces are finite, the existence of a solution to Bellman's equation is obvious. In a more general framework, when E is a countable infinite set, the existence can be shown by the application of Banach's fixed point theorem, given that L is a contracting operator in the Banach space V (see Puterman (1994)). A proof for the existence of a deterministic stationary policy, optimal for the problem, is given, for example, by Derman (1970). For a more general framework (compact state or action spaces), we can cite Puterman (1994) or Bertsekas (1995).

16.2.4.2. *Algorithms*

In this section, three classical algorithms are presented to solve the problem. We suppose that Assumptions 16.1 and 16.2 are satisfied.

16.2.4.2.1. Value iteration (VI) algorithm

One of the first authors to present this algorithm is Howard (1960). At each iteration, the value of the economic function v_n is enhanced (thus the name VI). The algorithm leads to an optimal policy with an error less than ϵ, hence the following definition (see Arapostathis et al. (1993, p. 287)).

DEFINITION 16.4.– *Let $\epsilon > 0$. The policy $\pi_\epsilon \in \Pi^H$ is said to be ϵ-optimal if*

$$||v_\gamma^{\pi_\epsilon} - v_\gamma^*|| \leq \epsilon.$$

The VI algorithm is defined as follows:

$$\begin{cases} v_0 \in V \text{ given} \\ v_{n+1}(i) = \max_{a \in K_i}\{w(i,a) + \sum_{j \in E} \gamma p_{ij}(a) v_n(j)\}, \quad i \in E. \end{cases}$$

Stop test (ϵ given):

$$||v_{n+1} - v_n|| < \frac{\epsilon(1-\gamma)}{2\gamma}.$$

Then, the deterministic stationary policy $\pi_\epsilon = (d_\epsilon)^\infty$ is obtained as follows: if we define

$$a_i^* = \text{argmax}_{a \in K_i}\{w(i,a) + \sum_{j \in E} \gamma p_{ij}(a) v_{n+1}(j)\}, \quad i \in E, \qquad [16.5]$$

then, d_ϵ is the unique decision rule satisfying:

$$d_\epsilon(a_i^* \mid i) = 1, \quad i \in E. \qquad [16.6]$$

THEOREM 16.2 (PUTERMAN 1994).– *The policy π_ϵ obtained by the VI algorithm is ϵ-optimal.*

The principle of the proof is to use the fact that L is a contracting operator in V. Consequently, by applying Banach's fixed point theorem, the VI algorithm converges to the solution of Bellman's equation.

In practice, the algorithm does not present good results as the convergence is linear. Moreover, the closer γ is to 1, the weaker the convergence. There are optimized versions of this algorithm. Puterman (1994) gives a review of efficient methods, for example the GSVI (Gauss Seidel Value Iteration), for which the convergence is still linear, but with a better speed.

16.2.4.2.2. Policy iteration (PI) algorithm

For this algorithm, Howard (1960) is again a precursor. Two sequences are simultaneously built: (d_n), a sequence of deterministic decision rules, and (v_n), a sequence representing the economic function values.

1) *Begin.* Let $n = 0$. Chose an arbitrary decision rule d_0 in D^*_M.

2) *Policy evaluation.* Obtain v_n by solving the equation (unknown v)

$$(I - \gamma P_{d_n})v = W_{d_n}.$$

3) *Policy improvement.* Chose d_{n+1} such as

$$d_{n+1} \in \text{argmax}_{d \in D^*_M} \{W_d + \gamma P_d v_n\}.$$

4) *Stop test.* If $d_{n+1} = d_n$, then end and define $d^* = d_n$. Otherwise, increment n for one unit and go to step 2.

THEOREM 16.3 (DERMAN 1970).– *The PI algorithm converges in a finite number of iterations. Moreover, the policy* $\pi^* = (d^*)^\infty$ *is optimal.*

The principle of the proof is to show that v_n is an increasing sequence. As state and action spaces are finite, V is finite. Thus, the sequence v_n takes only a finite number of values and converges.

Puterman (1994) gives a version of the algorithm when the state and action spaces are not finite. In the same book, he also gives optimized versions for this algorithm: the MPI (Modified Policy Iteration), which was introduced by Morton (1971).

Finally, the following algorithm presents good results and is easy to implement as it is the linear programming method.

16.2.4.2.3. Linear programming (LP) algorithm

Derman (1970) was one of the first to present this algorithm. Arbitrarily, we choose for $j \in E$, any $\alpha(j)$ satisfying

$$\alpha(j) > 0 \quad \text{and} \quad \sum_{i \in E} \alpha(i) = 1.$$

We define for $i \in E$ and $a \in K_i$, $x(i,a)$ the unknown of the problem, and $x = (x(i,a))$, the corresponding matrix.

The LP algorithm is defined as follows:

$$\max_x \textstyle\sum_{i \in E} \sum_{a \in K_i} w(i,a)x(i,a), \quad \text{subject to}$$
$$\begin{cases} \sum_{a \in K_j} x(j,a) - \sum_{i \in E} \sum_{a \in K_i} \gamma p_{ij}(a)x(i,a) = \alpha(j), & j \in E \\ x(i,a) \geq 0, \quad i \in E,\ a \in K_i. \end{cases}$$

Let x^* be an optimal solution for LP. Then, the policy $\pi^* = (d)^\infty$ is defined by

$$d(a \mid i) = \mathbf{P}(d(i) = a) = \frac{x^*(i,a)}{\sum_{a' \in A} x^*(i,a')}, \quad i \in E,\ a \in K_i. \qquad [16.7]$$

THEOREM 16.4 (DERMAN 1970).– *The policy π^* is optimal.*

The proof is based on the fact that the dual formulation of LP is such that the constraints correspond exactly to Bellman's equation. Consequently, applying the duality theorem, the economic function is equal to the optimal value.

When the space or state action space is infinite, there are also methods using the linear programming algorithms. Altman (1999) gives, for example, some methods (state space truncation and approximations).

In the case of large state and action spaces, the above three algorithms and their optimized variants become difficult to apply. To tackle this problem, methods combining simulation and stochastic approximation, namely reinforcement learning algorithms, can be used.

16.2.4.2.4. Reinforcement learning (RL)

For a good introduction, the following references can be cited: Kaelbling et al. (1996), Hagen and Kröse (1997) or Sutton and Barto (1998). In this section, we present only the most famous algorithm, i.e. the Q-learning introduced by Watkins (1989). We have a table $Q(i,a)$ for all possible states and actions. At step n of the algorithm, if the system is in the state x_n, the decision-maker chooses an action a_n. As a result, the system goes into the state y_n and an immediate reward r_n is earned. Typically, y_n is obtained by simulation from transition probabilities $p_{x_n,\cdot}(a_n)$ and $r_n = w(x_n, a_n)$. Thus, the sequence $< x_n, a_n, y_n, r_n >$ is used to update the table Q with a learning rate, α_n (thus the name Q-learning). After a sufficiently large number of iterations, the table becomes invariant and an optimal policy can be deduced.

The RL algorithm is as follows. Let $\{\alpha_n(x,a)\}$ be any positive reals.

1) *Begin.* Set $Q_0(i,a)$ to 0.

2) *Iterations for* $n = 1 \to \infty$.
- Chose x_n, a_n.
- Observe next state y_n and immediate reward r_n.
- Update table Q_n by

$$Q_{n+1}(x,a) = Q_n(s,a) + \alpha_n(x,a)e_n,$$

where

$$e_n = \begin{cases} r_n + \gamma \max_{b \in K_{y_n}} Q_n(y_n, b) - Q_n(x,a) & \text{if } (x,a) = (x_n, a_n) \\ 0 & \text{otherwise.} \end{cases}$$

THEOREM 16.5 (WATKINS 1989, 1992).– *If* $\sum_n \alpha_n(x,a) = \infty$, *if* $\sum_n \alpha_n^2(x,a) < \infty$ *and if all* (i,a) *are visited an infinite amount of times, then* Q_n *converges to* Q^*. *Moreover, an optimal deterministic stationary policy* $\pi^* = (d^*)^\infty$ *can be obtained by*

$$d^*(a \mid i) = \text{argmax}_{k \in K_i} Q^*(i,k), \quad i \in E,\ a \in K_i.$$

One possible proof is based on the stochastic approximation principle introduced by Robbins and Monro (1951) and developed, for example, by Benveniste et al. (1987). For a typical proof, see, for example, Jaakkola and Jordan (1994), Tsitsiklis (1994) or Borkar (1998). There is another type of reinforcement learning, such as TD(λ), introduced by Sutton (1998). Convergence properties are given, for example, by Sutton (1998) or Dayan (1992).

Finally, it should be noted that we can find methods when state and action spaces are very large (see, for example, Parr (1998) or Roy (1998)).

16.2.4.2.5. Other methods

By introducing several algorithms (VI, PI and LP), we have seen that the whole state space needs to be enumerated, which becomes impossible for very large models. To address this problem, reinforcement learning algorithms can use techniques to approximate the value function, in particular with neural networks (see Bertsekas (1996)), leading, for example, to the popular deep Q-learning approach, implemented today in Pytorch, one of the machine learning ecosystem's frameworks (see, for example, Sanghi (2021)).

In addition, there are methods developed in the mid-1990s that exploit the knowledge of the problem by representing state and action spaces with trees (see, for example, Boutilier et al. (1995)). Thus, rather than enumerating the whole space, the problem is reduced to tree explorations. One possible algorithm of this family is SPI (Structured Policy Iteration) based on the PI algorithm.

16.2.4.3. *Constraints on state–action frequencies*

As explained in the introduction, it is not obvious that new constraints will not have an impact on the search space, namely the stationary policy space as suggested by Theorem 16.1. Fortunately, there are some cases where we can stay in stationary policy space (see, for example, Altman (1992, 1999)), as, for example, with the following type of constraints:

$$D_\gamma^i(s,\pi) = (1-\gamma)\sum_{k=0}^{\infty} \gamma^k \mathbf{E}_s^\pi d^i(X_k, Y_k) \leq V^i, \quad i = 1, \ldots, I,$$

where s is the initial state, I is the number of constraints, $d^i(x,a)$ is the reward associated with state x and decision a, and V^i is a constant.

If the state and action spaces are finite, a stationary policy can be obtained by a linear programming algorithm by adding lines corresponding to the constraints. In this case, it can be proven that the policy is no longer deterministic and that the number of randomized actions depends on the number of constraints (see Altman (1992, 1999)).

Beutler and Ross (1985) give extensions where the action space is continuous and compact. In this case, additional assumptions have to be considered (see Altman (1992, 1999)).

16.2.5. *Infinite horizon expected average reward*

The objective is to find an optimal policy π that maximizes each component of g^π, i.e. to solve the following problem:

$$g^* = \max_{\pi \in \Pi^H} g^\pi. \qquad [16.8]$$

To present methods in a simple framework, we make the following assumption.

ASSUMPTION 16.3.– *For all decision rules $d \in D_M$, the matrix P_d is irreducible.*

This assumption leads to equality between all components of the optimal g^π that simplifies the study of the optimality equation. For reducible cases, see, for example, Puterman (1994).

In addition, we suppose that Assumptions 16.1 and 16.2 are satisfied.

16.2.5.1. *Optimality equation*

Let $\pi \in \Pi^{SD}$ be any deterministic stationary policy and consider the total reward on $[0, n]$ defined by

$$v_n(s) = \mathbf{E}_s^\pi \left[\sum_{k=0}^{n-1} w(X_k, Y_k) \right], \quad s \in E,$$

with v_n defined as the corresponding column vector, we have (see, for example, White (1993)):

$$v_n = \sum_{k=0}^{n-1} P_d^k W_d = W_d + P_d v_{n-1}. \qquad [16.9]$$

Moreover, under Assumption 16.3 when n tends to infinity, v_n can be written as (see, for example, White (1993))

$$v_n = ng\mathbf{1}_s + h + \epsilon_n, \qquad [16.10]$$

where 1_s is the column vector of size s (where $|E| = s$), all components are equal to 1, ϵ_n is such that $\lim_{n \to \infty} \epsilon_n = 0$, $g \in \mathbf{R}$ and $h \in \mathbf{R}^s$.

From expressions [16.9] and [16.10], we have

$$ng1_s + h + \epsilon_n = W_d + P_d((n-1)g1_s + h + \epsilon_{n-1}).$$

By using the fact that $P_d 1_s = 1_s$, we have

$$g1_s = W_d + (P_d - I)h + P_d\epsilon_{n-1} - \epsilon_n.$$

Thus, when n tends to infinity, we have the following equation:

$$W_d + (P_d - I)h - g1_s = 0. \qquad [16.11]$$

Similarly to the expected discounted reward, equation [16.11] gives us insight to find an optimal policy. Indeed, the principle is to find a pair (g, h) that maximizes the left-hand side. Thus, the following definition of the optimality equation.

DEFINITION 16.5.– *Let $(g, h) \in \mathbf{R} \times \mathbf{R}^s$. The pair (g, h) is said to be a solution of the optimality equation (EO) if*

$$\max_{d \in D_M^*} \{W_d - g1_s + (P_d - I)h\} = 0. \qquad [16.12]$$

THEOREM 16.6 (PUTERMAN 1994).– *There exists a solution $(g, h) \in \mathbf{R} \times \mathbf{R}^s$ for the optimality equation [16.12] and*

$$g1_s = g^*.$$

Moreover, there exists an optimal deterministic stationary policy $\pi^* = (d^*)^\infty \in \Pi^{SD}$, defined by

$$d^* = \text{argmax}_{d \in D_M^*} \{W_d + P_d h\}.$$

16.2.5.2. *Algorithms*

As for the expected discounted reward, there are four basic algorithms to obtain optimal policies: VI, PI, LP or reinforcement learning.

16.2.5.2.1. Value iteration (VI) algorithm

This algorithm appears in White (1963). The iteration is as follows:

$$\begin{cases} v_0 \quad \text{given} \\ v_{n+1}(i) = \max_{a \in K_i} \{w(i,a) + \sum_{j \in E} p_{ij}(a) v_n(j)\}, \quad i \in E. \end{cases}$$

For the stop test, we define α_n and β_n by

$$\alpha_n = \max_{i \in E}\{v_n(i) - v_{n-1}(i)\}, \quad \beta_n = \min_{i \in E}\{v_n(i) - v_{n-1}(i)\}.$$

For a given ϵ, the algorithm is stopped once we have

$$\alpha_n - \beta_n < \epsilon,$$

and we define the deterministic stationary policy π_ϵ similar to the VI algorithm for the expected discounted reward (expression [16.6] with $\gamma = 1$).

THEOREM 16.7 (WHITE 1993).– *The policy* $\pi_\epsilon = (d_\epsilon)^\infty$ *is* ϵ*-optimal for problem [16.8].*

REMARK 16.3.– *The sequences* α_n *and* β_n *tend to* g^* *(see White (1993, p. 78)).*

16.2.5.2.2. PI (Policy Iteration) algorithm

This algorithm was introduced by Howard (1960). Each iteration has two steps: policy evaluation and then policy improvement.

1) *Begin.* Set $n = 0$. Choose an arbitrary decision rule d_0 in D^*_M.

2) *(Policy evaluation).* Obtain g_n and a vector h_n (unknown g and h) by solving the following linear system

$$0 = W_{d_n} - g1_s + (P_{d_n} - I)h.$$

3) *(Policy improvement).* Choose $d_{n+1} \in D^*_M$ such as

$$d_{n+1} \in \operatorname{argmax}_{d \in D^*_M}\{W_d + P_d h_n\}$$

4) If $d_{n+1} \neq d_n$, let $d_{n+1} = d_n$ and increment one unit n, then go to step 2. Otherwise, terminate the algorithm and let $d^* = d_n$.

THEOREM 16.8 (WHITE 1993).– *The policy* $\pi^* = (d^*)^\infty$ *is optimal for problem [16.8].*

16.2.5.2.3. Linear programming (LP) algorithm

As for the expected discounted reward, we define for all $i \in E$ and $a \in K_i$, $x(i, a)$, the unknown of the linear programming problem. The problem to solve is the following:

$$\max_x \textstyle\sum_{i \in E} \sum_{a \in K_i} w(i,a)x(i,a), \quad \text{subject to}$$
$$\begin{cases} \sum_{a \in K_j} x(a,j) - \sum_{i \in E} \sum_{a \in K_i} p_{i,j}(a)x(i,a) = 0, & j \in E \\ \sum_{i \in E} \sum_{a \in K_i} x(i,a) = 1, \\ x(i,a) \geq 0, \quad i \in E, a \in K_i. \end{cases}$$

If we denote the optimal solution by x^*, we define the policy $\pi^* = (d)^\infty$ by equation [16.16].

THEOREM 16.9 (DERMAN 1970).– *The policy π^* is optimal for problem [16.8].*

16.2.5.2.4. Reinforcement learning

To solve the problem, adaptive algorithms can also be implemented (see Mahadevan (1996b) for a review of the methods). Schwartz (1994) developed an R-learning algorithm based on the same principles as the Q-learning algorithm. There are other algorithms in the literature (see, for example, Singh (1994) (variant of the Shwartz algorithm) or Bertsekas et al. (1996)).

16.2.5.3. *Constraints on state–action frequencies*

It is also possible to introduce constraints similar to those that we presented for the expected discounted reward case (see, for example, Beutler and Ross (1985), Altman and Shwartz (1991) and Altman (1999, 1992)). Again, in this case, we can restrict ourselves to search for an optimal stationary policy.

Indeed, let us consider the following function defined from $E \times \cup_{i \in E} K_i$ to $\mathbf{R}$, for each $s \in E$ and policy $\pi \in \Pi^H$:

$$c^\pi(s) = \limsup_{n \to \infty} \mathbf{E}_s^\pi \left[\frac{1}{n} \sum_{k=0}^{n-1} f(X_k, Y_k) \right]. \qquad [16.13]$$

Then, the constrained problem to solve will be the following, for all $s \in E$ and with β as a constant:

$$\begin{array}{ll} \max_{\pi \in \Pi^H} g^\pi(s), & \text{subject to} \\ c^\pi(s) \leq \beta. & \end{array} \qquad [16.14]$$

The following theorem characterizes the existence of solutions for problem [16.14] in the stationary policy set.

THEOREM 16.10 (PUTERMAN 1994).– *Under Assumptions 16.1, 16.2 and 16.3, there exists a stationary policy $\pi^* \in \Pi^S$, a solution for problem [16.14] for all initial states $s \in E$.*

REMARK 16.4.–

– *The Markov process X can contain some transient states.*

– *The optimal policy π^* does not depend on the initial state s. For all s, $g^{\pi^*}(s) = g^*$.*

To include the constraint and find an optimal policy, let us define the following linear programming algorithm (for all $i \in E$ and $a \in K_i$, $x(i,a)$ are again the unknown)

$$\max_x \sum_{i\in E}\sum_{a\in K_i} w(i,a)x(i,a), \quad \text{subject to}$$
$$\begin{cases} \sum_{a\in K_j} x(a,j) - \sum_{i\in E}\sum_{a\in K_i} p_{ij}(a)x(i,a) = 0, \; j \in E \\ \sum_{i\in E}\sum_{a\in K_i} x(i,a) = 1, \\ \sum_{i\in E}\sum_{a\in K_i} x(i,a)f(i,a) \leq \beta, \\ x(i,a) \geq 0, \quad i \in E, a \in K_i. \end{cases} \quad [16.15]$$

As for the expected discounted reward, from x^*, a solution to the above algorithm, we define the policy $\pi^* = (d)^\infty$ by

$$d(a \mid i) = \mathbf{P}(d(i) = a) = \frac{x^*(i,a)}{\sum_{a'\in A} x^*(i,a')}, \quad [16.16]$$

for all $i \in E$ and $a \in K_i$.

The following theorem states that π^* is the solution to the problem.

THEOREM 16.11 (PUTERMAN 1994).– *The policy π^* is optimal for the problem [16.14]. Moreover, the economic function of the LP for x^* is the optimal value of the expected average reward g^*.*

16.3. Markov decision process with a constraint on the asymptotic availability

16.3.1. *Introduction*

In many industrial applications, engineers are looking to optimize, in parallel, a system's design and its associated maintenance strategy to show compliance with availability requirements, generally defined in sustainment contracts.

For example, when considering a system (e.g. a vessel or a submarine) performing long-duration missions (e.g. patrols), maintenance activities, i.e. repairs, and supply elements have to be considered. The level of availability is an input among constraints on costs and technical feasibility (usually, during patrols, there is no external industrial support. As a result, maintenance tasks are required to be as simple as possible, such as replacing equipment using spares).

In this context, we propose a framework to obtain an optimal maintenance policy such that the asymptotic availability stays above a predetermined value, given a design definition, i.e. given an architecture in terms of serial and parallel redundancies. Then, our method can be applied iteratively as a decision support tool to find the best suitable design and associated logistic support system.

The following section presents this framework based on a particularization of cMDP presented earlier to obtain this optimal maintenance policy. To do so, we will see that the greatest effort will be to find a suitable definition of f (in equation [16.13]) involving the availability.

16.3.2. *Model*

We consider notations introduced in section 16.2.2. As we consider a constraint on the availability, we split the whole state space E into U, the up-states (in which the system is fully operational or may be affected by some acceptable failures), and D, the down-states (in which there are failures implying the unavailability of the system). Let $m = |U|$ be the number of up-states.

In many industrial applications involving long-duration missions, the asymptotic availability is considered and the client often specifies general formulae to be applied in logistic studies, such as $A = \mu/(\lambda + \mu)$, where λ is the failure rate and μ is the repair rate, in general given by $\mu = 1/T$, where T is the mean time to repair.

It is easy to show that this formula can be obtained by a Markov process for a single component if its failure rate and repair rate are constant. In this case, the instantaneous availability converges to a limit A as the times go to infinity (generally for highly reliable equipment, it converges quickly).

We keep the same principle, but extended to complex systems and not limited to elementary maintenance policies. More precisely, we define the asymptotic average availability as follows:

DEFINITION 16.6.– *For any policy $\pi \in \Pi^H$ and initial state $s \in U$, the asymptotic average availability is defined by:*

$$A^\pi(s) = \limsup_{n\to\infty} \frac{1}{n} \sum_{k=0}^{n-1} \mathbf{P}_s(X_n \in U). \qquad [16.17]$$

The idea is to apply the framework of cMDP to find an optimal maintenance strategy π such that the asymptotic average failure rate stays above a predetermined value $\tilde{A}$. That is, we have to solve the following problem:

$$\begin{array}{ll} \max_{\pi\in\Pi^H} g^\pi(s), & \text{subject to} \\ A^\pi(s) \geq \tilde{A}. & \end{array} \qquad [16.18]$$

The link between the generic problem and the problem involving the availability is not obvious. The following theorem is the focus point of this section.

THEOREM 16.12.– *Under Assumptions 16.1, 16.2 and 16.3, if for all* $i \in E$ *and* $a \in K_i$, *the function* f *is defined by:*

$$f(i,a) = \begin{cases} 1 \text{ if } & i \in D \\ 0 \text{ otherwise,} \end{cases}$$

and if $\beta = 1 - \tilde{A}$, *then* π^*, *the solution of problem [16.14] is also the solution of [16.18]. In addition,* $c^\pi(s)$ *is independent of* s, *i.e.* $c^\pi = c^\pi(s)$ *for all* $s \in U$.

As a result, algorithm [16.15] can be used to find an optimal policy, given the constraint on the asymptotic average availability.

PROOF OF THEOREM 16.12.–

With the particular definition of f, equation [16.13] gives

$$c^\pi(s) = \limsup_{n\to\infty} \left[\frac{1}{n} \sum_{k=0}^{n-1} \mathbf{P}_s^\pi (X_k \in D) \right] = 1 - A^\pi(s).$$

As a result, $c^\pi(s) \le \beta$ is equivalent to $A^\pi(s) \ge \tilde{A}$.

Under Assumption 16.3, for a fixed policy π, $X = (X_n)$ is a Markov chain for which the asymptotic average availability is

$$A^\pi(s) = \frac{1}{m} \sum_{i\in U} \beta(i),$$

where $\beta = (\beta_i)$ is the left normalized eigenvector associated with the Markov chain (X_n) transition matrix eigenvalue 1 (see Boussemart and Limnios (2004), Theorem A.3 applied to stochastic matrices). □

REMARK 16.5.– *Assumptions 16.1, 16.2, 16.3 are true, for example, when the Markov decision process has a finite state and action space, and where the underlying Markov chain (for any stationary policy) is irreducible (i.e. there is a path between each couple of states for all possible actions). However for this case, there can be periodic behavior of the availability (unless the Markov chain is ergodic), so the instantaneous availability does not necessarily converges to a limit. For this reason, the average availability is considered as it converges to a value.*

16.3.3. *Algorithm*

Given Theorem 16.12, algorithm [16.15] can be written as: find x^*, the solution of

$$\max_x \sum_{i\in E}\sum_{a\in K_i} w(i,a)x(i,a), \quad \text{subject to}$$
$$\begin{cases} \sum_{a\in K_j} x(a,j) - \sum_{i\in E}\sum_{a\in K_i} p_{ij}(a)x(i,a) = 0, \ j\in E \\ \sum_{i\in E}\sum_{a\in K_i} x(i,a) = 1, \\ \sum_{i\in D}\sum_{a\in K_i} x(i,a)f(i,a) \leq 1-\tilde{A}, \\ x(i,a)\geq 0, \quad i\in E, a\in K_i. \end{cases} \qquad [16.19]$$

Then, the optimal stationary policy $\pi^* = (d)^\infty$ can be easily obtained by equation [16.16]. It can be shown that (see, for example, Altman (1999)) there is one state for which the decision rule is not deterministic (the number of additional constraints correspond to the number of randomized decision rules).

16.3.4. *Application*

In this application, we apply the Lindqvist and Amundrustad model to describe the maintenance of a system performing missions, during which the degradation model is a continuous-time Markov process. This is a model that at first glance has no relation with MDP, but we will transform it into a Markov decision process.

16.3.4.1. *Lindqvist and Amundrustad's maintenance model (Lindqvist and Amundrustad 1998)*

Let us consider a system observed from time 0, with preventive maintenance (PM) possibly performed at discrete times $0, \tau, 2\tau, \ldots$

The sequence of events is as follows: a mission starts, the system changes state according to the underlying Markov process, the mission ends, the controller observes the system and performs preventive maintenance if necessary and then the next mission starts.

Let $\xi = (\xi_t)_{t\in\mathbf{R}}$ be the continuous-time homogeneous Markov process describing the state of the system at time t, with state space E, transition matrix T_t and initial distribution ρ.

We introduce for all $n\in\mathbf{N}$, the state of the system just before the n-th observation (see Figure 16.2):

$$X_n = \xi_{n\tau^-} = \lim_{t\uparrow n\tau} \xi_t.$$

We also define Z_n as a stochastic discrete-time process that represents the result of the maintenance just before the next mission. The effect of the maintenance at time

n is to change the state of the system from X_n to Z_n according to a transition matrix $R = (R_{ij})$, where

$$\mathbf{P}(Z_n = j \mid X_n = i) = R_{ij}, \quad \forall i, j \in E.$$

The matrix R describes the maintenance policy. The initial distribution of $X = (X_n)_{n \in \mathbf{N}}$ is given by

$$\mu(i) = \mathbf{P}(X_0 = i) = \mathbf{P}(\xi_0 = i) = \rho(i), \quad \forall i \in E.$$

Some assumptions have to be made when using Lindqvist's model:

ASSUMPTION 16.4.– *Given X_n, the process Z_n is independent of all the transitions of the system occurring before time $n\tau$.*

ASSUMPTION 16.5.– *Given its initial state Z_n, the Markov process ξ_t on $[n\tau, (n+1)\tau)$ is independent of all the transitions occurring before time $n\tau$.*

PROPOSITION 16.1 (LINDQVIST AND AMUNDRUSTAD 1998).– *Under Assumptions 16.4 and 16.5, $X = (X_n)_{n \in \mathbf{N}}$ is a Markov chain and its transition matrix P is given by*

$$P = RT_\tau. \quad [16.20]$$

In the next section, we show that there is a link between this model and the transition probabilities $\{p_{ij}(k)\}$ of an MDP.

16.3.4.2. *Definition of the Markov decision process*

Our objective is to replace the matrix R of the previous paragraph by the decision rule $d(a \mid i)$, for $i \in E$ and $a \in K_i$.

Let us define for all $i, k \in E$ and $a \in K_i$:

$$\beta_{ik}(a) = \mathbf{P}(Z_n = k \mid Y_n = a, X_n = i),$$

the probability of observing the state k immediately after the decision a was taken as the state i was observed.

PROPOSITION 16.2 (BICKARD ET AL. 2001).– *We have for all $i, j \in E$ and $a \in K_i$*

$$p_{ij}(a) = \sum_{k \in E} T_\tau(k, j)\beta_{ik}(a). \quad [16.21]$$

Generally, $\beta_{ik}(a) = 0$ or 1 (known values). Let us use some examples to illustrate several of the processes involved.

The matrix element $p_{ij}(a)$ is the probability of observing state j at the end of the next mission, given that state i was observed at the end of the last mission and decision a was made. During the mission, the system changes its states according to the degradation process (ξ_t).

If $i = 1$ is the full-operational state, there is no repair required. As a result, $\beta_{1k} = 1$ for $k = 1$, 0 otherwise and $p_{1j}(a) = T_\tau(1, j)$.

If $i > 1$ is a state with failure(s), $p_{ij}(a)$ will depend on decision a. For example, if a corresponds to a perfect repair, $p_{ij}(a) = T_\tau(1, j)$. If a corresponds to doing nothing, then $p_{ij}(a) = T_\tau(i, j)$.

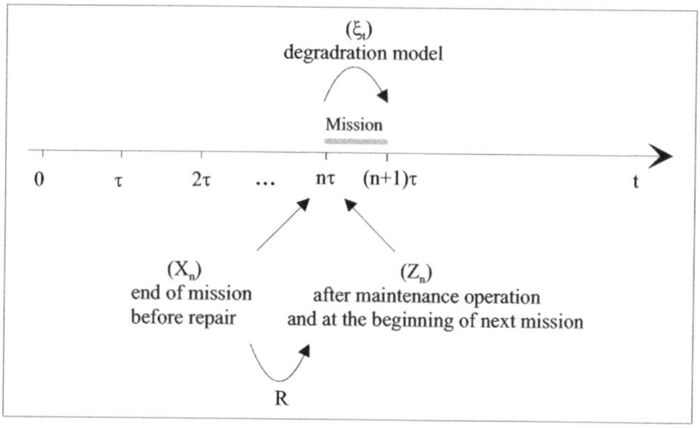

Figure 16.2. *Lindqvist and Amundrustad model. For a color version of this figure, see http://www.iste.co.uk/koroliouk/computational.zip*

16.3.4.3. *Example*

Let us consider a simple serial–parallel system with three different components c_1, c_2 and c_3 described by the reliability block diagram of Figure 16.3, with the component's failure rate, respectively, λ_1, λ_2 and λ_3.

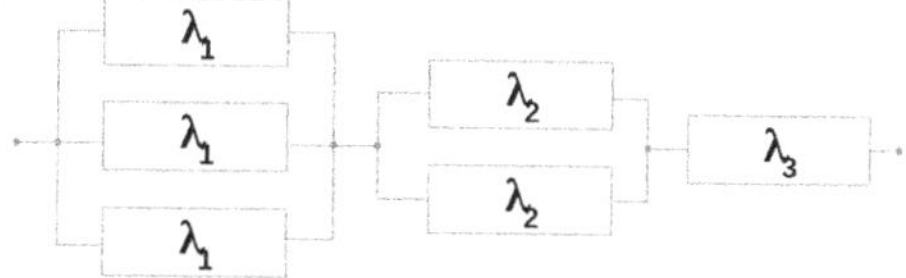

Figure 16.3. *Reliability block diagram*

The system accomplishes constant duration missions, during which there is no possible maintenance. At the end of each mission, we must decide whether to repair all or part of the failed components. In addition, a constraint on the average asymptotic availability, $\tilde{A}$, must be considered when optimizing the infinite horizon average maintenance cost. To address this maintenance problem, we propose the Markov decision process as defined in Figure 16.4, with state-space $E = \{1, 2, \ldots, 7\}$, up-states $U = \{1, 2, \ldots, 6\}$ and down-states $D = \{7\}$.

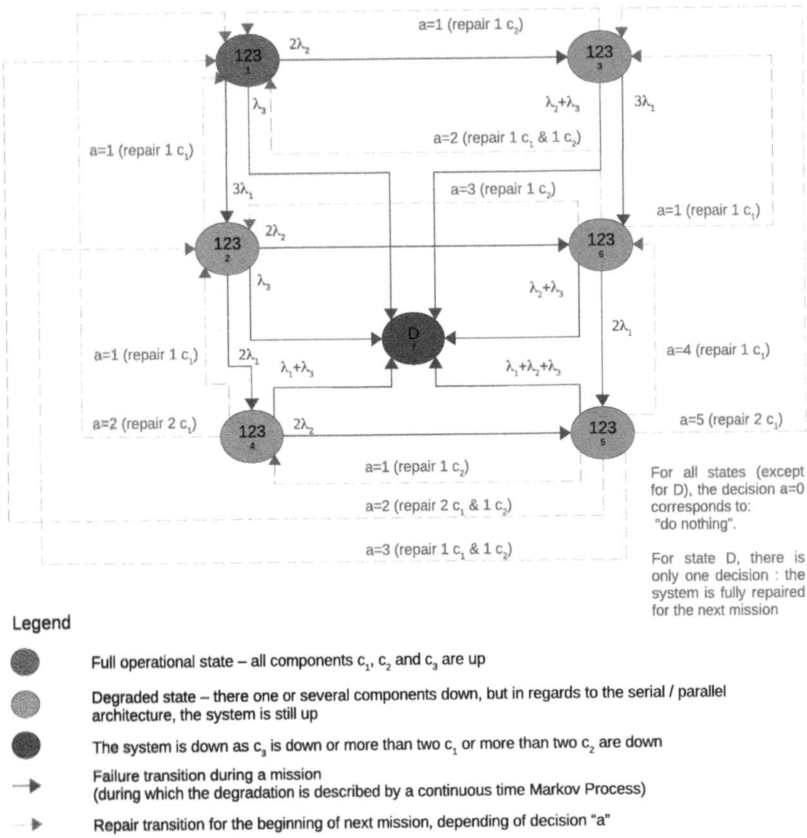

Figure 16.4. *Markov decision process model. For a color version of this figure, see http://www.iste.co.uk/korolioukcomputational.zip*

Let us define following parameters:

– c_1: component 1 replacement cost;

– c_2: component 2 replacement cost;

– c_3: component 3 replacement cost;

– c_4: average repair costs and impact of the unavailability when the system becomes unavailable during the mission;

– τ: duration of a mission.

From Figure 16.4, we define the action-spaces K_i, for $i = 1, \dots, 7$ as shown in Table 16.1.

K_i	Symbol	$a = 0$	$a = 1$	$a = 2$	$a = 3$	$a = 4$	$a = 5$
$i = 1$	123	Do nothing	.	.	.	.	.
$i = 2$	$\bar{1}23$	Do nothing	Repair 1 c_1	.	.	.	.
$i = 3$	$1\bar{2}3$	Do nothing	Repair 1 c_2	.	.	.	.
$i = 4$	$\bar{\bar{1}}23$	Do nothing	Repair 1 c_1	Repair 2 c_1	.	.	.
$i = 5$	$\bar{\bar{1}}\bar{2}3$	Do nothing	Repair 1 c_2	Repair all	Repair 1 c_1 and 1 c_2	Repair 1 c_1	Repair 2 c_1
$i = 6$	$\bar{1}\bar{2}3$	Do nothing	Repair 1 c_1	Repair all	Repair 1 c_2	.	.
$i = 7$	D	Repair all	.	.	.	.	.

Table 16.1. *Action-space K_i, for $i = 1, \dots, 7$*

MDP's rewards $\{W(i, a)\}$ must be defined as maintenance costs. As a result, we define negative rewards as shown in Table 16.2.

$w(i, a)$	$i = 1$	$i = 2$	$i = 3$	$i = 4$	$i = 5$	$i = 6$	$i = 7$
$a = 0$	0	0	0	0	0	0	$-c_4$
$a = 1$	.	$-c_1$	$-c_2$	$-c_1$	$-c_2$	$-c_1$	.
$a = 2$	.	.	.	$-2c_1$	$-2c_1 - c_2$	$-c_1 - c_2$	.
$a = 3$	.	.	.	.	$-c_1 - c_2$	$-c_2$	.
$a = 4$	.	.	.	.	$-c_1$	.	.
$a = 5$	.	.	.	.	$-2c_1$	.	.

Table 16.2. *Rewards $\{W(i, a)\}$*

To define MDP's $\{p_{ij}(a)\}$, equation [16.21] must be applied with $T_\tau = e^{-\tau A}$, where A is the Markov process generator as defined by the red transitions of Figure 16.4, i.e. the following matrix:

$$A = \begin{bmatrix} a_{11} & 3\lambda_1 & 2\lambda_2 & 0 & 0 & 0 & \lambda_3 \\ 0 & a_{22} & 0 & 2\lambda_1 & 0 & 2\lambda_2 & \lambda_3 \\ 0 & 0 & a_{33} & 0 & 0 & 3\lambda_1 & \lambda_3 \\ 0 & 0 & 0 & a_{44} & 2\lambda_2 & 0 & \lambda_1 + \lambda_3 \\ 0 & 0 & 0 & 0 & a_{55} & 0 & \lambda_1 + \lambda_2 + \lambda_3 \\ 0 & 0 & 0 & 0 & 2\lambda_1 & a_{66} & \lambda_2 + \lambda_3 \\ 0 & 0 & 0 & 0 & 0 & 0 & a_{77} \end{bmatrix},$$

where $a_{ii} = -\sum_{j \in E, j \neq i} a_{ij}$, for $i \in E$.

In equation [16.21], the $\{\beta_{ik}(a)\}$ make the link between the Lindqvist and Amundrustad model and the MDP. Following Figure 16.4, $\{\beta_{ik}(a)\}$ are defined in Table 16.3. Let us illustrate this with the following examples:

– At the end of a mission, if state $i = 1$ is observed, as it is the fully operational state, there is only one decision available: $a = 0$ (do nothing). As a result, the system stays in the same state at the beginning of the next mission: $k = 1$. Thus, $\beta_{ik}(a) = 1$ for $k = 1$ and $a = 0$ (for other values, $\beta_{ik}(a) = 0$). State $k = 1$ corresponds to the state at the beginning of the next mission, and consequently we have:

$$p_{1j}(0) = T_{\mathcal{T}}(1j),\ \forall j \in E.$$

– At the end of a mission, if state $i = 2$ is observed, there are two possible actions, $a = 0$ (do nothing) or $a = 1$ (repair failed component c_1). As a result, the state at the beginning of the next mission, k, is respectively 2 or 1. Thus, $\beta_{22}(0) = 1$, $\beta_{21}(1) = 1$ and:

$$p_{2j}(0) = T_{\mathcal{T}}(2j) \quad \text{and} \quad p_{2j}(1) = T_{\mathcal{T}}(1j),\ \forall j \in E.$$

– For $i = 4$, there are three possible actions, $a = 0$ (do nothing), $a = 1$ (repair one of the two failed components c_1) and $a = 2$ (repair the two failed components c_1). Thus, $\beta_{44}(0) = 1$, $\beta_{42}(1) = 1$, $\beta_{41}(2) = 1$ and:

$$p_{4j}(0) = T_{\mathcal{T}}(4j), \quad p_{4j}(1) = T_{\mathcal{T}}(2j) \quad \text{and} \quad p_{4j}(2) = T_{\mathcal{T}}(1j),\ \forall j \in E.$$

$\beta_{ik}(a)$	$i=1$	$i=2$	$i=3$	$i=4$	$i=5$	$i=6$	$i=7$
$a=0$	$k=1$	$k=2$	$k=3$	$k=4$	$k=5$	$k=6$	$k=1$
$a=1$	.	$k=1$	$k=1$	$k=2$	$k=4$	$k=3$	.
$a=2$	.	.	.	$k=1$	$k=1$	$k=1$	.
$a=3$	.	.	.	.	$k=2$	$k=2$	.
$a=4$	.	.	.	.	$k=6$	.	.
$a=5$	.	.	.	.	$k=3$	.	.

Table 16.3. $\beta_{ik}(a)$ *(values of* k *for which* $\beta_{ik}(a) = 1$, *for other values it is equal to 0)*

Now, let us consider the constraint on the availability [16.18]. It is clear that for some values of $\tilde{A}$, the problem becomes inconsistent. To avoid this, Theorem 16.13 gives bounds for admissible values.

THEOREM 16.13.– *For any policy* $\pi \in \Pi^S$, *if* $i = 1$ *denotes the fully operational state (the most reliable) and* $\mu(1) = 1$, *then the asymptotic average availability satisfies:*

$$1 - \frac{1}{[(I - Q_{\tau})^{-1}1_m]_1} \leq A^{\pi}(s) \leq 1 - \sum_{j\in D} T_{\tau}(1, j),$$

where Q_τ is the restriction of matrix T_τ on $U \times U$, I is the identity matrix, 1_m is the column vector of size $m = \mid U \mid$ with all components equal to 1 and $[\gamma]_1$ is the first component of the column vector γ.

PROOF OF THEOREM 16.13.–

Let $\pi \in \Pi^S$ be a policy such that the system is as good as new at the beginning of each mission (i.e. the availability at the end of each mission is constant and maximal). We have:

$$\begin{aligned}\mathbf{P}(X_{n+1} \in D) &= \sum_{i \in E, j \in D} \mathbf{P}(X_{n+1} = j \mid Z_n = i)\mathbf{P}(Z_n = i)\\ &= \sum_{j \in D} T_\tau(1, j) \quad (\text{as } \mathbf{P}(Z_n = 1) = 1),\end{aligned}$$

thus the right-hand side of the inequality, considering the expression of the asymptotic average availability.

The left-hand side can be obtained by applying the ergodicity theorem and applying hitting time formulas, as the Markov chain is irreducible. □

Considering the numerical data of Table 16.4, we have, for any stationary policy π:

$$0.8474 \leq A^\pi(s) \leq 0.9833. \quad [16.22]$$

λ_1	λ_2	λ_3	c_1	c_2	c_3	c_4	τ
$[h^{-1}]$	$[h^{-1}]$	$[h^{-1}]$	[$]	[$]	[$]	[$]	[h]
1.10^{-2}	5.10^{-3}	1.10^{-4}	1000.00	2000.00	3000.00	4000.00	24

Table 16.4. *Numerical data for the serial–parallel maintenance problem*

We propose to compare the results with several constraints on the asymptotic average availability.

– Constraint: $A^\pi(s) \geq 0$ (in fact, there is no constraint as the availability is always positive). The application of algorithm [16.19] leads to the optimal policy presented in Table 16.5. Without constraint, we observe in Figure 16.5 that the average availability converges to the minimum admissible (see inequality [16.22]).

– Constraint: $A^\pi(s) \geq 0.9833$. This is the upper admissible value (but it cannot be reached numerically, otherwise the linear programming algorithm becomes inconsistent). For this case, we observe the coherent optimal policy in Table 16.6:

when the system ends a mission with a number of failed components, all of them have to be repaired before the next mission. It is also coherent to see that there is one randomized decision (which tends asymptotically to 1 when the constraint tends to the upper admissible bound). In addition, Figure 16.6 shows that the asymptotic value is reached quickly. Finally, the maintenance cost is significantly higher than the previous case.

– Constraint: $A^{\pi}(s) \geq 0.9$ (see Figure 16.7 and Table 16.7). This case shows the advantage of using the model to optimize costs if there is a requirement about the availability. Indeed, it shows that it is not necessary to recover to the fully operational states all the time. Applying the optimal policy, it reduces costs by 40% compared to the maximum level of repair strategy.

	State			Decision	Probability
i	**Symbol**	**Description**	a	**Description**	$d(a \mid i)$
1	123	Fully operational state	0	Do nothing	1.000
2	123	1 c_1 down	0	No repair	1.000
3	123	1 c_2 down	0	No repair	1.000
4	123	2 c_1 down	0	No repair	1.000
5	123	2 c_1 down and 1 c_2 down	0	No repair	1.000
6	123	1 c_1 down and 1 c_2 down	0	No repair	1.000
7	D	3 c_1 down or 2 c_2 down or 1 c_3 down	0	Repair as good as new	1.000

Table 16.5. *Optimal policy for* $A^{\pi}(s) \geq 0$

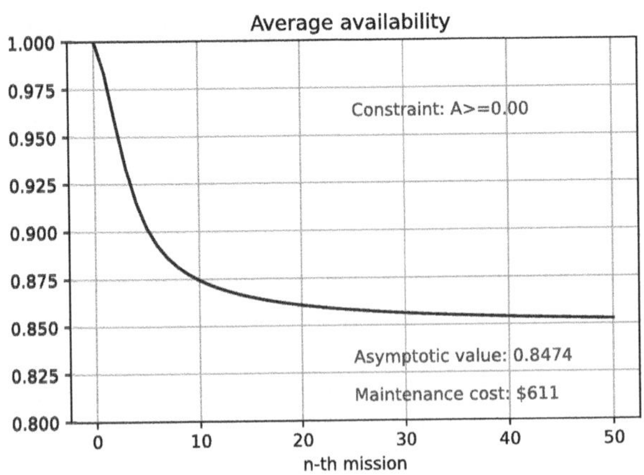

Figure 16.5. *Average availability for* π *optimal – no constraint. For a color version of this figure, see http://www.iste.co.uk/korolioukcomputational.zip*

		State		Decision	Probability
i	Symbol	Description	a	Description	$d(a \mid i)$
1	123	Fully operational state	0	Do nothing	1.000
2	$\bar{1}23$	1 c_1 down	1	Repair all	1.000
3	$1\bar{2}3$	1 c_2 down	1	Repair all	1.000
4	$\bar{\bar{1}}23$	2 c_1 down	2	Repair all	1.000
5	$\bar{\bar{1}}\bar{2}3$	2 c_1 down and 1 c_2 down	2	Repair all	0.998
5	$\bar{\bar{1}}\bar{2}3$	2 c_1 down and 1 c_2 down	5	Repair 2 c_1	0.002
6	$\bar{1}\bar{2}3$	1 c_1 down and 1 c_2 down	2	Repair all	1.000
7	D	3 c_1 down or 2 c_2 down or 1 c_3 down	0	Repair as good as new	1.000

Table 16.6. *Optimal policy for* $A^{\pi}(s) \geq 0.9833$

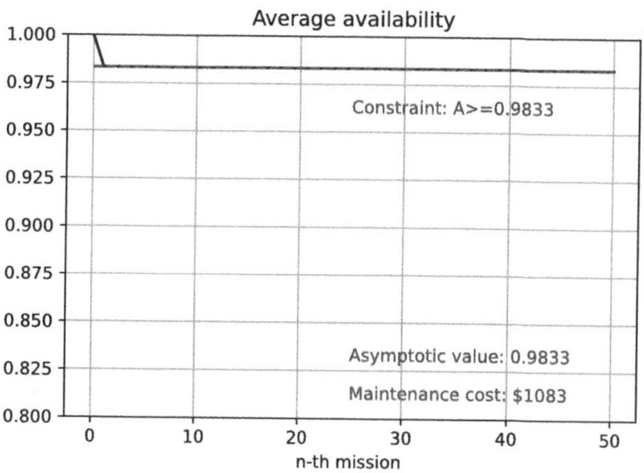

Figure 16.6. *Average availability for* π *optimal – constraint:* $A^{\pi}(s) \geq 0.9833$. *For a color version of this figure, see http://www.iste.co.uk/korolioukcomputational.zip*

		State		Decision	Probability
i	Symbol	Description	a	Description	$d(a \mid i)$
1	123	Fully operational state	0	Do nothing	1.000
2	$\bar{1}23$	1 c_1 down	0	No repair	1.000
3	$1\bar{2}3$	1 c_2 down	0	No repair	1.000
4	$\bar{\bar{1}}23$	2 c_1 down	1	Repair 1 c_1	1.000
5	$\bar{\bar{1}}\bar{2}3$	2 c_1 down and 1 c_2 down	0	No repair	1.000
6	$\bar{1}\bar{2}3$	1 c_1 down and 1 c_2 down	0	No repair	0.127
6	$\bar{1}\bar{2}3$	1 c_1 down and 1 c_2 down	0	Repair 1 c_1	0.873
7	D	3 c_1 down or 2 c_2 down or 1 c_3 down	0	Repair as good as new	1.000

Table 16.7. *Optimal policy for* $A^{\pi}(s) \geq 0.9$

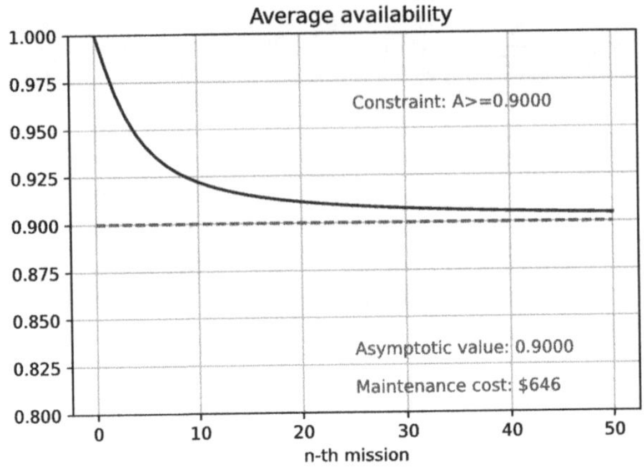

Figure 16.7. *Average availability for π optimal – constraint: $A^{\pi}(s) \geq 0.9$. For a color version of this figure, see http://www.iste.co.uk/koroliouk/computational.zip*

16.4. Markov decision process with a constraint on the asymptotic failure rate

16.4.1. *Introduction*

This section deals with the maintenance of full authority digital engine control (FADEC) systems for which the SAE International Group introduced a guideline for considering dispatch time for repairs in accordance with airworthiness rules (see SAE (1996)).

In this guideline, three main categories of dispatch time are considered: no dispach (ND), short-time dispatch (ST) and long-time dispatch (LT) (see, for example, Lu et al. (2018) for a recent application of this guideline). To show compliance with the security requirement, the model must demonstrate that the instantaneous failure rate of each multi-engined aircraft is less than 10^{-9} h^{-1}. But the FAA has also agreed that the use of a fleet-wide average (of many aircrafts) failure rate is acceptable when showing compliance with this requirement. Moreover, using a fleet-wide failure rate for an engine of 10^{-5} h^{-1} conforms to the FAA requirement for multi-engined aircraft. Finally, the FAA agrees with using a time-weighted average failure rate, instead of the fleet average failure rate. Indeed, a given engine may have more and less faults than expected, but for use in a large fleet of engines, the fleet average time of a fault will have the expected value.

In this context, the guideline SAE (1996) presents a methodology to compute this average failure rate, which can be considered as the asymptotic failure rate (as it appears in computation methods in this guideline).

In this section, the same principle of dispatch time for repairs is considered, but extended to more complex policies (instead of considering three classes). As a result, the goal is to find a policy for which the asymptotic failure rate is less than or equal to a prescribed value.

The idea is to use the cMDP framework presented in the previous sections. However, it can be easily shown that the asymptotic failure rate cannot be written as a function of state action frequencies. Consequently, the theory was adapted to consider this new constraint (see Bickard et al. (2001) and Boussemart and Limnios (2002, 2004)). The objective of the following section is to present these main adaptations, the innovation of which is based on the concept of quasi-stationary distributions. Similar to the standard approaches presented in the previous section, our main contribution is to show that an optimal policy can be found with a simple algorithm.

16.4.2. *Model*

16.4.2.1. *Problem definition*

As we are dealing with a maintenance problem, we still consider notations introduced in section 16.3.2, but now D is an absorbing set of states, i.e. we suppose that the following assumption is true:

ASSUMPTION 16.6.– *For all $\pi \in \Pi^H$, D is an absorbing set of states.*

As a result, for any policy $\pi \in \Pi^H$, the Markov process is not irreducible. However, we suppose that the restriction on $U \times U$ has the following irreducibility and aperiodicity properties:

ASSUMPTION 16.7.– *For any Markovian decision rule $d \in D_M$, if we define the matrix Q_d by:*

$$(Q_d)(i,j) = \sum_{a \in K_i} p_{ij}(a) d(i \mid a), \quad i,j \in U,$$

then, Q_d is primitive, i.e. an irreducible and aperiodic matrix, or precisely, such that

$$\exists k \in \mathbf{N}, \quad Q_d^k > 0.$$

Finally, for any Markovian decision rule $d \in D_M$, we define the column vector w_d as the restriction on $U \times U$ of W_d by:

$$w_d(i) = \sum_{a \in K_i} w(i,a) d(i \mid a), \quad i \in U.$$

In this section, we are concerned with performance measurements over a long time period when the system is in the up-states. The infinite horizon average expected reward presented in the previous section (or see, for example, Puterman (1994) and Altman (1999)) would not be an interesting criterion since from Assumption 16.6, it would only depend on the reward associated with the final (down) states. Since we consider highly reliable systems, we can assume that the up-time is large enough so that the system never enters the down-states. Thus, the economic performance that we propose is the infinite horizon expected reward, given that for each instant, the system is in an up-state, hence the following definition:

DEFINITION 16.7.– *For any initial distribution* μ *and* $\pi \in \Pi^H$, *we define*

1) the expected total reward on $[0, n]$ *before the failure by*

$$v_\mu^\pi(n) = \sum_{k=0}^{n} \mathbf{E}_\mu^\pi \left(w(X_k, Y_k) \mid T > k\right),$$

where $T = \inf\{n \geq 0 : \ X_n \in D\}$ *denotes the time of the first passage in the set* D, *with the convention* $\inf \emptyset = +\infty$;

2) the infinite horizon expected average reward before the failure by

$$g_\mu^\pi = \liminf_{n \to \infty} \frac{1}{n+1} v_\mu^\pi(n).$$

REMARK 16.6.– *From Assumption 16.6, we have* $\{T > k\} = \{X_k \in U\}$, $k \in \mathbf{N}$.

Now, let us define the security criterion.

DEFINITION 16.8.– *For any* $\pi \in \Pi^H$ *and any initial distribution* μ, *we define the (discrete) failure rate at time* $n \geq 0$ *by*

$$\lambda_\mu^\pi(n) = \mathbf{P}_\mu^\pi(T = n \mid T \geq n).$$

The security constraint is on the asymptotic upper values of the failure rate. If λ_c is the upper bound we want to impose, the objective is to solve the following problem:

$$g_\mu^* = \max_{\pi \in \Pi^H} g_\mu^\pi, \quad \text{subject to} \quad \limsup_{n \to \infty} \lambda_\mu^\pi(n) \leq \lambda_c. \qquad [16.23]$$

16.4.2.2. *Markov policies*

Before giving explicit forms for the economic criterion and the security criterion, let us recall that we can restrict the problem to searching only Markovian policies as stated by the following theorem:

THEOREM 16.14 (BOUSSEMART AND LIMNIOS 2002).– *For any* $\pi \in \Pi^H$ *and any initial distribution* μ, *there is a policy* $\pi' \in \Pi^M$ *such that*

$$g_\mu^\pi = g_\mu^{\pi'}.$$

16.4.2.3. *Explicit formulations for economic and security criteria*

THEOREM 16.15 (BOUSSEMART AND LIMNIOS 2002, PROPOSITIONS 16.2 AND 16.3).– *For any* $\pi = (d_1, d_2, \ldots) \in \Pi^M$ *and any initial distribution* μ*, we have:*

1)
$$v_\mu^\pi(n) = \sum_{k=0}^{n} \frac{\mu Q_{d_0} \cdots Q_{d_k} w_{d_{k+1}}}{\mu Q_{d_0} \cdots Q_{d_k} 1_m}, \quad n \in \mathbf{N}, \qquad [16.24]$$

where 1_m *is the column vector of size* m *with all entries equal to* 1*, where* $Q_{d_0} = I$ *(identity matrix), and* Q_{d_i} *is the restriction of the transition matrix* P_{d_i} *on* $U \times U$*;*

2)
$$\lambda_\mu^\pi(n) = 1 - \frac{\mu Q_{d_0} \cdots Q_{d_n} 1_m}{\mu Q_{d_0} \cdots Q_{d_{n-1}} 1_m}, \quad n \in \mathbf{N}. \qquad [16.25]$$

16.4.2.4. *Searching a stationary policy*

In this section, we provide a method to find an optimal policy for problem [16.23] when considering an equality constraint on the asymptotic failure rate and restricting to stationary policies. Then, we give a sufficient condition for this policy to be optimal in the class Π^H.

We assume that Assumptions 16.6 and 16.7 hold. Moreover, we set an initial distribution μ (obviously $\mu(D) = 0$).

THEOREM 16.16 (BICKARD ET AL. 2001).– *Under Assumption 16.7, for any* $\pi = (d)^\infty \in \Pi^S$*, the asymptotic failure rate is given by*

$$\tilde{\lambda}^\pi = \lim_{n\to\infty} \lambda_\mu^\pi(n) = 1 - \rho(Q_d), \qquad [16.26]$$

where $\rho(Q_d)$ *denotes the spectral radius of matrix* Q_d*.*

We will not directly solve problem [16.23]. Indeed, let us consider the following problem

$$\max_{\pi\in\Pi^S} g_\mu^\pi, \quad \text{subject to} \quad \lim_{n\to\infty} \lambda_\mu^\pi(n) = \lambda_c. \qquad [16.27]$$

First, we show that an optimal stationary policy for problem [16.27] exists and can be obtained by the following linear programing scheme (letting $r = 1 - \lambda_c$)

$$\max_x g(x) = \max_x \sum_{i\in U} \sum_{k\in C_i} x_{ik}\, w(i,k), \quad \text{subject to}$$
$$\begin{cases} \text{(i)} \;\; r\sum_{k\in K_j} x_{jk} - \sum_{i\in U}\sum_{k\in K_i} p_{ij}(k)\, x_{ik} = 0, \; j \in U \\ \text{(ii)} \; \sum_{i\in U}\sum_{k\in K_i} x_{ik} = 1 \\ \text{(iii)} \; x_{ik} \geq 0, \; i \in U, \; k \in K_i. \end{cases} \qquad [16.28]$$

THEOREM 16.17 (BICKARD ET AL. 2001).– *Under Assumption 16.7, the following two points hold time:*

1) Let $\pi = (d)^\infty \in \Pi^S$ *be such that* $\rho(Q_d) = 1 - r$ *and let* y *be the left Perron–Frobenius eigenvector of* Q_d *(i.e. satisfying* $yQ_d = ry$*) such that* $y1_m = 1$*. Then*

$$\forall i \in U, \quad \forall a \in K_i, \quad x_{ia} = d(a \mid i) y_i$$

is an admissible solution to problem [16.28].

2) Let x *be a solution to problem [16.28]. Then, the policy* $\pi = (d)^\infty$ *defined by*

$$\forall i \in U, \quad \forall a \in K_i, \quad d(a \mid i) = \frac{x_{ia}}{\sum_{k \in K_i} x_{ik}}$$

is a solution to problem [16.27] and $g(x) = \max_{\pi \in \Pi^S} g_\mu^\pi$*.*

Let us give some existence properties of problem [16.27].

PROPOSITION 16.3 (BICKARD ET AL. 2001).– *Let* $\mathcal{P}$ *be the set of feasible solutions* $x = (x_{jk})$ *of [16.28]. If* $\mathcal{P} \neq \emptyset$ *(see the remark below), then:*

1) there is an optimal stationary policy for problem [16.27];

2) there is an optimal stationary policy such that the decision rule is non-deterministic in at most one state.

REMARK 16.7.– *For some asymptotic failure rates imposed on the system, the constraints of the linear programming may become inconsistent (i.e.* $\mathcal{P} = \emptyset$*). In the following section, we will show that the possible values of* λ_c *(for which* $\mathcal{P} \neq \emptyset$*) belong to an interval depending on the transition probabilities* $\{p_{ij}(k)\}$*.*

16.4.2.5. *Optimality in the history-dependent set of policies* Π^H

In the following, we will give a sufficient condition (i.e. Theorem 16.18) for the policy $\pi = (d)^\infty$ obtained by Theorem 16.17(2) to be optimal for problem [16.23].

DEFINITION 16.9.– *A pair* $(\phi, h) \in \mathbf{R} \times \mathbf{R}^m$ *is called a solution to the average reward optimality equation (AROE) if*

$$h \geq 1_m \quad \text{and} \quad \max_{d \in D_M} \{w_d - \phi 1_m + (Q_d - rI)h)\} \leq 0, \qquad [16.29]$$

where $r = 1 - \lambda_c$*.*

Let us now recall the main theorem of the paper by Bickard et al. (2001).

THEOREM 16.18 (BICKARD ET AL. 2001).– *Let g be the optimal economic function of problem [16.28] and let $\pi = (d)^\infty$ be the corresponding policy obtained by Theorem 16.17(2). Let h be obtained by solving the dual linear programming problem (h corresponds to the m first Lagrange multipliers – the last multiplier corresponds to the last constraint (ii) of [16.28] and equals the economic function). If $h \geq 1_m$, then π is an optimal policy for problem [16.23].*

REMARK 16.8.– *The condition $h \geq 1_m$ is not a necessary condition for π to be optimal.*

16.4.2.6. *Extension to general reducible, periodic set of up-states*

Some readers may be interested in a more general case, as suggested by Assumption 16.7. Indeed, in discrete-time Markov decision process models that represent real applications, some periodic behavior can occur. This was precisely the case when we considered a real FADEC system subjected to periodic maintenance. In this context, our article Boussemart and Limnios (2004) presents this required extension.

16.4.3. *Algorithm*

To solve the problem, three steps are needed:

1) Select a value for $\lambda_c = 1 - r$, in the following range of admissible values (see Bickard et al. (2001)):

$$\min_{i\in U}\min_{k\in K_i}\sum_{j\in U} p_{ij}(k) \leq r \leq \max_{i\in U}\max_{k\in K_i}\sum_{j\in U} p_{ij}(k).$$

2) Solve problem [16.28] to obtain x, and apply Theorem 16.17(2) to obtain a stationary policy $\pi = (d)^\infty$. That is, optimal if we restrict ourselves to finding such policies.

3) Obtain h with the dual formulation of [16.28], i.e. find its Lagrange multipliers; these numbers are often given when applying any simplex algorithm.

4) If $h \geq 1_m$, it implies that the obtained policy π is optimal for problem [16.23], i.e. if the most general class of policies and the constraints on the failure rate are considered. Otherwise, we cannot conclude.

16.4.4. *Application*

We consider a FADEC, which is used to control aircraft's engines with a distributed architecture as described in Boussemart and Limnios (2004). The system consists of four processing units interconnected by high-speed point-to-point links. At least two processing units are necessary to perform the full control function. The continuous-time failure model $(\xi_t)_{t\in\mathbf{R}}$ is presented in Figure 16.8. Each state of the graph is a pair (n, m), where n is the number of links and m is the number of working processing units. A state corresponds to a set of equivalent configurations using the symmetry. Each transition of the graph represents a failure, either a processing unit (failure rate λ_S) or a link (failure rate λ_L). Some configurations, as a chain of four units, are forbidden (in this case, a failure of a link inside the chain would induce two separate sub-graphs, which is not allowed; as a result, it is considered as a loss of thrust control (LOTC) state).

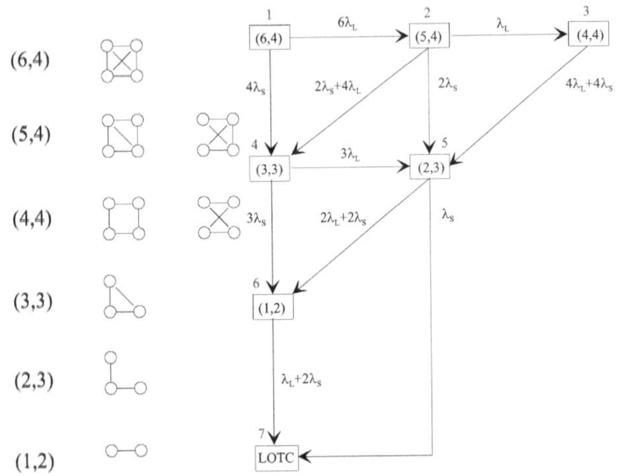

Figure 16.8. *Distributed architecture. For a color version of this figure, see http://www.iste.co.uk/koroliouk/computational.zip*

The system accomplishes constant duration missions, during which there is no possible maintenance. At the end of each mission, with regard to all possible states, we have to decide whether or not to repair the whole network. In addition, a constraint on the asymptotic failure rate λ_c must be considered when optimizing the infinite horizon excepted average maintenance cost. To address this maintenance problem, using the failure model presented in Figure 16.8, we propose to build the Markov decision process using the model in section 16.4.2 combined with the Lindqvist–Amundrustad model.

The state-space is $E = \{1, 2, \ldots, 7\}$, with up-states: $U = \{1, 2, \ldots, 6\}$ and down-state or LOTC state: $D = \{7\}$. The action space K_i, for $i = 1, \ldots, 6$ is defined in Table 16.8.

For states $i \in 2, \ldots 5$, the repair action is to bring the system to the full-up state ($i = 1$) for the beginning of the next mission. If we denote c_L and c_S as, respectively, the costs associated with a link repair and a processing unit repair, MDP's rewards $\{W(i, a)\}$ are defined in Table 16.9 (costs should be transformed into negative rewards).

K_i	State i	$a = 0$	$a = 1$	Comments
$i = 1$	$(6, 4)$	Do nothing	·	This is the full-up state (only one choice: do nothing)
$i = 2$	$(5, 4)$	Do nothing	Repair	Degraded state
$i = 3$	$(4, 4)$	Do nothing	Repair	Degraded state
$i = 4$	$(3, 3)$	Do nothing	Repair	Degraded state
$i = 5$	$(2, 3)$	Do nothing	Repair	Degraded state
$i = 6$	$(1, 2)$	Repair	·	This is a critical state, a repair is required (only one choice)

Table 16.8. *Action-space K_i, for $i = 1, \ldots, 6$*

$w(i, a)$	$i = 1$	$i = 2$	$i = 3$	$i = 4$	$i = 5$	$i = 6$
$a = 0$	0	0	0	0	0	$-2c_S - 5c_L$
$a = 1$	·	$-c_L$	$-2c_L$	$-c_S - 3c_L$	$-c_S - 4c_L$	·

Table 16.9. *Rewards $\{W(i, a)\}$*

To define the MDP's transition probabilities $\{p_{ij}(a)\}$, equation [16.21] must be applied with $T_\tau = e^{-\tau A}$, where A is the Markov process generator as defined by the transitions of Figure 16.8, i.e. the following matrix:

$$A = \begin{bmatrix} a_{11} & 6\lambda_L & 0 & 6\lambda_S & 0 & 0 & 0 \\ 0 & a_{22} & \lambda_L & 2\lambda_S + 4\lambda_L & 2\lambda_S & 0 & 0 \\ 0 & 0 & a_{33} & 0 & 4\lambda_S + 4\lambda_L & 0 & 0 \\ 0 & 0 & 0 & a_{44} & 3\lambda_L & 3\lambda_S & 0 \\ 0 & 0 & 0 & 0 & a_{55} & 2\lambda_S + 2\lambda_L & \lambda_S \\ 0 & 0 & 0 & 0 & 0 & a_{66} & 2\lambda_S + \lambda_L \\ 0 & 0 & 0 & 0 & 0 & 0 & a_{77} \end{bmatrix},$$

where $a_{ii} = -\sum_{j \in E, j \neq i} a_{ij}$, for $i \in E$.

In equation [16.21], the $\{\beta_{ik}(a)\}$ make the link between the Lindqvist and Amundrustad model and the MDP as shown in Table 16.10.

$\beta_{ik}(a)$	$i = 1$	$i = 2$	$i = 3$	$i = 4$	$i = 5$	$i = 6$	$i = 7$	Comments
$a = 0$	$k = 1$	$k = 2$	$k = 3$	$k = 4$	$k = 5$	$k = 1$	$k = 7$	Do nothing ($i \neq 6$) or repair ($i = 6$)
$a = 1$	·	$k = 1$	$k = 1$	$k = 1$	$k = 1$	·	·	Repair (bring the system in state 1)

Table 16.10. *$\beta_{ik}(a)$ (values of k for which $\beta_{ik}(a) = 1$, for other values it is equal to 0)*

The first step of the algorithm presented in section 16.4.3 is to define the constraint λ_c.

Considering the numerical data of Table 16.11, the admissible values for the constraint on the asymptotic failure rate are strictly included in the following interval:

$$0.001877 < \lambda_c < 0.077266. \qquad [16.30]$$

λ_S	λ_L	c_S	c_L	τ
$[h^{-1}]$	$[h^{-1}]$	[$]	[$]	[h]
3.10^{-3}	1.10^{-3}	5000.00	1000.00	24

Table 16.11. *Numerical data for the distributed architecture maintenance problem*

	State		Decision		Probability
i	$i = (m, n)$	Description	a	Description	$d(a \mid i)$
1	$(6, 4)$	Fully operational state	0	Do nothing	1.000
2	$(5, 4)$	Degraded state	1	Repair	1.000
3	$(4, 4)$	Degraded state	1	Repair	1.000
4	$(3, 3)$	Degraded state	1	Repair	1.000
5	$(2, 3)$	Degraded state	1	Repair	1.000
6	$(1, 2)$	Critical state	0	Repair	1.000

Table 16.12. *Optimal policy for* $\lambda_c \leq 0.001877$

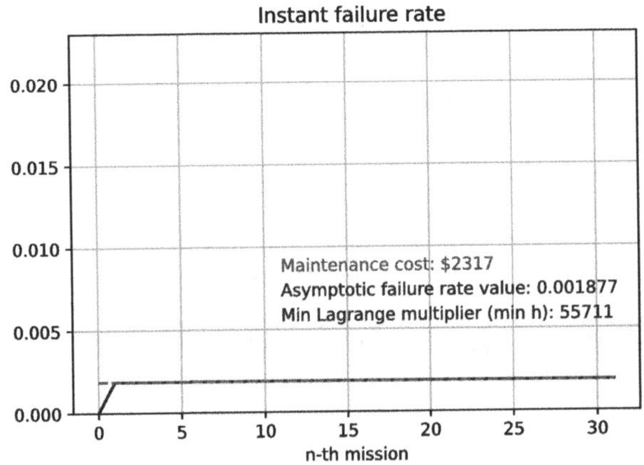

Figure 16.9. *Instant failure rate for* π *optimal – constraint:* $\lambda_c \leq 0.001877$. *For a color version of this figure, see http://www.iste.co.uk/koroliouk/computational.zip*

We propose to compare the results with two different constraints on the asymptotic failure rate.

– Constraint: $\lambda_c \leq 0.001877$ (which is close to the minimum value). As presented in section 16.4.3, the next step is to find x, the solution of problem [16.28] (and thus the solution for problem [16.27]). This can be done with a linear programming solver. Then, from Theorem 16.17(2), the corresponding stationary policy $\pi = (d)^\infty$ is computed as shown in Table 16.12. In Figure 16.9, we can observe that the instant failure rate converges very quickly to λ_c. In addition, the Lagrange multipliers (components of vector h) are greater than 1 (the minimum value is 55711): this means that the policy is also optimal in Π^H, as recalled in section 16.4.3, step (4).

– Constraint: $\lambda_c \leq 0.01$. As before, let us follow the steps of the algorithm presented in section 16.4.3:

Step (2). The optimal stationary policy $\pi = (d)^\infty$ is computed with a linear programming solver (see Table 16.13 and Figure 16.10).

Step (3). Compute the Lagrange multipliers (usually obtained with the linear programming solver or obtained with the dual formulation).

Step (4). As shown in Figure 16.10, all components of h are greater than 1 (the minimum value is 6937): this means that the policy is also optimal in Π^H.

For the second case, as $\lambda_c \leq 0.01$ is a less restrictive constraint than the first case, the policy shows that there is no need to repair at each instant; as a result, the maintenance cost is also reduced.

State			Decision		Probability
i	$i = (m, n)$	Description	a	Description	$d(a \mid i)$
1	$(6, 4)$	Fully operational state	0	Do nothing	1.000
2	$(5, 4)$	Degraded state	1	Repair	1.000
3	$(4, 4)$	Degraded state	1	Repair	1.000
4	$(3, 3)$	Degraded state	0	Do nothing	1.000
5	$(2, 3)$	Degraded state	1	Repair	0.813
5	$(2, 3)$	Degraded state	0	No repair	0.187
6	$(1, 2)$	Critical state	0	Repair	1.000

Table 16.13. *Optimal policy for* $\lambda_c \leq 0.01$

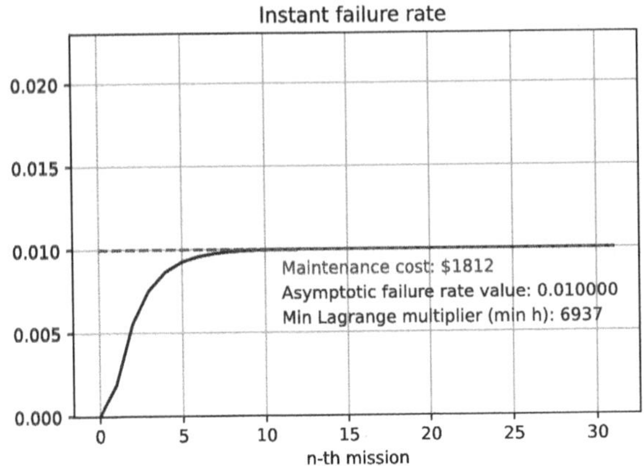

Figure 16.10. *Instant failure rate for π optimal – constraint: $\lambda_c \leq 0.01$. For a color version of this figure, see http://www.iste.co.uk/koroliouk/computational.zip*

16.5. Conclusion

In this chapter, foundations of Markov decision processes were presented, including classical algorithms to find optimal policies for two kinds of economic functions: first, the infinite horizon excepted discounted reward and, second, the infinite horizon expected average reward. Constraints on state–action frequencies were also presented and conditions were given to stay optimal in the most general class of policies, as it is not obvious in general. This framework was first applied to deal with a maintenance problem subject to an asymptotic average availability constraint, as this is the case for many industrial applications, especially for integrated logistics support (ILS) activities. Then, to consider safety constraints, in particular in the aeronautical industry, specifically for FADEC systems, the concept of asymptotic failure rate was introduced. As this quantity cannot be written as state–action frequencies, the authors proposed a new framework that was recalled in this chapter, with an application to a distributed architecture.

As presented in this chapter, the linear programming algorithm (the simplex) was considered to find optimal policies. Today, it is worth noting that efficient libraries can be found to deal with Markov decision processes, in particular reinforcement learning methods, which are able to tackle large models when combined with neural networks. In the literature, some reinforcement learning algorithms with constraints have been developed, for example, in robotics, to reach a target while avoiding obstacles (see, for example, Ge et al. (2019)). From a research perspective, it may be interesting to continue developing these kinds of methods in combination with deep learning for considering large models with general constraints while considering theoretical aspects, namely to remain in the largest class of policies.

16.6. References

Altman, E. (1992). Denumerable constrainded Markov decision problems and finite approximations. Research report, INRIA, 1568.

Altman, E. (1999). *Constrained Markov Decision Processes*. Chapman & Hall/CRC, Boca Raton, FL.

Altman, E. and Shwartz, A. (1991). Markov decision problems and state-action frequencies. *SIAM Journal on Control and Optimization*, 29(4), 786–809.

Arapostathis, A., Borkar, V., Ferandez-Gaucherand, E., Ghosh, M.K., Marcus, S.I. (1993). Discrete time controlled Markov processes with average cost criterion: A survey. *SIAM Journal on Control and Optimization*, 31(2), 282–344.

Barbu, V., Boussemart, M., Limnios, N. (2004). Discrete-time semi-Markov models for reliability and survival analysis. *Communications in Statistics – Theory and Methods*, 33(11), 2833–2868.

Bellman, R. (1957). A Markovian decision process. *Journal of Mathematics and Mechanics*, 6, 679–685.

Benveniste, A., Metivier, M., Priouret, P. (1987). *Algorithmes adaptatifs et approximations stochastiques*. Masson, Paris.

Bertsekas, D.P. (1995). *Dynaminc Programming and Optimal Control*, vol. 2. MIT Press, Cambridge, MA.

Bertsekas, D.P. and Tsitsiklis, J.N. (1996). *Neuro-Dynamic Programming*. MIT Press, Cambridge, MA.

Bertsekas, D.P., Abounadi, J., Borkar, V.S. (1996). Learning algorithms for Markov decision processes with average cost. Technical report LIDS-P-2434, MIT Laboratory for Information and Decision Systems, Cambridge, MA.

Beutler, F.J. and Ross, K.W. (1985). Optimal policies for controlled Markov chains with a constraint. *Journal of Mathematical Analysis and Applications*, 112, 236–252.

Bickard, T., Boussemart, M., Limnios, N. (2001). Markov decision processes with a constraint on the asymptotic failure rate. *Methodology and Computing in Applied Probability*, 3(2), 119–214.

Blackwell, D. (1962). Discrete dynamic programming. *Annals of Mathematical Statistics*, 33, 719–726.

Borkar, V.S. (1998). Asynchronous stochastic approximations. *SIAM Journal of Control and Optimization*, 36(3), 840–851.

Boussemart, M. and Limnios, N. (2004). Markov decision processes with asymptotic average failure rate constraint. *Communications in Statistics – Theory and Methods*, 7, 1689–1714.

Boussemart, M., Fillion, J.C., Limnios, N. (2002). Non ergodic Markov chain and decision processes with a constraint on the asymptotic failure rate: General class of policies. *Stochastic Models*, 18(1), 173–191.

Boutilier, C., Dearden, R., Goldszmidt, M. (1995). Exploiting structure in policy construction. *Proc. Fourteenth Inter. Conf. on AI (IJCAI-95)*, Montreal.

Dayan, P. (1992). The convergence of TD(λ) for general λ. *Machine Learning*, 8, 341–362.

Derman, C. (1970). *Finite State Markovian Decision Processes*. Academic Press, New York and London.

Ge, Y., Zhu, F., Ling, X., Liu, Q. (2019). Safe Q-learning method based on constrained Markov decision processes. *IEEE Access*, 7, 165007–165017.

Hagen, S.T. and Kröse, B. (1997). A short introduction to reinforcement learning. *Proc. BENELEARN-97, 7th Belgian-Dutch Conference on Machine Learning*, 7–12.

Howard, R.A. (1960). *Dynamic Programming and Markov Process*. John Wiley & Sons, New York.

Huang, Y. and Guo, X. (2011). Finite horizon semi-Markov decision processes with application to maintenance systems. *European Journal of Operational Research*, 212(1), 131–140.

Jaakkola, T. and Jordan, M.I. (1994). On the convergence of stochastic iterative dynamic programming algorithms. *Advances in Neural Information Processing Systems*, 6, 703–710.

Kaelbling, L.P., Littman, M.L., Moore, A.W. (1996). Reinforcement learning: A survey. *Journal of Artificial Intelligence Research*, 4, 237–285.

Limnios, N. and Oprişan, G. (2000). *Semi-Markov Processes and Reliability*. Birkhäuser, Boston, MA.

Lindqvist, B.H. and Amundrustad, H. (1998). Markov model for periodically tested components. *ESREL'98*, Trondheim.

Lippman, S.A. (1975). On dynamic programming with unbounded rewards. *Management Science*, 21, 1225–1233.

Lu, Z., Zhou, J., Li, X. (2018). Monte Carlo simulation based time limited dispatch analysis with the constraint of dispatch reliability for electronic engine control systems. *Aerospace Science and Technology*, 72, 397–408.

Mahadevan, S. (1996a). Optimality criteria in reinforcement learning. *AAAI Fall Symposium on Learning Complex Behaviors in Adaptative Intelligent Systems*. MIT, Boston, MA.

Mahadevan, S. (1996b). Average reward reinforcement learning: Foundations, algorithms, and empirical results. *Machine Learning*, 22, 159–196.

Morton, T.E. (1971). On the asymptotic convergence rate of cost differences for Markovian decision processes. *Operations Research*, 19, 244–248.

Parr, R.E. (1998). Hierarchical control and learning for Markov decision process. PhD Dissertation, Princeton University, NJ.

Puterman, M.L. (1994). *Markov Decision Processes: Discrete Stochastic Dynamic Programming*. John Wiley & Sons, New York.

List of Authors

Serhii BARANOVSKY
National University of Water and Environmental Engineering
Rivne
Ukraine

Andrii BOMBA
National University of Water and Environmental Engineering
Rivne
Ukraine

Michel BOUSSEMART
IoT Consultants
Adelaide
Australia

Dmytro CHERNIY
Taras Shevchenko
National University of Kyiv
Ukraine

Sergey V. DENISOV
Taras Shevchenko
National University of Kyiv
Ukraine

Ihor GAPYAK
Taras Shevchenko
National University of Kyiv
Ukraine

Yevhen HAVRYLKO
V.M. Glushkov Institute of Cybernetics
National Academy of Sciences of Ukraine
Kyiv
Ukraine

Vladislav HRYHORENKO
Taras Shevchenko
National University of Kyiv
Ukraine

Dmytro KLYUSHIN
Taras Shevchenko
National University of Kyiv
Ukraine

Dmitri KOROLIOUK
National Technical University of Ukraine
Igor Sikorsky Kyiv Polytechnic Institute
Ukraine
and
UNESCO Chair in Biotechnology
and Bioethics
University of Rome Tor Vergata
Italy

Pavlo KRYSENKO
National Technical University of Ukraine
Igor Sikorsky Kyiv Polytechnic Institute
Ukraine

Nikolaos LIMNIOS
Sorbonne University Alliance
Université de Technologie de Compiègne
Laboratoire de Mathématiques
Appliquées de Compiègne
France

Nataliia LYASHKO
V.M. Glushkov Institute of Cybernetics
National Academy of Sciences of Ukraine
Kyiv
Ukraine

Sergiy LYASHKO
Taras Shevchenko
National University of Kyiv
Ukraine

Ihor MOROZ
National University of Water and
Environmental Engineering
Rivne
Ukraine

Pavlo NEYEZHMAKOV
Ukrainian National Scientific Centre
Institute of Metrology
Kharkiv
Ukraine

Volodymyr PICHKUR
Taras Shevchenko
National University of Kyiv
Ukraine

Oksana PRYSHCHEPA
National University of Water and
Environmental Engineering
Rivne
Ukraine

Valerii SAMOILENKO
Taras Shevchenko
National University of Kyiv
Ukraine

Yuliia SAMOILENKO
National Technical University of Ukraine
Igor Sikorsky Kyiv Polytechnic Institute
Ukraine
and
Claude Bernard Lyon 1 University
France

Gennadiy V. SANDRAKOV
Taras Shevchenko
National University of Kyiv
Ukraine

Vladimir V. SEMENOV
Taras Shevchenko
National University of Kyiv
Ukraine

Valentyn SOBCHUK
Taras Shevchenko
National University of Kyiv
Ukraine

Mariana TURBAL
National University of Water and
Environmental Engineering
Rivne
Ukraine

Yurii TURBAL
National University of Water and
Environmental Engineering
Rivne
Ukraine

Vasyl USTIMENKO
Royal Holloway
University of London
United Kingdom
and
Maria Curie-Sklodowska University
Lublin
Poland

Evgen VAKAL
Taras Shevchenko
National University of Kyiv
Ukraine

Larysa VAKAL
V.M. Glushkov Institute of Cybernetics
National Academy of Sciences of Ukraine
Kyiv
Ukraine

Yuriy YAKYMENKO
National Technical University of Ukraine
Igor Sikorsky Kyiv Polytechnic Institute
Ukraine

Irina YALOVEGA
Simon Kuznets Kharkiv National
University of Economics
Ukraine

Maksym ZOZIUK
National Technical University of Ukraine
Igor Sikorsky Kyiv Polytechnic Institute
Ukraine

Stanislav ZUB
Ukrainian National Scientific Centre
Institute of Metrology
Kharkiv
Ukraine

Index

G, H

I, K

L, M

N, O

P, R

S, T

V, W